NEW GCSE SCIENCE

Additional Science B

For Specification Modules B3, B4, C3, C4, P3 and P4

OCR

Gateway Science

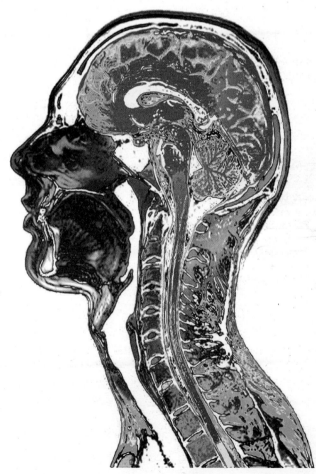

Series Editor: Chris Sherry

Authors: Colin Bell, Brian Cowie, Ann Daniels, Sandra Mitchell and Louise Smiles

Student Book

William Collins' dream of knowledge for all began with the publication of his first book in 1819. A self-educated mill worker, he not only enriched millions of lives, but also founded a flourishing publishing house. Today, staying true to this spirit, Collins books are packed with inspiration, innovation and practical expertise. They place you at the centre of a world of possibility and give you exactly what you need to explore it.

Collins. Freedom to teach

Published by Collins
An imprint of HarperCollinsPublishers
77–85 Fulham Palace Road
Hammersmith
London
W6 8JB

Browse the complete Collins catalogue at
www.collinseducation.com

10 9 8 7 6 5 4 3 2 1

ISBN-13 978-0-00-741531-1

British Library Cataloguing in Publication Data
A Catalogue record for this publication is available from the British Library

Commissioned by Letitia Luff
Project managed by Tammy Poggo and Hart McLeod
Production by Kerry Howie

Edited, proofread, indexed and designed by
Hart McLeod
Proofread by Lumière Chieh
New illustrations by Simon Tegg
Picture research by Caroline Green
Concept design by Anna Plucinska
Cover design by Julie Martin
'Bad Science' pages based on the work of Ben Goldacre

Printed and bound by L.E.G.O. S.p.A. Italy

Acknowledgements – see pages 319–320

Contents

Chemistry

C3 Chemical economics

C4 The periodic table

Physics

P3 Forces for transport

P4 Radiation for Life

How to use this book

Welcome to Collins New OCR Gateway Additional Science B

The main content

Each two-page lesson has three sections corresponding to the Gateway specification:

> For Foundation tier you should understand the work in the first and second sections.

> For Higher tier you need to understand the first section and then concentrate on the second and third sections.

Each section contains a set of level-appropriate questions that allow you to check and apply your knowledge.

Look for:

> 'You will find out' boxes

> Internet search terms (at the bottom of every page)

> 'Did you know' and 'Remember' boxes

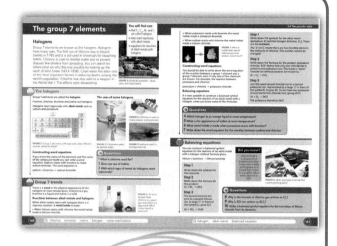

Remember!

To cover all the content of the OCR Gateway Science specification you should study the text and attempt the exam-style questions.

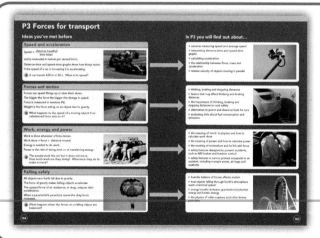

Module introductions

Each Module has a two-page introduction.

Link the science you will learn with your existing scientific knowledge to give you an overview before starting each Module.

Checklists

Each Module contains a checklist.

Summarise the key ideas that you have learned so far and look across the three columns to see how you can progress.

Refer back to the relevant pages in this book if you find any points you're not sure about.

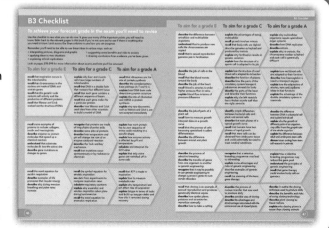

Exam-style questions

Every Module contains practice exam-style questions for both Foundation and Higher tier, labelled with the Assessment Objectives that it addresses.

Familiarise yourself with all the types of question that you might be asked.

Worked examples

Detailed worked examples with examiner comments show you how you can raise your grade. Here you will find tips on how to use accurate scientific vocabulary, avoid common exam errors, improve your Quality of Written Communication (QWC), and more.

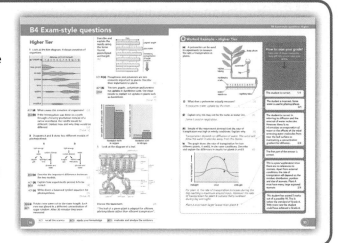

Preparing for assessment

Each Module contains preparing for assessment activities. These will help build the essential skills that you will need to succeed in your practical investigations and Controlled Assessment, and tackle the Assessment Objectives in your written exam.

Each type of preparing for assessment activity builds different skills.

> Applying your knowledge: Look at a familiar scientific concept in a new context.

> Research and collecting secondary data: Plan and carry out research using handy tips to guide you along the way.

> Planning and collecting primary data: Plan and carry out an investigation with support to keep you on the right track.

> Analysis and evaluation: Process data and draw conclusions from evidence. Use the hints to help you to achieve top marks.

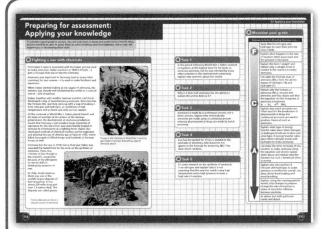

Bad Science

Based on *Bad Science* by Ben Goldacre, these activities give you the chance to be a 'science detective' and evaluate the scientific claims that you hear everyday in the media.

Assessment skills

A section at the end of the book guides you through your practical work, your Controlled Assessment tasks and your exam, with advice on: planning, carrying out and evaluating an experiment; using maths to analyse data; the language used in exam questions; and how best to approach your written exam.

B3 Living and growing

Ideas you've met before

Cells and reproduction

Eggs and sperm are specialised cells. When the nucleus of an egg joins with a sperm fertilisation takes place.

The fertilised egg cell divides into two new cells. The cells keep dividing to form a foetus. The foetus grows into a baby.

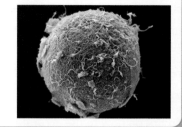

 How many sperm are needed to fertilise one egg?

Blood and transport

The circulatory system moves substances around the body.

Oxygen is taken from the lungs to the body.

Carbon dioxide is returned from the body to the lungs.

Blood is made up of red blood cells, white blood cells, plasma and platelets.

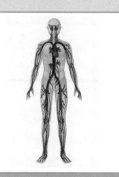

 Which part of the blood transports oxygen around the body?

Inheritance and selection

Characteristics such as hair and eye colour are inherited from our parents.

Characteristics are controlled by genes. Genes are found on chromosomes.

Selective breeding is when humans produce animals and plants that have the desired characteristics. For example, cows are bred to produce more milk.

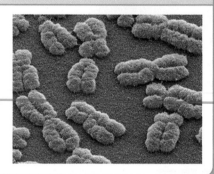

 Name the part of the cell that contains genes.

Respiration and enzymes

Respiration is one of the seven life processes. It takes place in all living cells.

The energy released in respiration is used for movement and keeping warm.

Enzymes are protein molecules that speed up chemical reactions.

There are enzymes in your digestive system that are used to break down food.

 Name three more of the seven life processes.

In B3 you will find out about...

> how eggs and sperm are adapted for fertilisation

> how cells divide by a process called mitosis

> how eggs and sperm are made by a special type of cell division called meiosis

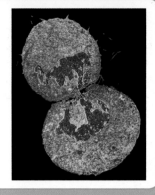

> how the blood cells are adapted to their function

> the different parts of the heart

> the role of haemoglobin in the transport of oxygen

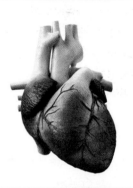

> how DNA from one organism can be transferred to another organism

> why selective breeding can be harmful

> how organisms like sheep can be cloned

> the two types of respiration

> how DNA is used to make a protein

> the effect of temperature on enzyme activity

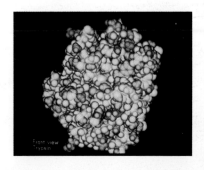

Molecules of life

You will find out:

> about the structure of cells

> about chromosomes and DNA

> about ribosomes

The secret of life

In 1953 two men walked into the Eagle pub in Cambridge and announced, 'We have found the secret of life.'

They were Francis Crick and James Watson and they had just discovered the structure of DNA (deoxyribonucleic acid). They had worked out that a DNA molecule is shaped like a double helix.

FIGURE 1: Crick and Watson with their model of part of a DNA molecule.

What's in a cell?

An animal cell has the following parts.

DNA

DNA is a molecule found tightly coiled inside the nucleus. It forms structures called **chromosomes**. A section of a chromosome is called a **gene**.

Francis Crick and James Watson discovered the structure of DNA.

cell **membrane** controls the movement of substances in and out of the cell

nucleus carries genetic information

cytoplasm where many chemical reactions happen

genetic information is carried on **chromosomes**

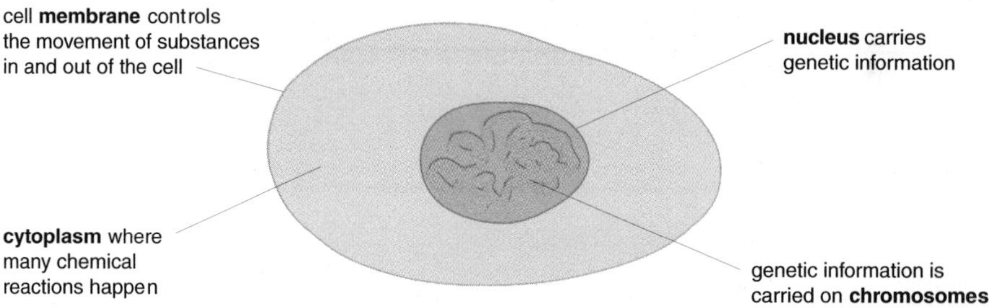

FIGURE 2: Look at the structure of an animal cell; which part of the cell contains chromosomes?

Mitochondria

Cell respiration is carried out inside **mitochondria**. During respiration energy is released from glucose in the presence of oxygen.

Mitochondria are very small. They cannot be seen using a school microscope.

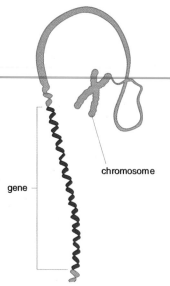

gene

chromosome

FIGURE 3: A chromosome – partly unravelled to show a gene. Which molecule are chromosomes made from?

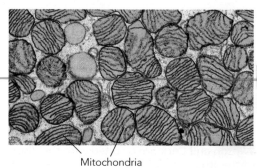

Mitochondria

FIGURE 4: Mitochondria seen using an electron microscope. Why are mitochondria needed in the body?

Did you know?

Jurassic Park could happen. Scientists have started to store DNA of animals facing extinction, so they can be recreated in the future. The project is called **The Frozen Ark.**

Questions

1 Write down the name of part of the cell that contains chromosomes.

2 What is a section of a chromosome called?

3 Which two scientists first worked out the structure of DNA?

More on mitochondria

Some types of cells have more mitochondria than others. Liver and muscle cells need large amounts of energy to carry out their functions. This is why they have higher numbers of mitochondria compared with most other cells.

DNA and chromosomes

Chromosomes are long coiled molecules of DNA divided up into regions called genes.

DNA molecules have two strands coiled to a double helix. The double helix has cross links made of chemicals called **bases**. There are four different bases in DNA. Each cross link contains two bases, which are known as base pairs.

Discovering DNA structure

Watson and Crick are the two scientists best known for discovering DNA. However, they could not have done so without the help of other scientists.

> Rosalind Franklin and Maurice Wilkins used X-ray crystallography to take pictures of DNA. The pictures showed DNA as two chains wound in a double helix.

> Erwin Chargaff discovered there are equal numbers of A and T bases and of G and C bases in DNA.

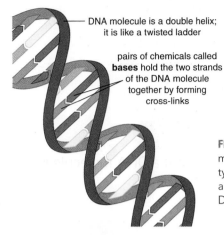

DNA molecule is a double helix; it is like a twisted ladder

pairs of chemicals called **bases** hold the two strands of the DNA molecule together by forming cross-links

FIGURE 5: How many different types of bases are found in DNA?

Questions

4 Explain why muscle cells have a lot of mitochondria.

5 What are the cross links in DNA made from?

6 Which evidence was used to show DNA was a double helix?

7 Who discovered that there were equal numbers of A and T bases in DNA?

Ribosomes

Cells have lots of structures inside them that cannot be seen using a light microscope. **Ribosomes**, which are found in the cytoplasm, are one example. Ribosomes are used in the cell when proteins are made.

More on the discovery of DNA

Watson and Crick are credited with the discovery of the structure of DNA in 1953. However, without the work of other scientists they may have got it wrong. Originally they thought DNA was a triple helix. They needed an x-ray photograph taken by Rosalind Franklin to confirm it was a double helix.

When scientists make new discoveries it is important that their results are repeated by other scientists. This will help make the results valid.

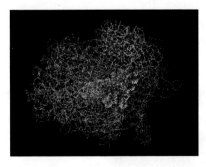

FIGURE 6: Ribosomes are needed in the body for repair of damaged tissue. Explain why.

Questions

8 Suggest why ribosomes can't be seen using a light microscope.

9 Why was Rosalind Franklin's work so important in the discovery of DNA?

10 Why is it important that scientists repeat the work of other scientists?

Genetic code

The genes in your body make up your genetic code. Your genetic code controls the activities of your cells. It also controls your different characteristics like the colour of your hair.

Genes are made of DNA. The DNA is a code for making proteins. Without DNA you could not make the proteins needed for growth and repair of cells.

Did you know?

50% of our DNA is shared with the DNA of a banana.

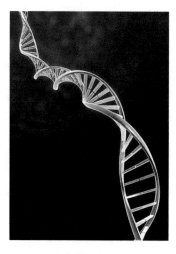

FIGURE 7: What is a section of DNA called?

Questions

11 Which part of the nucleus carries your genetic code?

12 Why does a cell need to make proteins?

13 Which **molecule** in the cell is a code for making proteins?

DNA and proteins

Each gene in DNA has a different sequence of bases.

The sequence of bases code for a particular protein.

Your body has a code for every protein it needs to make. The different codes give the proteins different shapes. To make the protein a copy of the gene is made as the gene cannot leave the nucleus. The copy leaves the nucleus so that the protein can be made on the ribosomes within the cytoplasm.

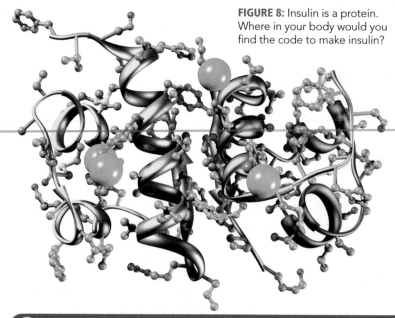

FIGURE 8: Insulin is a protein. Where in your body would you find the code to make insulin?

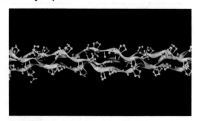

FIGURE 9: Collagen is a different shape to insulin. What does that tell you about the DNA codes for collagen and insulin?

Questions

14 Finish the sentence. The order of bases in DNA is called the base ....

15 Some people cannot make the protein insulin. Use ideas about DNA to suggest a reason why.

DNA and proteins

The four bases in DNA are shown by their initials A, T, C and G. A only links with T and C only links with G. This is important when DNA is copied to make proteins.

The order of bases found in DNA is called the base code. Each three bases code for an amino acid. For example the sequence CAA codes for the amino acid called valine.

By coding for proteins DNA controls the functions of the cell. For example DNA codes for the enzymes involved in respiration. Without these enzymes respiration would not take place.

Protein synthesis

> To make proteins the DNA code must be copied (this is called transcription).

> The copied code is called **messenger RNA** (mRNA).

> The mRNA leaves the nucleus and travels to the ribosomes.

At the ribosomes the code is used to put the amino acids together in the right order and form the protein (this is called translation).

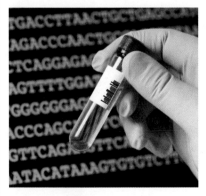

FIGURE 11: Can you spot the code for valine in this base code?

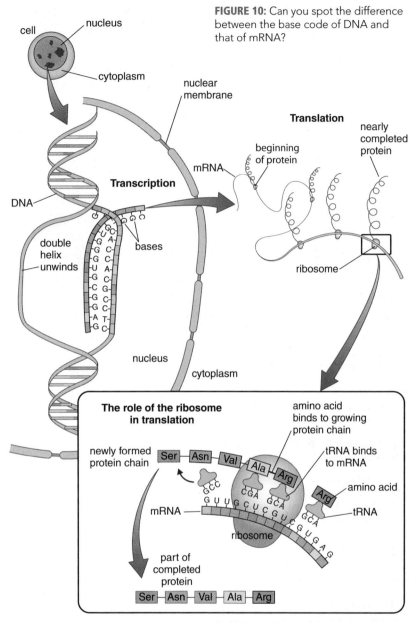

FIGURE 10: Can you spot the difference between the base code of DNA and that of mRNA?

The role of the ribosome in translation

Questions

16 Which base always joins to C in DNA?

17 How many amino acids are coded for in the following section of DNA?
AAATATCTCCCCTCAACCGGGCGGTAAATG

18 Write down the complementary base sequence for the section of DNA in Q17.

19 Describe the role of mRNA in protein synthesis.

Proteins and mutations

You will find out:

> about different types of proteins

> about the functions of different proteins

> why some proteins are not made by all the cells

Sickle cells

Red blood cells carry oxygen around the body. The oxygen is attached to a protein called haemoglobin.

In some people the haemoglobin is the wrong shape. This makes their red blood cells look different. It also means they can't carry as much oxygen around the body.

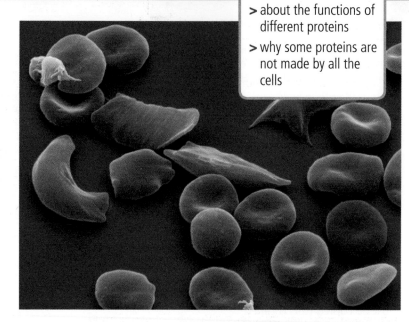

FIGURE 1: Can you spot the sickle cells?

 Proteins

Proteins are important chemicals found in your body. There are many different types all with a different job to do.

> **Collagen** – this protein can be found in the walls of arteries. It makes the walls stronger.

> **Insulin** – this is a hormone used to control blood glucose levels.

> **Haemoglobin** – this used to carry oxygen around the body.

The cells in your body do not make every protein. Insulin is only made in the cells of your pancreas. Not all organisms make the same proteins. For example, plants do not make collagen because they do not need it.

Mutations

Genes code for proteins. Sometimes a gene code can change. This is called a **mutation**. Mutations to genes can cause the shape of the protein to change so it can no longer do its job in a cell.

 Questions

1 Write down the name of the protein that makes your blood red.

2 Why is collagen needed in artery walls?

3 Write down the term used to describe a change to the gene code.

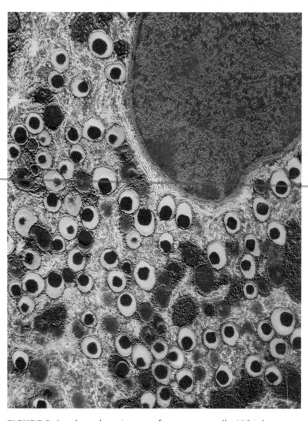

FIGURE 2: Look at the picture of pancreas cells. Which protein is made only by these cells?

Q proteins   enzymes   mutations

Protein function

Proteins are made from long chains of **amino acids**. Different types of proteins have different functions.

Some proteins form part of the structure of your body.

Collagen can be found in bones, tendons and cartilage.

Hormones are proteins. Hormones help control many of your body functions. For example insulin controls the level of glucose in your blood.

Haemoglobin is a carrier molecule. It carries oxygen around your body.

Enzymes are proteins. Enzymes control many activities in the body like digestion.

More on mutations

Gene mutations can occur spontaneously or be caused by

> radiation

> chemicals such as tar in cigarettes.

A mutated gene may code for a different shaped protein. The new protein can be harmful and even cause death. Haemophilia is caused by a mutated gene. The blood of a person with haemophilia is unable to clot, which can be dangerous, as they may not be able to stop bleeding. This is because one of the proteins they need to clot their blood cannot be made by the mutated gene.

Sometimes a mutation can be beneficial. For example, the mutation that causes sickle cell anaemia can mean that people with sickle cell anaemia are less likely to die of malaria. Some mutations have no effect at all.

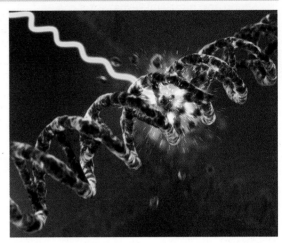

FIGURE 3: Radiation can change DNA. What is this change called?

Questions

4 Which protein forms part of bone and cartilage?

5 Write down one use of enzymes in the body.

6 Write down two causes of mutations.

Amino acid sequence

Each protein has its own number and sequence of amino acids. This is called its primary structure. The primary structure of the protein is determined by the sequence of bases in the gene that codes for the protein. The primary structure is then folded into different shapes. This means that each type of protein molecule has a different shape, which is the ideal one to carry out its function. Collagen has a long fibrous shape whereas haemoglobin is globular.

If a mutation changes the sequences of bases in the DNA, the amino acid sequence could change. This would result in a different shaped protein that could not do its job.

Switched on or off

Each cell in your body has a complete set of genes. So each cell could make all the proteins your body needs. However, genes can be switched on or off. Only cells in your pancreas switch on the insulin gene. All your other cells have this gene switched off.

Questions

7 Why is the sequence of bases in DNA important?

8 Why don't the cells in your eye make insulin?

9 What caused the shape of sickle cell haemoglobin to be different to normal haemoglobin?

Making cheese

Simon and his friend Nisha visit a cheese-maker. Simon knows that cheese can be made from sour milk. When milk turns sour it separates into a solid and a liquid. The solid part is called **curds**. It is the curds that form cheese.

The chemical reaction that turns milk sour is slow. The cheese-makers tell Simon and Nisha that an **enzyme** called rennin is added to the milk. Enzymes are molecules that speed up chemical reactions and work best at particular temperatures.

Part of an enzyme is called the **active site**. The **substrate** molecules fit into the active site when a reaction takes place. When milk turns sour the substrate are the molecules in the milk which fit into the active site of the enzyme rennin.

FIGURE 4: Making cheese. What is added to milk to speed up the souring reaction?

Questions

10 Name **one** example of an enzyme.

11 What do enzymes do to chemical reactions?

Did you know?

There are enzymes under your kitchen sink! Household products, such as washing powder, use enzymes to speed up the breaking down of stains.

Enzymes and rate of reaction

An enzyme is a **biological catalyst**. It is a protein that speeds up a biological reaction. Enzymes catalyse most chemical reactions occurring within cells, such as respiration, photosynthesis and protein synthesis. Each enzyme is **specific** to one **substrate**.

In an enzyme-catalysed reaction, **substrate** molecules are changed into **product** molecules.

Enzyme controlled reactions are affected by:

> pH

> temperature.

Enzymes have an optimum pH. Above and below this pH the rate of reaction will slow down.

pH

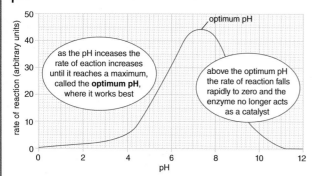

FIGURE 5: Graph to show how pH affects the rate of an enzyme-catalysed reaction. What is the optimum pH for this reaction?

temperature

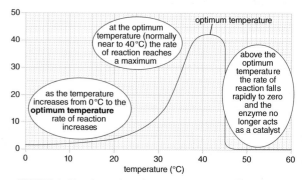

FIGURE 6: Graph to show how temperature affects the rate of an enzyme-catalysed reaction. Suggest why an enzyme ceases to function above approximately 40 °C.

Q enzymes   lock and key theory   catalysts

Temperature

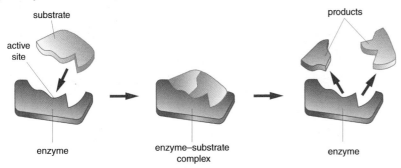

FIGURE 7: The lock and key theory of enzymes. How does the theory explain the specificity of enzymes?

How enzymes work – 'the lock and key' theory

Each enzyme has a unique sequence of amino acids. This results in each enzyme having a different shape. Within this shape is a structure called an **active site**. Only one type of substrate can fit into the active site making an enzyme specific to a reaction. Once the substrate is attached to the active site it is turned into a product. The enzyme is like a lock and the substrate a key. Only one key can open a lock and only one enzyme can change a particular substrate.

Questions

12 What other name is given to enzymes?

13 Describe how changing the temperature changes the rate of an enzyme-catalysed reaction.

14 Pepsin is an enzyme found in the stomach. It breaks down proteins into amino acids. Draw a diagram to explain why pepsin cannot break down starch.

Denaturing enzymes

If the shape of an enzyme changes, it can no longer catalyse the reaction. The enzyme has become **denatured**. This is an irreversible change which means the enzyme will no longer function. The substrate can no longer fit into the active site. An enzyme can be denatured by the following.

> Extremes of pH. Every enzyme has an optimum pH at which it works most efficiently. At this pH, the active site and the substrate molecule are a perfect fit. The further away from this pH, the more the enzyme molecule distorts and the less perfect the fit is. Eventually the substrate can no longer fit the active site and the enzyme stops working.

> High temperatures. As the temperature increases, the molecules gain more energy. More collisions occur and the rate of reaction increases. Above the optimum temperature, the enzyme denatures and the reaction stops.

Q_{10}

Temperature coefficient or Q_{10} is the effect of temperature on the rate of a chemical reaction. It is calculated using the formula

$$Q_{10} = \frac{\text{rate at higher temperature}}{\text{rate at lower temperature}}$$

Many enzymes have a Q_{10} of around 2. This means that increasing the temperature of the reaction by 10 °C should double the rate of the reaction.

Questions

15 Biological washing powders contain enzymes. Explain why the washing powder may become inefficient at high temperatures.

16 An enzyme controlled reaction at 20 °C and a rate of 5 arbitrary units. When the reaction was repeated at 30 °C the rate changed to 12 arbitrary units.

Calculate the Q_{10} for this reaction.

Respiration

You will find out:
> why animals and plants need energy
> about respiration
> about RQ values

Burning sugar

If you set fire to sugar it releases a lot of heat energy. Your body can release energy in sugar. However it does this in a very controlled way. If your body tried to release all the energy at once you would burst into flames.

FIGURE 1: Which type of energy is released when sugar burns?

Did you know?

Each day you:

> breathe about 25 000 times

> breathe about 6 litres of air every minute.

Energy and respiration

Plants and animals need energy to carry out life processes such as movement.

To release energy, animal and plant cells need glucose and oxygen. Releasing the energy in cells is called **respiration**.

Aerobic respiration

Aerobic respiration uses oxygen to release energy. The word equation for aerobic respiration is

glucose + oxygen → carbon dioxide + water

The energy from aerobic respiration is needed for many different things in the body. Mammals like humans use the energy to maintain body temperature. Energy is also needed to make proteins and contract muscles.

Remember!

Breathing is simply getting air into and out of the body. Respiration is the release of energy in cells from food.

FIGURE 2: Plants need energy to grow. Write down the name of the main process that releases the energy in plant cells.

Questions

1 Which gas is used in aerobic respiration?

2 Which gas is produced during aerobic respiration?

3 Write down one life process that needs energy.

Q aerobic respiration   using a respirometer

Measuring the rate of respiration

The balanced symbol equation for respiration is

$$C_6H_{12}O_6 + 6O_2 \rightarrow 6CO_2 + 6H_2O + energy$$

The rate of respiration in an organism can be measured using a respirometer. The faster the organism respires the more oxygen it takes in.

Respiration rate can also be measured by recording the amount of carbon dioxide made.

Once the volumes have been found the **respiratory quotient (RQ)** can be calculated using the formula

$$RQ = \frac{carbon\ dioxide\ produced}{oxygen\ used}$$

For aerobic respiration that uses glucose the RQ is always 1.

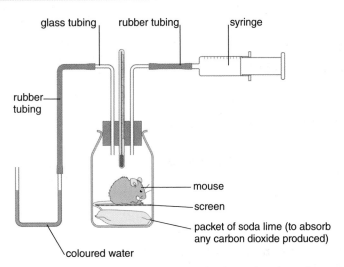

FIGURE 3: As the mouse uses oxygen, the dye in the glass tubing moves toward the mouse. If the mouse runs around, the dye moves faster. Explain why?

Questions

4 Write down the term used to describe respiration that uses oxygen.

5 Look at the picture of the respirometer it is used to measure respiration rate. 2.0 cm³ of oxygen was used by the mouse and 1.4 cm³ of carbon dioxide was released. Calculate the RQ value.

More about respiration

Respiration releases energy, which is then stored in **ATP** (adenosine triphosphate) molecules. The ATP molecule is used as an energy source for many processes inside the cell.

The process of respiration is controlled by enzymes. Changes to pH and temperature can affect enzymes. This is why the rate of respiration is also dependent of pH and temperature.

Metabolic rate

Your **metabolic rate** is the amount of energy your body needs. Aerobic respiration provides your body with energy. This means that metabolic rate can be estimated by measuring oxygen uptake.

Questions

6 What is the role of ATP in cells?

7 Explain why metabolic rate can be estimated by measuring oxygen uptake.

8 Snakes need to warm their bodies before they can become active. Explain why using ideas about enzymes and respiration.

Q ATP   RQ values

Feeling tired

You will find out:

> why exercise changes breathing and pulse rate

> about the difference between aerobic respiration and anaerobic respiration

> about oxygen debt

It has been a long hard game of football, and David feels tired.

When the cells in David's body do more work, they need more oxygen and glucose. His breathing rate and pulse rate increase.

This helps to get rid of the waste carbon dioxide produced by his cells as quickly as possible. Carbon dioxide is breathed out of David's lungs.

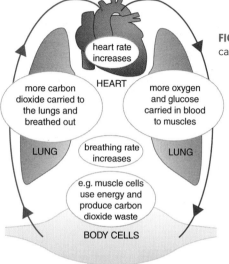

FIGURE 4: How does David get rid of carbon dioxide from his body?

Did you know?

You can find out how fit you are by measuring your pulse rate. Measure your pulse rate before and after exercise. If you are fit it will soon return to normal.

Measuring recovery rate

David can measure how quickly he recovers from playing football.

He sits down and counts his pulse before he plays. He does this by placing a finger on his pulse and counting the beats. After exercise he counts his pulse again. He does this every minute until his pulse rate levels off. The time it takes to return to resting pulse is recovery time.

Questions

9 What happens to David's breathing rate when he plays football?

10 Which waste gas does David need to breathe out?

11 Which cells in David's legs need more oxygen when he plays football?

Anaerobic respiration

David goes to training sessions to get fit.

During hard exercise anaerobic and aerobic respiration takes place in muscles. **Anaerobic respiration** releases energy without the use of oxygen and produces lactic acid as waste.

glucose → lactic acid

🔍 anaerobic respiration   muscle fatigue

Anaerobic respiration releases much less energy per glucose molecule than aerobic respiration.

The training coach tells David that after hard exercise lactic acid collects in his muscles. Lactic acid stops muscles working temporarily, a condition called muscle fatigue. This is also known as cramp and can be very painful. This explains why David's muscles hurt after a game of football.

When David is walking his body uses aerobic respiration. When David sprints, his muscle cells cannot get oxygen quickly enough to release all the energy he needs. His muscle cells therefore resort to obtaining some energy by anaerobic respiration.

FIGURE 5: David's cells are getting enough oxygen.

FIGURE 6: David is working so hard his cells are now not getting enough oxygen. Why is David using anaerobic and aerobic respiration?

Questions

12 What type of respiration does David use when he sprints?

13 What chemical is produced during anaerobic respiration?

14 Why do David's muscles hurt after a hard game of football?

15 Explain the difference between aerobic and anaerobic respiration.

What is an oxygen debt?

The energy released by anaerobic respiration is less than that from aerobic respiration because glucose is only partly broken down. Lactic acid also builds up. Extra oxygen is needed to remove this lactic acid, so at the end of the sprint:

> David's heart continues to beat faster than normal so his blood can carry lactic acid away from his muscles to be broken down in his liver. This requires oxygen.

> David breathes heavily for a few minutes to obtain extra oxygen. The extra oxygen required to remove the lactic acid is called the **oxygen debt**.

> Panting – shallow rapid breathing after exercise – also replaces oxygen to allow aerobic respiration to take place.

Did you know?

Sprinters running in a 100 m race do not usually breathe during the race.

Questions

16 Why does panting occur after exercise?

17 How does David's body recover from hard exercise?

Q lactic acid

Cell division

You will find out:
> why cells divide
> about advantages of being multicellular
> how cells divide by mitosis
> how DNA replicates

Growing human tissue

Did you know it is possible to grow human tissue outside the body? Burns victims can now have their damaged skin repaired using sheets of new skin formed from skin cells grown in the laboratory.

Using a plastic framework under the skin of a mouse to provide a growing frame and a blood supply, scientists can also 'grow' human cartilage and skin cells into the shape required. The ability to grow something that looks like an ear can now help accident victims who need cosmetic surgery.

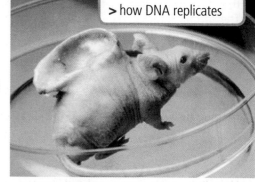

FIGURE 1: Why would a scientist want to grow an ear on a mouse?

Making new cells

Every day thousands of dead skin cells are lost from the skin. They are continuously shed by our bodies. All of these cells need to be replaced. To make new cells the body carries out cell division. Cells divide whenever the body needs to:

> grow

> replace worn out cells

> repair damaged tissue.

Body cells contain matching pairs of chromosomes. When cells divide these chromosomes are copied. This means all the new cells will have the same chromosomes.

The more cells the better

Amoeba is a **unicellular organism**. Its whole body is a single cell. It is microscopic in size. To reproduce, the amoeba cell just divides to form two new cells. This is an example of **asexual reproduction**. To grow larger than an amoeba, an organism has to become **multicellular**.

FIGURE 2: Why do skin cells need to divide?

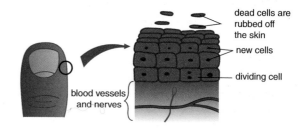

dead cells are rubbed off the skin

new cells

dividing cell

blood vessels and nerves

Questions

1 How does the body replace lost skin cells?

2 Write down three reasons why the body needs to make new cells.

Advantages of being multicellular

Humans are multicellular. This gives them many advantages.

> Multicellular organisms can grow large.

> **Cell differentiation** takes place. Cells become different shapes or sizes to carry out specialised jobs.

> Organisms become more complex. Multicellular organisms can develop different organ systems. All the systems work together to make up the organism.

FIGURE 3: Twenty-three homologous pairs of chromosomes from a human cell. How many single chromosomes are there in a human cell?

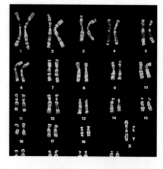

cell division   growing skin cells

Dividing cells

Inside the nucleus of a human body cell there are 46 chromosomes. They are all different shapes and sizes. Each chromosome can be matched up with another of the same size. Humans have 23 pairs of chromosomes. The chromosomes in a pair look the same and carry similar information. They are called homologous pairs. When a cell has pairs of chromosomes it is called a **diploid** cell. In all mammals body cells are diploid. During growth a type of cell division called **mitosis** makes new cells. The new cells are genetically identical. This is because the DNA (chromosomes) in the nucleus was replicated before the cell divided.

Questions

3 Suggest one advantage of being multicellular.

4 Which type of human cell does not have chromosomes?

5 a Explain what is meant by the term diploid.

b How many pairs of chromosomes are there in a diploid human body cell?

Size matters

Unicellular organisms can absorb oxygen and nutrients across their membrane. Multicellular organisms have a large volume compared to their surface area. This means that nutrients and oxygen cannot be absorbed across their skin and reach all their cells. Instead they have specialised organ systems to ensure that all their cells get enough nutrients and oxygen. These include:

> nervous systems to communicate between cells

> transport systems to supply cells with nutrients

> respiratory systems for exchange of gases with the environment.

Mitosis and chromosomes

During mitosis, genetically identical cells are made. Each new cell contains identical chromosomes to the original cell.

DNA replication

Before cells can divide DNA needs to be copied

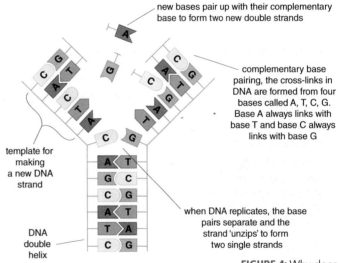

new bases pair up with their complementary base to form two new double strands

complementary base pairing, the cross-links in DNA are formed from four bases called A, T, C, G. Base A always links with base T and base C always links with base G

template for making a new DNA strand

when DNA replicates, the base pairs separate and the strand 'unzips' to form two single strands

DNA double helix

FIGURE 4: Why does the DNA unzip?

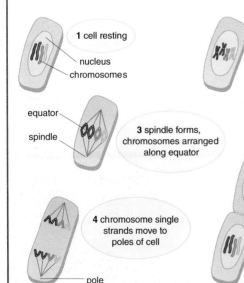

1 cell resting

nucleus

chromosomes

2 each chromosome is copied, the single strand forms double-stranded 'X' shape

equator

spindle

3 spindle forms, chromosomes arranged along equator

4 chromosome single strands move to poles of cell

pole

5 two genetically identical cells are produced

FIGURE 5: Mitosis. At what stage in mitosis are chromosomes copied?

Questions

6 Explain why multicellular organisms need a respiratory surface.

7 Describe the process of DNA replication.

8 Describe the stages involved when the body makes a new layer of skin.

Eggs and fertilisation

You will find out:

> about eggs and sperm
> about fertilisation
> how gametes are made by meiosis

Mohammed likes eating an egg for breakfast. He knows a hen's egg has a hard outer shell for protection. Cracking open the egg, Mohammed finds the albumen (white) and the yolk (yellow). The hen's egg has not been fertilised by sperm from a male so cannot develop into an embryo.

Sexual reproduction

Eggs and sperm are called sex cells or reproductive **gametes**. Gametes join during **fertilisation**. To increase the chance of one sperm reaching the egg a male releases millions of sperm at a time.

The gametes have half the number of chromosomes of a body cell. So when they join the new cell has the full number of chromosomes.

The individual that develops from the cell will have inherited genes from both parents. This means they have their own unique set of genes.

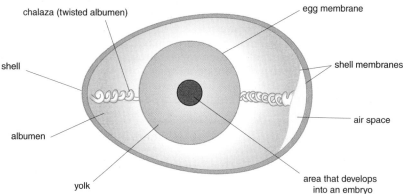

chalaza (twisted albumen)

egg membrane

shell

shell membranes

albumen

air space

yolk

area that develops into an embryo

FIGURE 6: A hen's egg. What needs to happen for the egg to develop into an embryo?

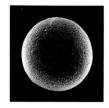

FIGURE 7: Human egg. Why do egg cells only have half the number of chromosomes found in body cells?

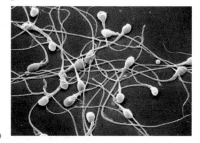

FIGURE 8: Human sperm. Why do males release millions of sperm?

Questions

9 What is another name for a male gamete?

10 What is it called when a sperm and an egg join?

11 Why does the new individual have their own unique set of genes?

Making gametes

Human body cells have 23 pairs of chromosomes. This full set of chromosomes is called the diploid number. The gametes (egg and sperm) contain only one chromosome from each pair. So gametes have half a set of chromosomes, called the **haploid** number. During fertilisation the gametes join to form a **zygote**.

The zygote is diploid and can develop into an embryo.

Meiosis is a special type of cell division that produces gametes. When fertilisation takes place gametes from a male and female join. The resulting offspring have genes from both parents. This new set of genes controls the characteristics of the zygote. The characteristics will be a combination of the parents so the new individual will be different to the parents. Reproduction using meiosis results in a lot of genetic variation within a species. Animals such as amoeba use mitosis for reproduction so all their offspring are genetically identical.

FIGURE 9: What process during sexual reproduction introduces genetic variation?

Q sperm   fertilisation

Moving sperm

Sperm have to travel a long way and then get inside an egg. They are specially adapted to do this.

> A sperm has large numbers of mitochondria to release energy for motion.

> A structure called an **acrosome** on the sperm head releases enzymes that digest the cell membrane of an egg allowing the sperm inside.

acrosome contains enzymes to digest the egg cell membrane

many mitochondria in cells release energy

nucleus containing chromosomes

FIGURE 10: A sperm. Why are there a lot of mitochondria in some of its cells?

Remember!
Body cells are diploid. Gametes are haploid. They only have half the number of chromosomes.

Questions

12 Where in the human body does meiosis take place?

13 If sperm and eggs were diploid how many chromosomes would the human zygote have?

14 Explain the function of the following parts of a sperm cell:

a mitochondria

b acrosome.

Meiosis and chromosomes

Gametes are made when diploid cells divide by meiosis to produce haploid cells. Meiosis involves two divisions:

> first the pairs of chromosomes separate

> then the chromosomes divide in the same way as in mitosis.

Remember!
Your body cells divide by mitosis. When making eggs or sperm, cells divide by meiosis.

diploid cells – the single strands are copied to make x-shaped chromosomes

4 four new haploid cells form, all genetically different from each other

1 homologous chromosomes pair up

pole

2 one from each pair moves to each pole

3 the strands of each chromosome are pulled apart to opposite poles

Questions

15 Describe the stages involved when sperm is made inside the testes.

16 Describe two ways in which meiosis is different to mitosis.

FIGURE 11: Meiosis. At what stage do chromosome pairs move to the poles of the cell?

 meiosis

Preparing for assessment: Applying your knowledge

To achieve a good grade in science, you not only have to know and understand scientific ideas, but you need to be able to apply them to other situations and investigations. These tasks will support you in developing these skills.

✳ Genetic disorders in the royal family

Imagine what would happen if, when you cut yourself, the wound did not heal. This condition is called haemophilia and it is a disorder of the blood.

A person suffering from haemophilia lacks a clotting agent in their blood. If they are cut their wounds continue to bleed; there can be bleeding inside the body and their body bruises more easily.

Haemophilia is an inherited condition; it is not contagious which means it cannot be caught by being near someone who has it.

Haemophilia is a condition that is sex-linked – it can only be carried on an X chromosome. A woman has two X chromosomes. A man has one X chromosome and one Y chromosome. Haemophilia is a recessive characteristic, which means that if a woman has inherited a 'healthy' X chromosome from one parent and a 'damaged' (i.e. haemophilia) chromosome from the other parent, the healthy chromosome dominates the damaged one and the woman does not suffer from haemophilia – though she does carry it and may pass it on. If this is the case she is called a 'carrier'. However, a man only has one X chromosome (from his mother) so if it is damaged he will suffer from haemophilia.

Queen Victoria, who was queen of England for a large part of the 19th Century, was a carrier of haemophilia. She did not suffer from the disease herself but she passed on the affected chromosome to some of her nine children. Sons who inherited the affected chromosome suffered from haemophilia (Prince Leopold died at the age of 31) and daughters who inherited the affected chromosome carried it on to the next generation. Two of Victoria's grandsons suffered from haemophilia and seven of her great grandsons.

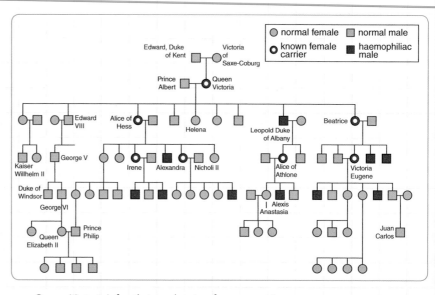

Queen Victoria's family tree showing four generations.

Task 1

Queen Victoria was the first in her family to carry the haemophilia gene. It is thought that the gene was the result of a mutation. What is a mutation?

Task 2

Genes are found on chromosomes. Which molecule are chromosomes made of?

Genes are codes. What substance in the body do they code for?

Task 3

Queen Victoria had the haemophilia gene but did not have haemophilia. Explain why.

Use ideas about X and Y chromosomes to explain why it is mostly boys that have haemophilia.

Task 4

Use ideas about meiosis to explain why not all Queen Victoria's sons had haemophilia.

Is it possible for a daughter to inherit haemophilia? Explain your answer.

Task 5

Explain how the mutation results in a lack of clotting agent in the blood.

✳ Maximise your grade

Answer includes showing that you can...

Describe mutations as changes to genes and that genes code for proteins.
Recall that chromosomes are made of DNA.
Explain that you have two copies of each genes and if you have the correct gene then you will not have haemophilia.
Explain that the gene for the clotting agent is on the X chromosome not on the Y so boys only get one chance at inheriting the correct gene.
Show how girls can inherit the haemophilia gene from their fathers.
Describe the stages in meiosis to show how eggs can have different genes.
Explain how mutations change the base sequence, which results in a protein being made that is different to the clotting agent.
As above, but with particular clarity and detail.

The circulatory system

You will find out:
> about blood cells and what they do
> about different types of blood vessels
> how blood vessels are adapted to their functions

Why your heart is so important

Every day your heart will beat about 100 000 times.

On average your heart pumps 1 500 gallons of blood each day. This makes it the hardest working muscle in the body. All this hard work goes on without you even thinking about it. However if your blood stopped moving your body would stop getting the oxygen it needs to stay alive. That is why your heart is so important.

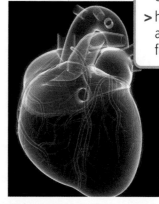

FIGURE 1: About how many times has your heart beat so far in your life?

Is blood really red?

Blood looks like a red liquid. But looked at through a microscope blood is very different. It consists of a yellow liquid called **plasma**. Blood looks red because it contains tiny **red blood cells**. These cells transport oxygen around the body. There are also **white blood cells** in blood. White blood cells help to defend the body against disease. Blood also contains tiny cell fragments called **platelets**. Platelets help to clot the blood if we cut ourselves.

Blood vessels

Blood is carried around the body in three different types of blood vessels:

> **artery**

> **vein**

> **capillary**.

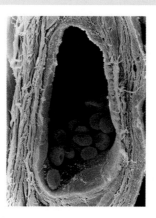

FIGURE 2: Blood flowing through a vein. What types of cells can you see?

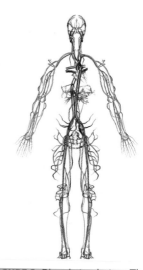

FIGURE 3: Blood circulation. The arteries are shown in red and the veins in blue. Which blood vessels are not shown in the picture?

Did you know?

Red blood cells are so small, 5 000 000 000 will fit into 1 cm^3.

Remember!

White blood cells engulf microbes. They don't fight or kill microbes.

Questions

1 Match up each part of the blood to the job it does.

Part of the blood	Job
red blood cell	defends against disease
white blood cell	helps blood to clot
platelet	transports oxygen

2 Name three types of blood vessels.

blood plasma   blood vessels   haemoglobin   oxyhaemoglobin

Doing their job

Blood has many important functions. Each part of the blood is adapted to carry out its function.

Red blood cells

Red blood cells are adapted to carry as much oxygen as possible.

> Their red colour comes from a chemical called haemoglobin. Oxygen joins to haemoglobin, which allows it to be transported around the body.

> They do not have a nucleus. This leaves more room to carry oxygen.

> They are disc-shaped and have a dent on both sides. This helps them to absorb a lot of oxygen.

> They are very small so they can carry oxygen to all parts of the body.

Plasma

Liquid plasma is adapted to transport water and dissolved substances such as food, hormones, antibodies and waste products around the body.

The transport vessels

> Arteries transport blood away from the heart.

> Veins transport blood to the heart.

> Capillaries join arteries to veins and carry blood through tissues. Materials such as oxygen are exchanged between the capillaries and the body tissue.

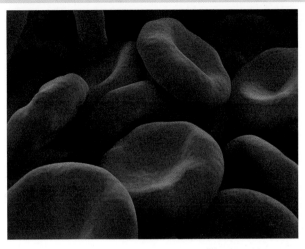

FIGURE 4: Red blood cells. Why are they disc-shaped?

Questions

3 Explain one way in which a red blood cell is adapted to its function.

4 Write down one substance transported in blood plasma.

5 Why is plasma a liquid?

6 Which type of blood vessel transports blood away from the heart?

More on red blood cells

The shape and small size of a red blood cell means it has a large surface area compared to its volume. This enables it to absorb a lot of oxygen.

Haemoglobin is a very special chemical. In the lungs it reacts with oxygen to form oxyhaemoglobin. When it reaches tissue it separates into haemoglobin and oxygen. The oxygen diffuses into the tissue cells and the red blood cells return to the lungs to pick up more oxygen.

Adaptation of blood vessels

Blood vessel	Structure	Adaptation
artery		thick muscular and elastic wall to help it withstand high blood pressure as the blood leaves the heart
vein		large lumen to help blood flow at low pressure; valves stop blood from flowing the wrong way
capillary		thin, permeable wall to allow exchange of material with body tissue

Questions

7 Why is oxyhaemoglobin such an important molecule and where does it form?

8 Explain the differences between veins and arteries relative to their functions.

9 Why are the walls of capillaries permeable?

Q platelet   red blood cell   white blood cell

What does the heart do?

The heart pumps blood around the body. There are two sides to a heart.

> The right side pumps blood to the lungs.

> The left side pumps blood to the rest of the body.

> The blood leaves the heart in arteries where the pressure is high.

> The blood returns to the heart at low pressure in veins.

Higher blood pressure in the arteries allows the blood to flow from the arteries into capillaries. The lower blood pressure in the veins allows the blood to flow from the capillaries into the veins and back to the heart.

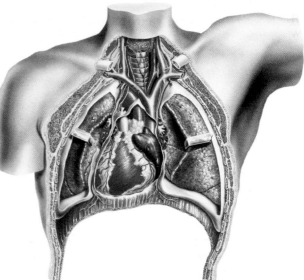

FIGURE 5: The heart and lungs. Which side of the heart pumps blood to the lungs?

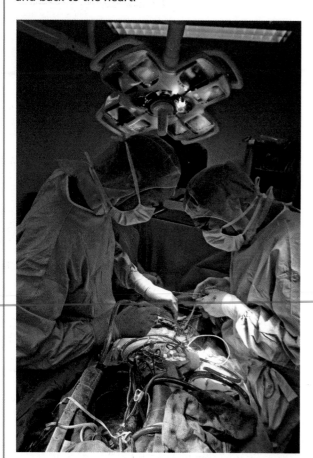

FIGURE 6: Why do some people need a heart transplant?

Did you know?

Even though a human heart weighs only 300 g it beats about 2 500 million times in a lifetime.

Questions

10 Which side of the heart pumps blood to the lungs?

11 Which side of the heart pumps blood to the rest of the body?

12 Which type of blood vessel contains blood at high pressure?

Q atrium   bicuspid valve   double circulatory system

Structure and function of the heart

There are four parts to the heart called chambers.

> The right and left atria receive blood from veins.

> The right and left ventricles pump blood into arteries.

There are also valves in the heart. They prevent the blood flowing backwards when the heart relaxes and so maintain blood pressure.

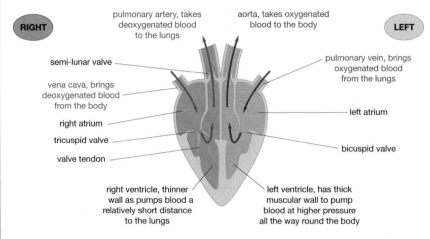

RIGHT

pulmonary artery, takes deoxygenated blood to the lungs

aorta, takes oxygenated blood to the body

LEFT

semi-lunar valve

vena cava, brings deoxygenated blood from the body

right atrium

tricuspid valve

valve tendon

pulmonary vein, brings oxygenated blood from the lungs

left atrium

bicuspid valve

right ventricle, thinner wall as pumps blood a relatively short distance to the lungs

left ventricle, has thick muscular wall to pump blood at higher pressure all the way round the body

FIGURE 7: Blood flow in the heart. Which blood vessel brings blood to the heart from the body?

Questions

13 List the structures blood passes through as it travels from the vena cava to the aorta.

14 Explain why the left ventricle has a thicker muscle wall than the right ventricle.

15 Name three different valves found in the heart and describe the function of each.

Double circulation

Humans have a **double circulatory system** of arteries, veins and capillaries.

> One circuit links the heart and lungs.

> One circuit links the heart and the body.

The heart is made up of two pumps. The advantage of this is that blood going to the body can be pumped at a much higher pressure than blood going to the lungs. This provides a greater rate of flow to all the body tissue.

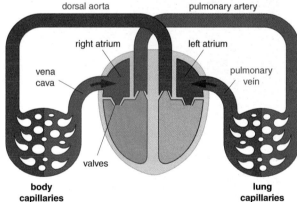

dorsal aorta

pulmonary artery

right atrium

left atrium

vena cava

pulmonary vein

valves

body capillaries

lung capillaries

FIGURE 8: Double circulatory system. In which circuit does the blood travel furthest?

Questions

16 Fish only have a single circulatory system. After the blood leaves the heart it travels to the gills and then round the body.

a Describe one difference between double and single circulatory systems.

b Describe two advantages double circulatory systems have compared to single circulatory systems.

17 Some babies are born with a hole between the two ventricles in their heart. Suggest why their muscles would receive less oxygen.

Q human heart   ventricles

Growth and development

You will find out:
> about plant cells
> about bacterial cells
> how cells can change as they grow
> about stem cells

Chicken, nerves and human stem cells

If a person damages their nerve cells their body cannot make new ones. However, scientists have found a way to grow human nerve cells in chickens.

They have taken stem cells from human bone marrow and put them into chicken embryos. Stem cells are special cells that have the ability to form all the different cells. Normally stem cells from bone marrow turn into blood cells. Scientists have found that these stem cells can be turned into nerve cells in chicken embryos.

However, scientists are still a long way off using this to replace damaged nerve cells in the human body.

FIGURE 1: How can chickens be used to grow human nerve cells?

Plant cells

A plant cell has the following parts.

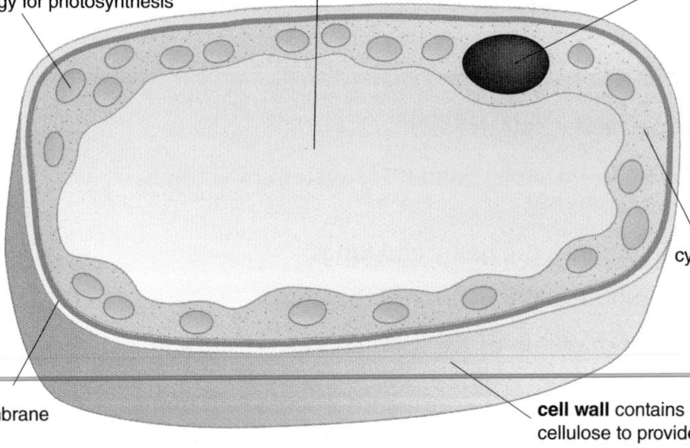

Bacterial cells

Bacteria are usually made from one cell. Bacterial cells are much smaller and simpler than plant and animal cells. They still have DNA but no nucleus.

Changing cells

When plants and animal grow, cells need to divide and change into specialised cells. The specialised cells can carry out different jobs. Some cells turn into nerve cells. Others change into bone cells. When a cell changes to become specialised the process is called **cell differentiation**.

Seeing cells

Cells can be seen using a microscope. Some of the easiest cells to see using a microscope are onion cells. To prepare a slide sample:

> peel the epidermis (skin) from the inside of an onion

> lay the piece of epidermis flat on a microscope slide

> cover the epidermis with iodine solution to stain the cells

> lower a cover slip onto the slide

> place the slide onto the stage of the microscope

> use the focus knob to get a clear image of the cells.

chloroplasts absorb light energy for photosynthesis

vacuole contains cell sap and provides support

nucleus

cytoplasm

cell membrane

cell wall contains cellulose to provide support

FIGURE 2: A plant cell. Which part of the cell contains DNA?

Questions

1 Match up each part of a cell with the job it does.

Part of cell	Job
vacuole	absorb light energy for photosynthesis
cell wall	contains cell sap and provides support
chloroplast	provides support

2 Which part of a plant cell contains cellulose?

Q bacterial cell   cell differentiation   cell wall   chloroplast

Similar but different

Both animal and plant cells have a nucleus, cytoplasm and a cell membrane. This makes them similar. However there are ways in which they are very different as shown in the table.

Plant cell	Animal cell
cellulose cell wall for support	no cell wall
most have chloroplasts for photosynthesis	no chloroplasts
large vacuole containing cell sap	may have a small vacuole but no cell sap

More about bacterial cells

The cells of bacteria do not have a 'true' nucleus. Instead bacterial DNA is found in the cytoplasm. Mitochondria and chloroplasts are also missing from bacterial cells.

Stem cells

A few days after an egg is fertilized it contains a group of undifferentiated cells called **stem cells**. These all have the same simple cell structure. They divide and then change to form all the different specialised cells in the body. As the embryo grows all the specialised cells form tissues and organs.

Stem cells can be taken from embryos and used to treat medical conditions. However, many people object to stem cell research for ethical reasons because it can involve human embryos.

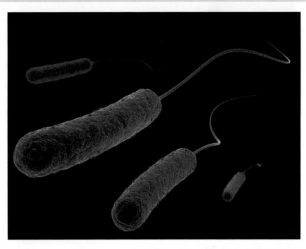

FIGURE 3: How are bacterial cells different to animal cells?

 Questions

3 Name three structures found in a plant cell but not in an animal cell.

4 Name three structures found in both an animal cell and a plant cell.

5 Write down three different specialised cells in a human body.

6 Should scientists be allowed to use embryo stem cells? Give one reason for your answer.

Bacterial DNA and stem cells

Plants and animals have several chromosomes made of DNA held within the nucleus of their cells. Bacteria are different they do not have chromosomes or a nucleus instead they have a single circular strand of DNA in the cytoplasm.

Stem cell research

Scientists have found ways of making stem cells develop into other specialised cells in the hope of replacing damaged cells. They can use either adult or embryonic stem cells. However, embryonic stem cells have the ability to specialise into most cells of the body. Adult stem cells are only able to differentiate into a few different types of cells.

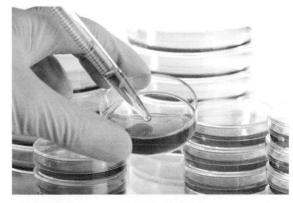

FIGURE 4: Why do some people object to stem cell culture?

 Questions

7 Explain the difference between embryonic and adult stem cells.

organ   stem cell   tissue   vacuole

Growth

Animals only grow in the early stages of their life but plants keep on growing. However, in animals the whole animal grows, but in plants only specialised parts of a plant can continue to grow.

Measuring growth

Growth can be measured by recording increases in height or mass (wet mass or dry mass).

The picture shows the growth curve of a plant. Like most growth curves it has a 'S' shape.

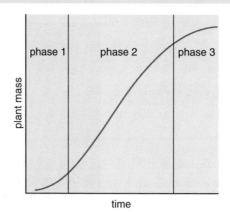

FIGURE 5: Which letter of the alphabet is the same shape as a growth curve.

FIGURE 6: How can plant growth be measured?

Questions

8 Use this graph of growth in boys and girls to answer the questions.

a Between which ages do boys grow fastest?

b Describe the change in the graph when girls reach adulthood at 18.

c Between which ages do girls grow faster than boys?

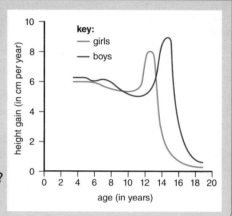

Animal and plant growth

The cells of animals and plants cause them to grow in different ways.

Animals tend to grow to a certain size and then stop. Plants can continue to grow. When plants grow cell division take place in special areas called meristems. **Meristems** are found at the tips of shoots and roots.

Plants increase in height because their cells get larger rather than just divide. Also many plant cells retain the ability to differentiate unlike animal cells.

Mass and growth

The growth of animals and plants can be measured by measuring increase in mass. This can be done in two ways. Measuring wet mass or measuring dry mass.

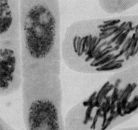

FIGURE 7: Cell division in the tip of a root of a germinating plant.

Q dry mass   growth curve

> Wet mass is the mass of the whole organisms and can be measured when the organism is alive.

> Dry mass is the mass of the organism after all the water has been removed. This can only be measured on dead organisms.

Measuring dry mass is more accurate than measuring wet mass. This is because water content can change depending on conditions. A plant may contain less water if the weather has been hot and dry. Measuring dry mass of plants shows that their growth has three phases. Growth starts slow, speeds up and then slows down again (see figure 5). Most organisms will have the same growth pattern.

Growth of humans

Measuring growth of plants shows that their growth has three phases. Humans are different they normally have two phases of rapid growth; one phase just after birth and the other during adolescence.

Did you know?

Blue whales grow very fast. A calf can have a mass of 26 tonnes before it is a year old. That is about the weight of two double-decker buses!

Questions

9 Use figure 5 to answer these questions.

a When is the rate of growth fastest, phase 1, 2 or 3?

b Suggest one reason why growth of a plant might slow down?

10 Human growth cannot be measured using dry mass. Explain why.

11 Describe one way the pattern in human growth is different to that of an apple tree.

More on measuring mass

There are advantages and disadvantages to the different methods of measuring mass.

method	advantage	disadvantage
length	quick and easy method	organism may increase in size but length may not change
wet mass	live organism can be measure	water content can change
dry mass	more accurate method	cannot measure live organism

Growth of a baby

When an organism grows some parts of the organism may grow faster than others. When a human foetus develops in the uterus the head grows faster than the rest of the body. This enables the brain to develop quickly so that it can coordinate the complex human structure and chemical activity.

foetus at 2 months   foetus at 3 months   newborn

FIGURE 8: Suggest which part of a human foetus grows the fastest during the first 2 months.

Questions

12 Explain the disadvantage of using length to measure the growth of an apple?

13 Explain why measuring dry mass is more accurate than measuring wet mass.

14 Suggest how you would find the dry mass of an apple.

Q human growth   foetal development

New genes for old

You will find out:

> how plants and animals can be changed by selective breeding
> about advantages and disadvantages of selective breeding
> how mutations happen

An apple a day could prevent tooth decay

One day tooth decay may be prevented by eating apples. Scientists have identified a protein that stops tooth-rotting bacteria sticking to teeth. They hope to place the gene that codes for the protein into apples. The apple would then make the protein and eating an apple would stop the bacteria sticking to your teeth!

Selecting the best

Farmers are always trying to make their animals and plants produce more. They choose animals and plants with the characteristics they want. For example, they choose apple trees with the tastiest apples on. Then, they breed them to produce offspring that have the characteristics. This is known as selective breeding.

Lettuces

Chickens

Cows

FIGURE 1: Write down one characteristic farmers would want in each of the animals and plants shown in the pictures.

Designing a better cow

To breed a cow that has a high yield (quantity) of creamy milk farmers carry out a **selective breeding** programme.

> Choose a Jersey cow that produces the creamiest milk.

> Choose a Friesian cow with the highest milk yield.

> Cross-breed these cows by mating a Friesian cow with a Jersey bull and a Jersey cow with a Friesian bull.

> Select the best cross-breeds, i.e. the offspring that produce large quantities of creamy milk.

> Repeat the selection and breeding process for a number of generations.

Q characteristic   cross-breed   gene pool   inbred

FIGURE 2: Jersey cows (left) produce small amounts of very creamy milk and Friesian cows (below, left) produce large amounts of slightly creamy milk.

Questions

1 Why do farmers want to plant wheat crops that grow quickly?

2 Why do farmers want pigs that produce a lot of baby pigs?

3 Name one characteristic of strawberry plants that is useful to gardeners.

4 Describe how a farmer could produce a plant with large sweet strawberries if she starts with a plant that has small sweet berries and one with large non-sweet berries.

Problems with a 'designer animal'

Selective breeding often involves animals that are closely related. This is called **inbreeding**. Inbreeding often leads to health problems within a species.

FIGURE 3: Bulldogs have been bred with large folds of skin on their faces. This is a recessive characteristic that affects their sight. Suggest reasons why a bulldog would not survive in the wild.

Questions

5 Why is variation in inbreeding a problem to some species?

Did you know?

There are over 300 different dog breeds in the world. All these different breeds exist because of selective breeding.

Inbreeding and the gene pool

Inbreeding causes a reduction in the **gene pool** (the different alleles available in a species). With a smaller gene pool there is less variation. For example, cows could lose alleles of genes that could help them survive a new disease.

Some animals are bred to show in competitions. The more an animal is selectively bred, the more chance there is of harmful recessive genes being expressed.

Questions

6 Farmers are now starting to keep many different breeds of sheep instead of just one type. Explain the advantage of keeping more than one breed.

Q selective breeding   variation   yield

Transferring genes

Scientists can take genes from one organism and put them into a different organism and get the organism to express them.

This transferring of genes is called **genetic engineering** or genetic modification (GM). The organisms made by genetic engineering have different characteristics to the original animals.

Scientists are using genetic engineering to treat people with genetic disorders like cystic fibrosis. To do this they would need to change the person's genes.

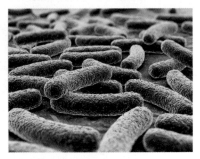

FIGURE 5: Would bacteria modified to produce insulin be useful?

Did you know?

Scientists have put a jellyfish gene into mice. This makes the mice glow in the dark!

You will find out:

> how genetic engineering is done

> about some examples of genetic engineering

> about the advantages and disadvantages of genetic engineering

FIGURE 4: Would a square tomato be useful?

Questions

7 Suggest how the following may be useful.

a Maize that is resistant to weedkiller.

b Bacteria that produce insulin.

8 Copy and complete the sentence using some of the following words.

cells   genes   genetic engineering   genetic patchwork

When a scientist takes _____ from one organism and puts them into another it is called _____ _____.

What happens in genetic engineering?

Genetic engineering involves adding a gene to the DNA of an organism. The altered organism is described as genetically modified (GM). The GM organism makes the protein the new gene codes for.

> Insulin. Some people with diabetes cannot produce enough insulin to control the level of glucose in their blood. These people rely on daily injections of insulin. Large supplies of insulin are needed. A bacterium called *E. coli* has been genetically engineered to make human insulin. The bacteria reproduce rapidly to make large quantities of insulin.

> Vitamin A. Rice is the main diet for people living in Asian countries. It does not contain vitamin A, which is needed to prevent 'night

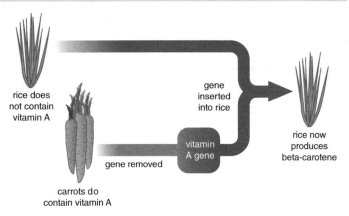

rice does not contain vitamin A

gene removed

carrots do contain vitamin A

gene inserted into rice

vitamin A gene

rice now produces beta-carotene

FIGURE 6: Genetically modified rice. What benefit does this rice bring to people living in Asian countries?

 gene therapy   genetic engineering

blindness'. The eyes of people lacking vitamin A do not adjust to dim light. Scientists have taken the gene to make beta-carotene from carrots and have put it into rice plants. Humans eating the rice can then convert the beta-carotene into vitamin A.

Gene therapy

Scientists are trying to cure genetic disorders by changing a person's genes. This is called gene therapy. One example is treating cystic fibrosis a genetic disorder that affects the lungs. People with cystic fibrosis have a faulty gene; scientists have tried to place the correct gene into the lung cells. This has helped reduce the symptoms of cystic fibrosis.

Advantages and disadvantages

Changing an organism by genetic engineering can have many uses. Scientists are able to develop crops quickly that are resistant to herbicides, frost and disease. However, these methods could cause unexpected problems. For example, some people are concerned about how our bodies will react after eating GM foods for several years.

Genetic engineering: right or wrong?

Genetic engineering has many advantages and is said to be no different from ancient selective breeding methods, but this is inaccurate. Selective breeding involves using the same or similar species. In contrast, genetic engineering allows us to combine fish, mouse, human and insect genes in the same person or animal. It is for this reason that many people have concerns about 'playing God' with nature. What do you think?

Did you know?

Make sure you understand the basic ideas about chromosomes, DNA and genes.

Questions

9 Which organisms are being used to make insulin?

10 Why would people living in Asian countries find it useful to grow rice that makes vitamin A?

11 Suggest one possible disadvantage of GM crops.

12 Would you eat GM crops? Explain your answer.

How genetic engineering works

Modifying DNA by genetic engineering follows these basic steps:

> select the characteristic

> identify and isolate the gene

> insert the gene into the chromosome of a different organism

> replicate (copy) the gene in the organism and produce the protein.

Enzymes are used to cut a gene out of an original chromosome and to splice it into a new chromosome.

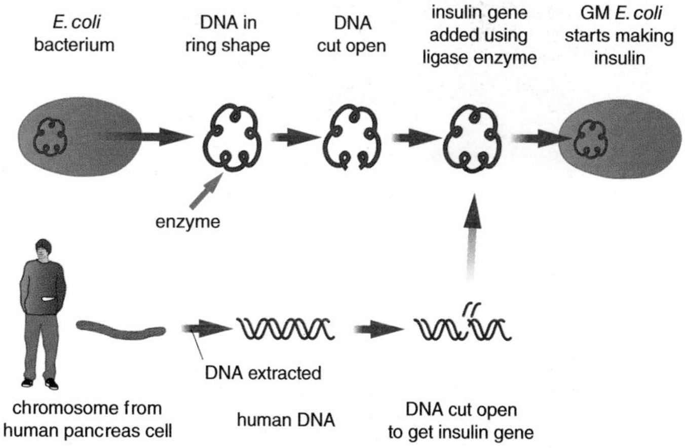

Two types of gene therapy

Genetic disorders can be treated using **gene therapy** in two different ways. Changing the genes of body cells or changing the genes of gametes before fertilisation take place. Many people are concerned about gene therapy using gametes. Can you suggest some reasons why?

FIGURE 7: Genetic engineering. Which organism is used to make human insulin?

Questions

13 Describe how genetic engineering is used to make insulin. Include in your answer ideas about isolating the gene, insertion and replication.

14 What is the difference in the cells used for the two types of gene therapy?

Cloning

You will find out:
> about asexual reproduction
> how cloning produces genetically identical copies
> how cows can be cloned
> how cloning was used to make Dolly the sheep

Meet 'Snuppy' the cloned puppy

Dolly the sheep was the first mammal to be cloned from an adult. Now the world's first cloned dog has been created. He is an Afghan hound called Snuppy. Some pet owners hope that the technique can be used to clone dead pets but scientists stress their research is for medical purposes.

FIGURE 1: What is unusual about Snuppy the dog?

Making copies

The process of cloning is used to make copies of animals and plants. The copies are called **clones**. Clones are genetically identical. They all have the same DNA as the original animal or plant.

Cloning involves only one parent. It is an example of asexual reproduction.

Natural clones

Clones are genetically identical organisms. Sometimes clones are produced naturally. Human twins can be genetically identical. They are called natural clones.

FIGURE 2: Identical twins (left) are natural clones as they have the same DNA. Why are non-identical twins (right) not clones?

Questions

1 Why is cloning an example of asexual reproduction?
2 What was the first mammal to be cloned from an adult?
3 What is special about the DNA of a clone and its parent?
4 Why are identical twins called natural clones?

Q asexual reproduction   clone cloning   dolly the sheep   natural clone

Making Dolly

Dolly was cloned using the DNA in the nucleus of udder cells. The process is called **nuclear transfer**. It involved taking a body cell from a sheep, removing the nucleus and transferring it to an egg cell.

Importance of cloning

Cloning of animals has very important uses. Cloning could result in large numbers of animals with the right characteristics, such as cows that could live in drought conditions but still provide plenty of milk.

Scientists are also hoping to clone pigs to supply organs for transplants into humans.

Human embryos could also be cloned to provide stem cells. Stem cells could be transplanted into people suffering from diabetes so that they could make their own insulin. However, people are concerned that this would be unethical because the embryo is a living thing. Some people are also afraid that scientists will eventually be able to clone adult humans.

FIGURE 3: Dolly the cloned sheep. What is the term used to describe the placing of nucleus from a body cell into an egg cell?

Questions

5 Describe one advantage of cloning large numbers of animals.

6 Why do scientists hope to clone pigs?

7 Explain why some people are concerned about human cloning.

Cloning sheep

Dolly was genetically identical to her single parent. However, this was not the sheep that gave birth to her. The diagram explains why.

Risks involved in cloning

> There is a low rate of success. There were over 200 attempts to clone a sheep before Dolly was born.

> Research into human cloning raises many moral and ethical issues about creating life and then using it to help others.

> Dolly died of conditions linked to old age, yet she was only 7 years old. Her DNA may have already been old before she was born.

Benefits of cloning

> Cloned pigs could make up for a shortage in transplant organs and patients needing a transplant would not have to wait for someone to die.

> Diseases could be cured using embryonic stem cells.

egg cell taken from sheep A and nucleus removed

cells taken from the udder of sheep B and the nucleus removed

nucleus from sheep B is put into egg of sheep A

after being given an electric shock to make it divide, the egg cell is put into a surrogate mother sheep to grow

FIGURE 4: Nuclear transfer in sheep. Why is the lamb produced a clone of sheep B and not of sheep A?

Questions

8 Explain why Dolly was not related to the sheep that gave birth to her.

9 If you needed a new heart would you object to one from a cloned pig? Give one reason for your answer.

10 A friend objects to the use of embryonic stem cells. Write a short passage to persuade them they are wrong.

Q genetically identical   nuclear transfer   stem cell   surrogate

Asexual reproduction in plants

You will find out:

> how asexual reproduction takes place in plants

> about the advantages and disadvantages of cloned plants

> about plants cloned using tissue culture

Many plants reproduce by asexual reproduction. This process produces new plants very quickly. In asexual reproduction there is no fertilisation between male and female gametes. New plants are produced using mitotic cell division only. The new plants are clones of the parent plant.

> The part of the potato we eat is the tuber. Left long enough, it will grow shoots and roots from the 'eye' (bud).

> Strawberries grow stems called runners. The runners spread over the ground and have buds that grow into tiny strawberry plants called **plantlets**. These plantlets put down roots and grow into adult plants.

> Spider plants grow new plants on their stems. The new plants are called plantlets. If the plantlets are cut off the parent plant and planted in soil they grow into adult plants.

New plants from old

Plants grown from cuttings or tissue culture are clones.

Gardeners can clone plants by taking cuttings.

> A short stem is cut off the parent plant with a sharp knife.

> The end of the stem is dipped into plant hormone rooting powder. The hormone helps the plant to grow roots.

> The cutting is put into a pot containing sandy soil.

> A polythene bag is then put over the plant to keep the moisture in.

FIGURE 5: How do potatoes reproduce? What is the process called?

Did you know?

Bananas have been cloned for so many years they have little variation. A new disease could destroy the banana species.

Questions

11 Describe how a strawberry plant reproduces asexually.

12 Describe how a new plant can be made from a cutting.

13 Describe how you could produce five clones from a potato tuber.

FIGURE 6: Taking a cutting. Why is the bag placed over the cutting?

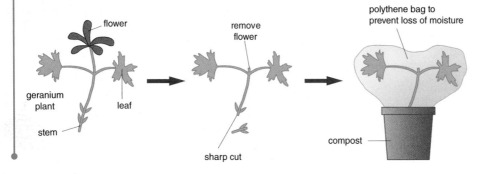

flower

geranium plant

leaf

stem

remove flower

sharp cut

polythene bag to prevent loss of moisture

compost

Q aseptic technique   cell division   cutting

Why clone plants?

Advantages

> All the plants are genetically identical. All the cuttings taken from a red rose grow into red roses.

> Plants can take a long time to grow from seeds. Cloning produces lots of identical plants more quickly.

> Cloning enables growers to produce plants that are difficult to grow from seed such as bananas.

> Can be used to preserve good characteristics.

Disadvantages

> The plants are all genetically identical. If the environment changes or a new disease breaks out all the plants would be affected in the same way. This could mean they all die. With variation some could survive because they have resistant genes.

> Cloning plants over many years has resulted in little genetic variation because there is no mixing of genes.

FIGURE 7: Gardeners are trying to grow old varieties to increase genetic variation in vegetables. Why is variation in a species beneficial?

 Questions

14 Describe one advantage of cloning plants.

15 Describe one disadvantage of cloning plants.

Tissue culture

Small sections of plant tissue can be cloned using **tissue culture**. Tissue culture must be carried out using aseptic technique (everything has to be sterile).

> Plants with the desired characteristics are chosen.

> A large number of small pieces of tissue are taken from the parent plant.

> They are put into sterile test tubes that contain growth medium.

> The tissue pieces are left in suitable conditions to grow into plants.

Animal or plant clones?

Humans have been cloning plants for hundreds of years. Animals have only been cloned over the last few years. Why is this?

> Many plant cells retain the ability to differentiate into different cells. Root cells used in tissue culture have to change into all the different types of cells found in a plant.

> Most animal cells have lost the ability to differentiate.

FIGURE 8: Tissue culture. Suggest why it is important that aseptic technique is used.

Questions

16 Explain the term aseptic technique.

17 Suggest two suitable conditions needed for tissue cultures to grow into plants.

18 Explain why strawberry plants are easier to clone than sheep.

Preparing for assessment: Research and collecting secondary data

To achieve a good grade in science, you not only have to know and understand scientific ideas, but you need to be able to apply them to other situations and investigations. These tasks will support you in developing these skills.

 Task

What factors affect recovery time after exercise?

 Context

When we exercise our heart and breathing rates increase. This is because our muscles need more oxygen for respiration.

There are two types of respiration, aerobic and anaerobic. When muscles have enough oxygen they release energy from glucose using aerobic respiration.

During exercise muscles may not get enough oxygen so they break down the glucose using anaerobic respiration. This produces a chemical called lactic acid.

After exercise our bodies need to break down the lactic acid. Oxygen is needed to do this. The amount of oxygen needed is called the oxygen debt.

When we stop exercising, it takes time for our heart and breathing rates to return to normal. This is because we are taking in extra oxygen to repay the oxygen debt. The time taken for the heart and breathing rates to return to normal is called the recovery time.

People have different recovery times. This could because they do more exercise or because they smoke. There are also many other reasons.

Matthew and Charlotte are discussing recovery times.

Matthew thinks that males all have better recovery times than females. He explains that it is because males tend to have larger lung capacities. This means they can get more oxygen into their blood.

Charlotte tells Matthew that it depends on how much exercise you do. She explains that the more exercise you do the more efficient your lungs are.

You need to do some research to find out who is correct.

 Planning

Plan how you are going to collect this information. You need to:

1. Write down how you found the information.

2. Write down a list of all the sources of your information.

3. Clearly present the information so it could be used to plan an actual investigation.

✹ General Rules

1. You may work with other students but your written work should be done on your own.

2. You cannot get detailed help from your teacher.

3. You are not allowed to redraft your work.

4. Your work can be handwritten or word processed.

5. It is expected that you complete this task in two hours.

6. You are allowed to do this research outside the laboratory.

7. You must be aware of and mention any health and safety issues.

✹ Research and collecting secondary data

To research:

- the effect of exercise on heart and breathing rates
- the use of anaerobic respiration by muscles
- oxygen debt and recovery time
- effects of gender and smoking

Information sources:

- science textbooks in the school library
- general books on exercise
- the internet

Research found:

- Recovery times for different people
- The effect of smoking on recovery times

Record:

- Use a notebook to write down the information and where it came from.
- Use a computer to record images, conversations and diagrams.

> What do you need to research?

> Where would you find this information?

> Find out about heart and breathing rates, anaerobic respiration, oxygen debt.

> Use the search terms **exercise recovery time, smoking and exercise**.

> What did you find out about?

> How do you record all this information?

> Put together or print off a final account of your research and hand it to your teacher.

B3 Checklist

To achieve your forecast grade in the exam you'll need to revise

Use this checklist to see what you can do now. It gives you many of the important points you will need to know. Refer back to the relevant pages in this book if you're not sure and to see if there is anything else you need to know. Look across the three columns to see how you can progress.

Remember you'll need to be able to use these ideas in various ways, such as:

> interpreting pictures, diagrams and graphs
> applying ideas to new situations
> explaining ethical implications
> suggesting some benefits and risks to society
> drawing conclusions from evidence you've been given.

Look at pages 278–299 for more information about exams and how you'll be assessed.

To aim for a grade E

recall that respiration occurs in the mitochondria
recall that chromosomes in the nucleus are made of DNA and contain genes
recall that the genetic code controls cell activity and the production of different proteins
recall that Watson and Crick worked out the structure of DNA

recall some examples of proteins to include: collagen, insulin and haemoglobin
describe enzymes as protein molecules that speed up a chemical reaction
understand that substrate molecules fit into the active site
describe gene mutations as changes to genes

recall the word equation for aerobic respiration
describe examples of life processes that require energy
describe why during exercise breathing and pulse rates increase

To aim for a grade C

explain why liver and muscle cells have large numbers of mitochondria
recall that DNA is a double helix that contains four different bases
recall that each gene contains a different sequence of bases
recall that each gene codes for a particular protein
describe how Watson and Crick used data from other scientists to build a model of DNA

recognise that proteins are made of long chains of amino acids
describe some jobs of proteins
describe how temperature and pH will change the rate of an enzyme-catalysed reaction
describe the 'lock and key' mechanism
recall that mutations occur spontaneously or by radiation or chemicals

recall the symbol equation for aerobic respiration
use data from experiments to compare respiration rates
calculate respiratory quotient
explain why anaerobic and aerobic respiration takes place during hard exercise
recall the word equation for anaerobic respiration

To aim for a grade A

recall that ribosomes are the site of protein synthesis
describe the complementary base pairings: A–T and G–C
explain how DNA base code determines protein structure
explain the role of ribosomes and mRNA in protein synthesis
explain why new discoveries, such as Watson and Crick's, are not accepted immediately

explain how each protein has its own sequence of amino acids resulting in a specific shape
explain how enzyme activity is affected by pH and temperature
calculate and interpret the Q_{10} value
explain that only some genes are switched off in some cells

recall that ATP is made in respiration
explain how to measure metabolic rate
explain why temperature and pH affect rate of respiration
explain fatigue in terms of lactic acid build up (oxygen debt) and how this is removed during recovery

To aim for a grade E

describe the difference between unicellular and multicellular organisms

understand that to produce new cells the chromosomes are copied

recall that in sexual reproduction gametes join in fertilisation

describe the job of cells in the blood

recall that the blood moves around the body

describe the job of the heart

recall blood in arteries is under higher pressure than in veins

explain blood flow in terms of pressure difference

describe the job of parts of a plant cell

recall how to measure growth

interpret data on a growth curve

recall that the process of cells becoming specialised is called differentiation

describe the difference between animal and plant growth

describe the process of selective breeding

describe the transfer of genes from one organism to another as genetic engineering

recognise that it may be possible to use genetic engineering to change a person's genes to cure certain disorders

recall that cloning is an example of asexual reproduction and produces genetically identical copies

describe how spider plants, potatoes and strawberries reproduce asexually

describe how to take a cutting

To aim for a grade C

explain the advantages of being multicellular

recall growth involves mitosis

recall that body cells are diploid

describe gametes as haploid and produced by meiosis

explain why fertilisation results in genetic variation

explain how the structure of a sperm cell is adapted to its job

explain how the structure of a red blood cell is adapted to its function

describe the function of plasma

describe how the parts of the circulatory system transport substances around the body

identify the parts of the heart and describe their functions

explain why the left ventricle has a thicker muscle wall than the right ventricle

identify simple differences between bacterial cells and plant and animal cells

describe the main phases of a typical growth curve

recall that humans have two phases of rapid growth

recall that stem cells can be obtained from embryonic tissue and could potentially be used to treat medical conditions

recognise that a selective breeding programme may lead to inbreeding

explain some advantages and risks of genetic engineering

describe examples of genetic engineering

recall the meaning of the term gene therapy

describe the process of nuclear transfer that was used to produce Dolly

describe possible uses of cloning

describe the advantages and disadvantages associated with the commercial use of cloned plants

To aim for a grade A

explain why multicellular organisms require specialised organ systems

describe how DNA replicates

describe mitosis

explain why, in meiosis, the chromosome number is halved and each cell is genetically different

explain how red blood cells are adapted to their function

describe how haemoglobin is used to transport oxygen

explain how the adaptations of arteries, veins and capillaries relate to their functions

explain the advantage of the double circulatory system in mammals

describe the difference between DNA in a bacterial cell and a plant/animal cell

explain why the growth of different parts of an organism may differ from the growth rate of the whole organism

explain the difference between adult and embryonic stem cells

explain how a selective breeding programme may reduce the gene pool

understand the principles of genetic engineering

recall that gene therapy could involve body cells or gametes

discuss issues arising from gene therapy involving gametes

describe in outline the cloning technique used to produce Dolly

describe the benefits and risks of using cloning technology

describe plant cloning by tissue culture

explain why cloning plants is easier than cloning animals

B3 Exam-style questions

Foundation Tier

1 Zainab and James investigate exercise and pulse rate. They measure Zainab's resting pulse and then she exercises for two minutes. After exercise they measure her pulse again. When Zainab's pulse rate returns to normal they repeat the process but this time she exercises longer.

Look at the table; it shows their results.

length of exercise in minutes	pulse rate in beats per minute
0	78
2	84
4	96
6	105
8	116

AO2 **(a) (i)** Describe the pattern in the results. [1]

AO2 **(ii)** Use ideas about oxygen and carbon dioxide to explain the results. [3]

AO2 **(b)** James thinks the experiment went well but Zainab thinks it could be improved. Suggest one way they could improve the method to make the results more valid [1]

[Total: 5]

2 Look at the picture of a **red** geranium plant.

AO1 **(a) (i)** Describe how you could take a cutting from the geranium. [3]

AO2 **(ii)** What colour flowers would all your cuttings produce? Explain your answer. [2]

AO1 **(b)** Taking cuttings is a method of cloning Describe one advantage and one disadvantage of cloning plants. [2]

[Total: 7]

3 Sue investigates plant cells. She starts by making a microscope slide of some onion cells.

AO1 **(a)** Describe how Sue could make a microscope slide of some onion cells. [3]

AO1 **(b)** Sue draws some of the cells that she sees on her slide. Look at the picture of her drawing.

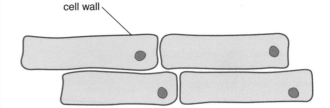

cell wall

AO1 **(i)** Describe the job of the cell wall [1]

AO2 **(ii)** Plant cells normally have chloroplasts. Suggest why they are not in onion cells. [1]

AO1 **(c)** Sue also investigates bacterial cells. She finds out that bacterial cells are much smaller than onion cells. Describe **one other** way onion cells are different to bacterial cells. [1]

[Total: 6]

AO1 **4 (a)** Kevin is a horse breeder. He breeds horses for racing. He tries to breed the fastest horses. The first step is to breed a fast female horse with a fast male horse. Describe the next stages. [2]

AO1 **(b)** Kevin tries to make sure that the male and female horses he breeds together do not have the same parents. Explain why. [2]

[Total: 4]

AO1 recall the science   AO2 apply your knowledge   AO3 evaluate and analyse the evidence

✳ Worked Example – Foundation Tier

Look at the pictures of a baby and a seedling.

The baby and the seedling will both grow into adults. The growth of animals and plants are similar in some ways and different in others.

Describe the how growth the baby and the seedling is similar and how it is different.

The quality of written communication will be assessed in your answer to this question. [6]

*Both animals and plants grow when their **cells divide**.*

After the cells divide they become change into different types of cell. This means they change into cells that do certain jobs like blood cells or leaf cells.

*To grow they both need **proteins**. The plants and the animals make proteins using the **DNA** code.*

The baby will grow until it reaches adult age then it will stop growing in height. The seedling will continue to grow until it dies.

Every part of the baby grows but the seedling only grows at special points.

The baby will also have growth spurts when there will be a sudden increase in growth. With plants they tend to grow more when the conditions are better. For example, when it is warmer in the summer.

Remember!

This question is about growth but don't just write about facts from the Growth and Development section. Growth involves cell division so extend your ideas to include information from Cell Division and Molecules of Life.

How to raise your grade!

Take note of these comments – they will help you to raise your grade.

Cells dividing to make new cells is an E grade statement. A grade C student should mention mitosis.

This is describing the term cell differentiation. Using the term would help improve your communication mark.

At grade C it should be mentioned that in an adult plant the most of the cells can still become specialised but not in an adult animal.

Explaining that these special points are called meristems and that they are found at the root and shoot tips would extend this idea to a grade C answer.

This part brings in ideas from other parts of the specification. 3/6

These longer 6 mark answers usually have marks awarded for the Quality of Written Communication shown by this symbol ✐ so answers need planning, and care is needed with spelling, punctuation and grammar. (See the example banded mark scheme on page 297.)

For the most part the information is relevant and presented in a structured and coherent format.

Specialist terms (shown in bold) are used although there are only a few; cell differentiation, specialised cells, mitosis could have been used. There are occasional errors in grammar, punctuation and spelling.

This student has scored 3 marks out of a possible 6. This is below the standard of Grade C. With a little more care the student could have achieved a Grade C.

B3 Exam-style questions

Higher Tier

1 Chloe and Muhammad investigate the enzyme amylase. They add 15cm³ of starch solution to a boiling tube containing 1cm³ of amylase solution at 20 °C. They then add drops of iodine solution to a spotting tile. Chloe and Muhammad time how long it takes for the starch to breakdown. They do this by taking a drop of the mixture and adding it to the iodine solution in the spotting tile. This is done every minute until the iodine no longer turns blue-black.

Look at the table; it shows their results.

temperature in °C	time for starch to breakdown in minutes			
	first attempt	second attempt	third attempt	mean time
20	12	11	12	11.7
30	7	6	9	7.3
40	4	3	4	
50	9	8	5	7.3
60	19	21	17	19.0

AO2 **(a) (i)** Calculate the mean time for 40 °C [1]

AO2 **(ii)** Draw a graph to show the mean times against temperature. [3]

AO2 **(b) (i)** Describe the pattern in the results. [1]

AO2 **(ii)** Use ideas about enzymes to explain the results. [3]

(c) Chloe thinks the experiment went well but Muhammad thinks they have an anomalous result.

AO3 **(i)** Which result is an anomaly? [1]

AO3 **(ii)** Suggest what might have caused the anomaly [1]

[Total: 10]

2 Look at the picture of a bacterial cell.

AO1 **(a)** Write down one structure found in plant and animal cells that is not found in a bacterial cell. [1]

AO1 **(b)** Bacteria reproduce by dividing. The bacterial DNA has to replicate before the cell can divide. Describe how DNA replicates. [2]

AO2 **(c)** Bacteria can be genetically engineered to produce human insulin. The gene for human insulin is placed inside the bacterial DNA. Explain how the bacteria can use this gene to make human insulin [4]

[Total: 7]

3 Look at the picture of a fish heart.

AO2 **(a)** Describe how the fish heart is different to a human heart. [2]

AO1 **(b)** Fish hearts have valves. What is the function of the valve? [1]

AO2 **(c)** Fish have a single circulatory instead of a double circulatory system found in humans. This means the rate of blood flow to the tissue is less. Explain why the rate of blood flow to the tissue is less in a single circulatory system. [1]

ventricle

atrium

[Total: 4]

4 Organisms can use different types of food in aerobic respiration.
Look at the balanced symbol equation using a fat.

$$C_{25}H_{100}O_6 + 77O_2 \rightarrow 55CO_2 + 50H_2O$$

AO2 **(a) (i)** Use the following equation to calculate the RQ value for respiration of this fat. [2]

$$RQ = \frac{\text{carbon dioxide produced}}{\text{oxygen used}}$$

AO2 **(ii)** Suggest how the RQ value calculated in part (a) would change if it was aerobic respiration of glucose. [1]

AO2 **(iii)** Explain your answer. [1]

[Total: 4]

AO1 recall the science AO2 apply your knowledge AO3 evaluate and analyse the evidence

✱ Worked Example – Higher Tier

The diagram shows what happens to urine concentration when too much water is lost from the body by sweating.

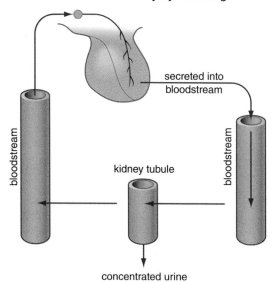

secreted into bloodstream

bloodstream

bloodstream

kidney tubule

concentrated urine

(a) (i) Which hormone is involved in this process and where is it produced? [2]

Anti-diuretic hormone (ADH) is involved. It is produced in the brain.

> The student has correctly named the hormone but it is produced by the pituitary gland. 1/2

(ii) Use the diagram to help you explain how the water concentration on the blood is restored. [4]

Changes in the blood concentration are detected by the brain. ADH is secreted and is carried by the blood to the kidney. ADH increases the permeability of the kidney. Therefore more water is reabsorbed back into the blood.

> The student's answer is too vague. The answer should have referred to a high blood concentration, i.e. too little water present.

> The student has referred to the kidney rather than the kidney tubules.

> This part is correct. 2/4

(b) A dialysis machine is used when a patient has kidney failure.

Explain how the machine removes waste from the blood. [3]

The artificial kidney removes urea from the blood. The urea molecules diffuse out of the blood and into fluid in the machine.

> The student has correctly stated that the waste is urea and that it diffuses out of the blood. However, this is not a full answer. The answer should have included information about partially permeable membranes allowing the diffusion of small molecules such as urea. 2/3

> This student has scored 5 marks out of a possible 9. This is below the standard of Grade A. With more care the student could have achieved a Grade A. (See example banded mark scheme on page 297.)

B4 It's a green world

Ideas you've met before

Organisation of cells

Plant cells make up tissues, e.g. storage tissue.

Different tissues form an organ, e.g. a leaf.

Different organs work together in a plant system, e.g. a water transport system.

 Name three main parts of a plant.

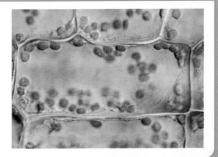

Variation

All living things show variation and are interdependent, interacting with each other and the environment.

Living organisms have similarities because they are alive.

Living organisms show differences even though they are the same species. This is variation.

 Suggest two ways in which sunflowers can be different from each other.

Behaviour

Behaviour is influenced by internal and external factors.

All organisms respond to stimuli.

Some reactions are by instinct, e.g. reflexes.

Some actions can be learnt, some actions can be conditioned.

 What other types of weather conditions can affect plants?

The environment, Earth and Universe

Human activity and natural processes can lead to changes in the environment.

 Name two other examples of human activity that can change the environment.

In B4 you will find out about...

> the parts of a plant cell

> the names and location of different cells and tissues in a leaf and stem

> how the cellular structure of a leaf is adapted for efficient photosynthesis

> the effects of the uptake and loss of water on plant and animal cells

> the use of keys to identify plants and animals

> the variety of organisms in a small area

> the biodiversity of natural and artificial ecosystems

> the advantages and disadvantages of biological decay

> why plant leaves have different arrangements to reduce excessive water loss

> how the distribution of organisms is affected by other living organisms

> how to map the distribution of organisms in a habitat

> why plants wilt when they are short of water

> about intensive farming methods

> how intensive farming can cause harm to the environment

> the use of fertilisers and pesticides in farming

> overall gaseous exchange in plants during the day and night

Ecology in the local environment

You will find out:
> about ecosystems
> kite diagrams
> about biodiversity

How many red squirrels are there in the UK?

In January 2006, the UK Government introduced a culling programme to control the grey squirrel. It is an introduced species and is rapidly replacing our native red squirrel. It is estimated that there are only about 140 000 red squirrels left in the UK.

But exactly how many are there?

FIGURE 1: Why are the numbers of red squirrels in the UK decreasing?

Biodiversity

Living organisms are found where conditions are best for them. Their distribution is affected by:

> the presence of other living organisms such as predators or herbivores, which eat them

> physical factors such as soil type and water availability.

The variety of different species living in a habitat is called **biodiversity**.

Ecosystems

Ecosystems can be:

> natural, such as lakes and native woodland

> or artificial, such as fish farms and forestry plantations.

A native woodland contains plants that have grown there naturally and have not been introduced by humans. Artificial ecosystems have been made by humans, often using plants and animals from other countries.

FIGURE 2: Natural woodland and a forestry plantation. How can you tell which is which?

Did you know?

The Amazon rainforest is one of the most important ecosystems in the world. Yet, an area the size of Wales is being destroyed each year by man.

Questions

1 Write down three examples of: **a** natural ecosystems and **b** artificial ecosystems.

2 Name three animals that can live in a woodland ecosystem.

Ecosystems

A garden is an example of an ecosystem. All the living things in the garden and their surroundings make up the ecosystem. Where a plant or animal lives is called its **habitat**. In the garden, the worm's habitat is the soil. All the animals and plants living in the garden make up the **community**.

The number of a particular organism present in the community is called its **population**.

Distribution of organisms

The distribution of organisms in a habitat can be mapped using string as a **transect** line. At regular intervals along the line the number of organisms can be counted (for animals) and percentage cover estimated (for plants) on each side of the line using a five point scale. The results are plotted to give **kite diagrams**.

Biodiversity

Natural ecosystems such as a natural woodland and lake will both contain a large variety of different plants and animals, and have a large **biodiversity**.

Artificial ecosystems such as forestry plantations and fish farms have a very small biodiversity.

> A forestry plantation will often contain only one species of tree and a limited number of animals.

> A fish farm will contain only one type of fish.

FIGURE 3: Kite diagram showing distribution of plants and animals near a woodland path. Which plants did not grow near the path?

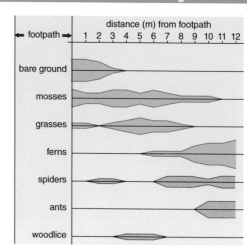

Questions

3 Explain the difference between an ecosystem and a habitat.

4 For animals, what is measured to produce a kite diagram?

5 Which plants increased their percentage cover away from the path?

Remember!
An ecosystem describes living things and their environment. A habitat describes the place where the living things live.

Self supporting systems

Natural woodlands and lakes have a large biodiversity. The variety of trees and shrubs in woodland and aquatic plants in lakes will provide:

> a wide variety of food

> food throughout the year

> shelter at different heights above ground or depth in water

> a natural environment, without artificial pesticides and fertilisers, which could encourage some organisms but harms others.

A forestry plantation has a small biodiversity. Only a small number of different tree species are planted so food supplies are limited and not varied enough, resulting in a small number of animal species.

Small shrubs and plants such as grasses cannot compete for light. Fish farms also have a small biodiversity since they farm only one type of fish, discouraging any other living organisms.

However, all ecosystems are self supporting, apart from an energy source which is usually the Sun. The Sun's energy is used by green plants in photosynthesis to make food chains. Food chains and food webs show that all animals directly or indirectly depend on plants for food. Plants depend on animals for pollination and seed dispersal. Elements such as nitrogen and carbon are also recycled. Plants and animals are also dependent on each other for supplies of oxygen and carbon dioxide.

Distribution of organisms

The kite diagram in Figure 3 shows the **zonation** of plants and animals across a habitat. The zonation is caused by abiotic (not living) factors such as availability of water, exposure and pH. Simple, small plants such as mosses can withstand dry and poor conditions near a footpath. Ferns prefer wetter and more protected conditions away from the path. The gradual change in abiotic factors causes this zonation.

FIGURE 4: A fish farm. The fish are kept in submerged cages. Why do they grow faster than fish in the wild?

Questions

6 Availability of water, exposure and pH are abiotic factors.

Name one other abiotic factor which affects plants and describe its effect.

7 Explain why a forestry plantation of pine trees will have a low biodiversity.

🔍 transect line   kite diagrams   biology

Collecting living things

Sam wanted to find out what animals lived on the school field. He and his classmates tried to catch the animals using:

> pooters

> nets

> pitfall traps.

FIGURE 5: Animals must be handled carefully and released back into their habitat at the end of the experiment.

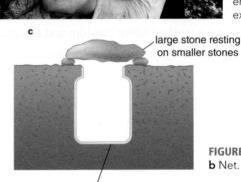

a   b   c

large stone resting on smaller stones

jam jar buried in ground

FIGURE 6: a Pooter. **b** Net. **c** Pitfall trap.

Pooter – a small re-sealable jar used for collecting insects. The lid contains two tubes. The end of one tube is covered with a fine mesh, with the other end placed in the user's mouth. The second tube is placed over the insect, the user then sucks on the first tube drawing the insects into the jar.

Nets – commonly used to collect butterflies and other airborne insects. This is a form of active collecting.

Pitfall trap – a small jar is buried within the ground and lightly covered with leaves or grass. Small insects, amphibians and reptiles fall in and cannot escape. This is called passive collecting.

A   B   C

FIGURE 7: These are the animals that Sam and his classmates found.

He identified the animals using an identification key.

To find out which plants were growing, Sam used a square frame called a quadrat. He threw the quadrat over his shoulder so it landed at random. Where it landed he counted and recorded the number of different types of plants and the numbers of each type. He also used a pooter to collect the small animals in the quadrat for identification.

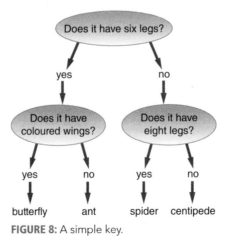

Does it have six legs?

yes   no

Does it have coloured wings?   Does it have eight legs?

yes   no   yes   no

butterfly   ant   spider   centipede

FIGURE 8: A simple key.

FIGURE 9: What did Sam do to make sure his sample was random?

Header at top right

Questions

8 Use the key to identify the three animals Sam caught.

9 Which of Sam's catching methods do you think caught the butterfly?

10 Ants crawl along the ground. Which method do you think Sam used to catch the ant?

Counting animals

Counting one type of animal in a habitat is difficult, as they do not stand still. To estimate the population of a species, scientists can count the number of organisms in a small area, such as 1m², then scale up to the actual size by multiplication.

Scientists can also use a method called **'capture and recapture'**.

> The animals are trapped, recorded and marked in some harmless way.

> They are then released and the traps are set again a few days later. The animals with and without markings are recorded.

To estimate a population size the following formula is used.

$$\text{population size} = \frac{\text{number in 1}^{st}\text{ sample} \times \text{number in 2}^{nd}\text{ sample}}{\text{number in 2}^{nd}\text{ sample previously marked}}$$

This can only be an estimate as not all the animals are counted.

FIGURE 10: Why is this moth trap hanging above the ground?

Remember!
A population is a group of animals or plants of the same species.
A community is lots of different species living in the same ecosystem.

Questions

11 Scientists want to know the number of moths living in a wood. They set traps and catch 200 moths. They mark the moths' under-wings with a harmless spot of red paint and release them. Two days later they set the traps again and catch 150 moths. Fifty of them have paint on their under-wing.

a Suggest why the paint is not put on the upper surface of the wing.

b Estimate the population of moths.

12 Suggest two ways in which a population may increase overnight.

13 Explain the difference between a population and a community.

Counting plants

Scientists have to remember that their samples may be unrepresentative of the population as a whole. Population sizes are always changing because:

> animals are being born and others are dying

> there is movement of animals in and out of an ecosystem.

A small error in a small sample will result in a large error when multiplied to the actual area being investigated.

To increase the accuracy of an estimate the process is repeated several times and the sample size is as large as possible. When such techniques as capture and recapture are used it is assumed that:

> there are no deaths, immigration or emigration

> identical sampling methods are used each time so comparisons can be made

> that marking animals does not affect their survival by damaging their bodies or affecting their camouflage.

Questions

14 When estimating the population of plants in a habitat, suggest why it is unreliable to place just one quadrat.

15 Suggest why it is important to place quadrats in a random way rather than choose an area.

Q sampling   population

Photosynthesis

You will find out:

> how plants make food using photosynthesis

> what plants need for photosynthesis

> how plants use the products of photosynthesis

Killer trees

One of the most dangerous trees in the world grows in the Caribbean. It is called the Manchineel tree. It looks like an apple tree but its small green fruit can kill.

Contact with the sap causes an eruption of blisters and can cause blindness if it gets into the eyes. The sap was used by the Carib Indians to poison their arrow tips for hunting.

However, like all trees, the Manchineel tree does a useful job, it photosynthesises.

FIGURE 1: The Manchineel tree.

Plants and photosynthesis

Ben likes to grow carrots to enter in the local garden show. He knows that to grow well, his carrots need to carry out lots of **photosynthesis**. Photosynthesis means using light (photo-) to make food (-synthesis).

The carrots' leaves carry out photosynthesis using light energy from the Sun. They contain a green pigment called **chlorophyll**.

Photosynthesis can be described using the following word equation:

$$\text{carbon dioxide} + \text{water} \xrightarrow[\text{(chlorophyll)}]{\text{(light energy)}} \text{glucose} + \text{oxygen}$$

Glucose and sucrose are **soluble** sugars. Soluble sugars are transported to all parts of the plant where they are needed. Glucose can be converted into starch for storage. Starch is better than glucose for storage because it is **insoluble**. Ben's carrots store the starch in their large orange roots.

Both glucose and starch can be converted into other substances. These can then be used for energy, growth and other storage products.

A plant also produces oxygen as a waste product of photosynthesis. This oxygen is very important to both plants and animals. They need it for **respiration**.

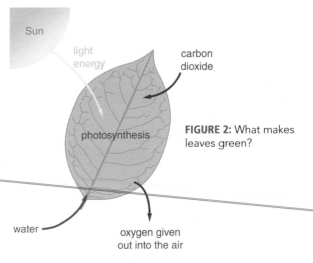

FIGURE 2: What makes leaves green?

Did you know?

Centuries ago, cultivated carrots were purple. Breeding experiments have produced the popular orange colour.

Questions

1 Copy and complete the following sentence.

Plants need c _ _ _ _ _ dioxide, l _ _ _ _ and w _ _ _ _ for photosynthesis.

2 Name the two things plants make during photosynthesis.

3 What do plants do to glucose so that they can store it?

Balancing the photosynthesis equation

The chemical symbol for glucose is $C_6H_{12}O_6$.

Therefore the overall balanced chemical equation for photosynthesis is:

$$6CO_2 + 6H_2O \xrightarrow[\text{(chlorophyll)}]{\text{(light energy)}} C_6H_{12}O_6 + 6O_2$$

Products of photosynthesis

The products of photosynthesis are glucose and oxygen. The glucose can be used by the plant as energy. The energy can be released from, which may use up some of the oxygen. The rest of the oxygen is released into the air and used by other organisms.

FIGURE 3: Harvesting olives.

Other uses of glucose

Not all of the oxygen and glucose is used up in respiration. Glucose and starch are converted into:

> cellulose to make new cell walls

> proteins for growth and repair

> fats and oils are stored for future use.

This is how we get vegetable oil, used in cooking.

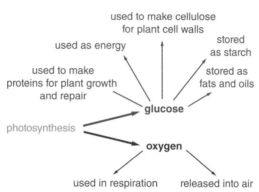

FIGURE 4: What are the two direct products of photosynthesis?

Questions

4 Why can sugars such as sucrose be carried around the plant easily?

5 Apart from starch, name three substances that glucose can be converted to by plants.

More about photosynthesis

Glucose ($C_6H_{12}O_6$) is a small, soluble molecule unsuitable for storage. This is because it easily dissolves and can be lost from the cell. It also increases the concentration of the cell contents, which could damage the cell. For storage, the glucose molecules are joined together to make larger molecules of starch. Starch:

> is insoluble and therefore not easily lost from the cell in solution

> is not very reactive making it a good storage molecule

> does not affect the water concentration inside cells.

Photosynthesis is a two stage process. Experiments using isotopes have shown that the light energy is used to split water, not carbon dioxide as originally thought. The water is split up into oxygen gas and hydrogen ions. Isotopes are different forms of the same element and easy to detect.

A single celled green algae called *Chlorella* was used in experiments on photosynthesis. *Chlorella* was supplied with water whose oxygen atom was replaced by an isotope of oxygen, ^{18}O. The oxygen isotope was in the oxygen given off by the plant. When *Chlorella* was supplied with the oxygen isotope as

part of carbon dioxide instead of being part of the water, none of the oxygen isotope was given off. This experiment proves that the plant splits the water and not the carbon dioxide during photosynthesis.

In the second stage of photosynthesis, carbon dioxide gas combines with the hydrogen from the split up water, eventually forming glucose.

starch is insoluble and is made up of many glucose molecules joined together

glucose molecules are soluble

FIGURE 5: Starch is a large molecule and glucose is a small molecule.

Questions

6 Explain why being insoluble makes starch a better storage molecule than glucose.

7 Name the elements present in a molecule of starch.

You will find out:

> what factors increase the rate of photosynthesis
> what factors limit the rate of photosynthesis
> when plants carry out respiration

Summer growth

Ben also grows peas in his garden. He likes to harvest the peas in June, so he needs to plant the seeds in February.

February can be very cold so the pea seeds have to be covered by glass to keep them warm. Ben does not see much growth at first.

When spring and summer arrive the pea plants get more light for photosynthesis. The extra light and warmth help the peas grow faster, ready for picking in June.

The higher temperature increases the rate of reaction of enzymes involved in photosynthesis.

FIGURE 6: Why are dull cold days in winter not good for growing plants?

FIGURE 7: Why are bright warm days good for growing plants?

Remember!

Plants grow faster when they can photosynthesise faster. Plants photosynthesise faster on hot sunny days.

Questions

8 Give two reasons why plants grow faster in the spring and summer months.

9 Why do the pea plants need to be covered by glass during February?

A photosynthesis time line

Greek scientists believed that plants gained mass by taking minerals from the soil.

1600: Van Helmont grew a willow tree in a plant pot for 5 years. It showed a large increase in mass but there was little change in the mass of the soil. The increase could not be solely due to soil minerals.

1771: Joseph Priestly put a part of a mint plant in a transparent closed space with a burning candle (which used up the oxygen). After 27 days he was able to relight the candle. The plant had produced a gas (oxygen).

Conditions that speed up photosynthesis

Ben has a greenhouse in his garden. Plants grow better in a greenhouse because the conditions are better for photosynthesis.

The following actions increase the rate of photosynthesis:

> keeping the greenhouse warm
> providing the greenhouse with extra carbon dioxide
> increasing the amount of light in the greenhouse.

Plants and respiration

In respiration, oxygen is used to release energy from glucose. At the same time it releases carbon dioxide and water. This makes the exchange of gases in respiration the reverse of photosynthesis.

Unlike photosynthesis, respiration never stops because the plant needs the constant supply of energy it releases to remain alive.

FIGURE 8: Van Helmont's experiment. How much mass did the plant gain and how much did the soil lose?

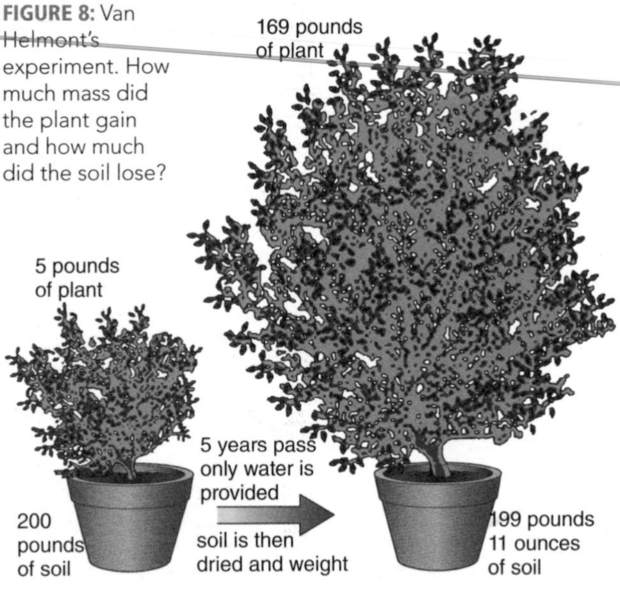
169 pounds of plant
5 pounds of plant
5 years pass only water is provided
200 pounds of soil
soil is then dried and weight
199 pounds 11 ounces of soil

Limiting factors in photosynthesis rate of photosynthesis

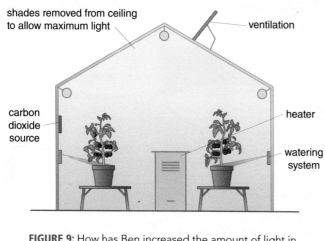

shades removed from ceiling to allow maximum light

ventilation

carbon dioxide source

heater

watering system

FIGURE 9: How has Ben increased the amount of light in his greenhouse?

Questions

10 Paraffin burns to release carbon dioxide. Give two reasons why a paraffin heater would be useful in a greenhouse.

11 What would happen to a plant if it stopped carrying out respiration?

Factors that limit photosynthesis

Experiments can be done to find out what speeds up photosynthesis and what limits it.

Some results are shown in the graph.

Scientists found that:

> the rate of photosynthesis increases as the light intensity increases until point B is reached. Up to this point, light is the limiting factor.

> at point B, the rate of photosynthesis stays the same, because something else is limiting it

> the **limiting factor** could be carbon dioxide level or temperature

> carbon dioxide level, light intensity and temperature are all limiting factors of photosynthesis.

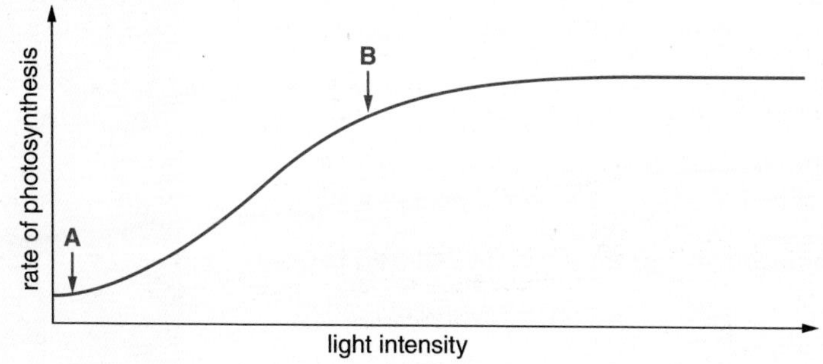

Gas exchange

As long as a plant is photosynthesising it needs to take in carbon dioxide through its leaves. At the same time it will release oxygen.

At night the plant still needs oxygen for respiration. It takes in oxygen from the air and releases carbon dioxide.

FIGURE 10: Why does the curve in the graph flatten out?

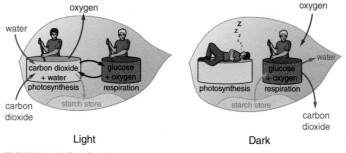

Light

Dark

FIGURE 11: Why do plants not photosynthesise at night?

Questions

12 Name three factors that could limit photosynthesis.

13 Copy the graph showing the effect of light intensity on photosynthesis. Sketch a line to show the results you would expect if the experiment was repeated at a higher carbon dioxide level.

14 A plant produces carbon dioxide from respiration 24 hours a day. Explain why it releases more carbon dioxide in the dark.

Leaves and photosynthesis

You will find out:

> about the contents of plant cells
> about the cells in leaves
> why leaves are green

Giant plants

The Giant Amazon Water Lily has the largest leaves of any known plant. Each leaf can grow to 2.5 m across and floats on the water surface.

The leaves can support the weight of some animals – even a human baby.

FIGURE 1: These leaves would not support an adult so don't try to walk on them.

Green plants

Sean is a gardener. He likes to grow lots of different plants in his garden to show lots of different leaf shapes.

Most plant leaves and some plant stems are green.

The green colour of plants is caused by green **chloroplasts** inside some cells in the stems and leaves. Cells in plant roots do not contain chloroplasts because they are not exposed to light.

The chloroplasts contain green pigments called chlorophyll.

Reactions inside the chloroplasts make sugars and starch. To provide energy for these reactions the chlorophyll pigments in the chloroplasts absorb light energy which is usually from the Sun and artificial light.

The process is called **photosynthesis** because light ('photo-') is used to make ('synthesis') molecules such as glucose.

FIGURE 2: There are lots of different leaf shapes. What do all the leaves have in common?

Questions

1 Which parts of a plant can be green?

2 Which parts of a plant cell are green?

3 What process takes place in chloroplasts?

4 What type of energy is used by chloroplasts?

cell wall, gives support

green chloroplasts

FIGURE 3: The structure of a plant cell. Which part of a plant cell provides support?

vacuole, holds water, gives shape

nucleus, controls cell functions

plant stomata   plant cells

The structure of a leaf

The whole of a leaf looks green but only certain leaf cells contain chloroplasts.

These cells are in the **palisade** and **spongy mesophyll** layers. Guard cells also contain chloroplasts. Chloroplasts contain green pigments called **chlorophyll**.

Guard cells are specialised cells. They surround small pores called **stomata** in the leaf surfaces. They occur mostly on the underside of a leaf.

The vascular bundles are also made up of specialised cells. They are adapted to carry water and food.

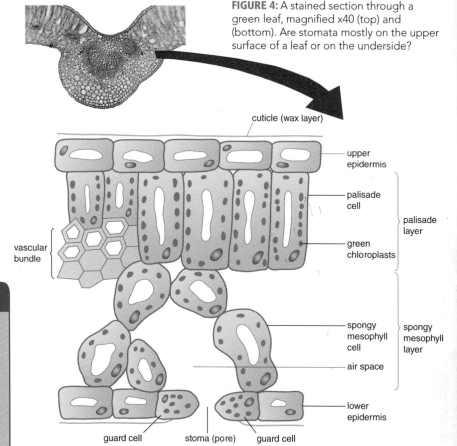

FIGURE 4: A stained section through a green leaf, magnified x40 (top) and (bottom). Are stomata mostly on the upper surface of a leaf or on the underside?

cuticle (wax layer)

upper epidermis

palisade cell

palisade layer

green chloroplasts

vascular bundle

spongy mesophyll cell

spongy mesophyll layer

air space

lower epidermis

guard cell   stoma (pore)   guard cell

Questions

5 Name three types of cells that photosynthesise.

6 How can a cell be tested for photosynthesis?

7 What are the jobs of the cells in vascular bundles?

8 What are stomata?

How a leaf is adapted to photosynthesise

The main function of a leaf is photosynthesis. The structure and arrangement of cells are adapted for maximum efficiency.

> The epidermis is thin and transparent, allowing light through to inner cells.

> The palisade cells contain large numbers of chloroplasts to absorb a lot of light energy. The top layers of cells contain the most chloroplasts.

> The chloroplasts are arranged mainly down the sides of the palisade cells, allowing some light to reach the mesophyll cells.

> Air spaces between mesophyll cells allow gases to diffuse easily and reach all cells.

> The mesophyll cells are small and irregular. This increases their surface area to volume ratio so large amounts of gases can enter and exit the cells.

Questions

9 Scientists are trying to develop new strains of crop plants containing many more chloroplasts in their leaf cells.

Explain why.

10 Suggest why most plant leaves only have stomata in their lower surfaces.

Q types of leaves   chloroplasts

What materials are needed for photosynthesis?

You will find out:

> how materials needed for photosynthesis enter leaves

> how leaves are adapted for photosynthesis

The following three things are needed for photosynthesis to take place:

> Light from the Sun. The broader the leaf, the more surface area it has so more sunlight can be absorbed.

> Carbon dioxide, which enters the leaf through **stomata** on the underside of leaves.

Even though the air contains only 0.04 per cent of carbon dioxide, this is enough for photosynthesis.

> Water, which enters the plant through the root hairs. This is why Sean waters the soil around his plants and not the leaves.

Oxygen is made by photosynthesis. Some of the oxygen that plants make is released into the air through stomata in the leaves.

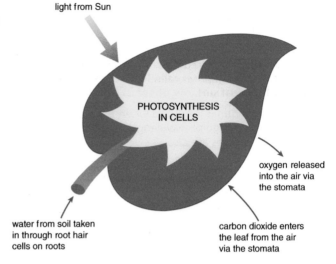

FIGURE 5: The process of photosynthesis. What three things have to be present in a leaf for photosynthesis to take place?

light from Sun

PHOTOSYNTHESIS IN CELLS

oxygen released into the air via the stomata

water from soil taken in through root hair cells on roots

carbon dioxide enters the leaf from the air via the stomata

Questions

11 Which gas is used in photosynthesis?

12 Which part of the plant takes in this gas?

13 Why does Sean water the soil around a plant and not its leaves?

14 Which gas is made in photosynthesis?

Autumn leaves

Leaves on **deciduous** trees change colour in autumn as the trees absorb the green pigment from the chloroplasts. The colours of other pigments can now be seen.

During the winter it is too cold for photosynthesis and too much water is lost from leaves. Because leaves are specially adapted to carry out photosynthesis many cannot survive winter frosts and they drop off the trees.

A leaf is adapted for photosynthesis by:

> being broad so it has a large surface area to absorb light

> being thin so gases do not have far to diffuse and light is able to reach all the way through it

> having a number of chlorophyll pigments, each of which absorb light from different parts of the visible spectrum, in most of its cells

> having a network of specialised cells in **vascular bundles** to support it and carry water and food, such as sugars, to different parts of the plant

> having stomata to allow carbon dioxide to diffuse into it and oxygen to diffuse out of it

> having guard cells which control the size of the stomata during the day and night.

FIGURE 6: Autumn leaves on deciduous trees lose their green colour. Why?

FIGURE 7: Cabbage plants growing in a market garden. The cabbage leaves have a large surface area to aid photosynthesis. Can you suggest why they are a good food source?

Q chloroplasts in cells   palisade cells

Questions

15 Why is it important for leaves to have a large surface area?

16 Which part of a leaf carries sugars to other parts of the plant?

17 Holly leaves have a thick waxy covering. Suggest why.

Absorption of light

White sunlight is actually a rainbow of colours called the visible spectrum. Individual colours have different wavelengths. A green leaf appears green because the leaf absorbs a range of colours but reflects green light.

An action spectrum is produced when the rate of photosynthesis is plotted against wavelength of light. The wide range of wavelengths used ensures efficient photosynthesis in different lighting conditions.

Chloroplasts contain a number of different pigments such as chlorophyll a, chlorophyll b and carotenoids. The carotenoids include the pigments **carotene** and **xanthophyll**. They each absorb different wavelengths of light so a wide range of wavelengths can be used in photosynthesis.

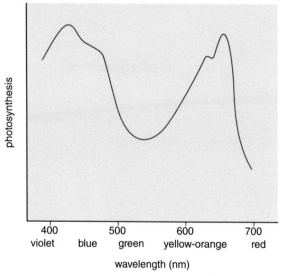

FIGURE 8: An action spectrum of photosynthesis. Which colours are absorbed in photosynthesis?

Did you know?

Seaweeds can be green, red or brown. Green seaweeds are found in shallow water and brown seaweeds are found in deep water.

The varying amount of light they receive determines their colour. Different pigments use different wavelengths of light that penetrate different depths of water.

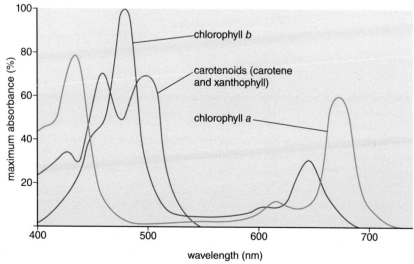

FIGURE 9: Absorption by different pigments. Which wavelengths are not important in photosynthesis?

Questions

18 Suggest why seaweeds are different colours and why this determines where they are found.

Diffusion and osmosis

You will find out:
> how substances enter and exit plants
> about diffusion and osmosis
> why plants wilt

Diffusion

If someone lets off a stink bomb at one end of a corridor why does the whole corridor smell? Simple – it's called diffusion. The smelly gas molecules spread out into all the space available.

Diffusion is defined as the movement of a substance from a region of high concentration to a region of low concentration.

Diffusion helps to move substances in and out of cells.

Diffusion and gases

Diffusion and cells

All living things are made of cells. In plant cells the contents are surrounded by a **cell membrane** and a cell wall. Substances such as oxygen and carbon dioxide pass through these by **diffusion**.

Diffusion and the leaf

Plants use up carbon dioxide during photosynthesis. The concentration of carbon dioxide inside a leaf is therefore low during photosynthesis. The higher concentration of carbon dioxide in the air around the plant causes the gas to move into the leaf by diffusion. It diffuses through small pores called stomata. The stomata are mainly in the lower epidermis (the underneath surface) of leaves.

During photosynthesis, oxygen levels inside the leaf increase. This causes oxygen to diffuse out of the leaf into the air.

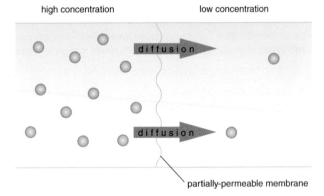

FIGURE 1: What is happening to the particles?

Questions

1 Describe how oxygen escapes from leaves.
2 Why is there more diffusion from the lower rather than the upper surface of leaves?
3 Explain why diffusion of gases does not take place through a metal container.

Diffusion

When molecules in a liquid or a gas spread out they move in all directions. This is known as random movement. Because the particles move randomly the overall result will be that they will spread from where there are many, to where there are few. Using these ideas, diffusion can be defined as:

> the net movement of particles from an area of high concentration to an area of lower concentration due to the random movement of individual particles.

If no particles are removed, the concentrations will eventually be equal. However the particles will still continue to move randomly.

 partially-permeable membrane

Leaf adaptations for diffusion

Leaf cells need carbon dioxide for photosynthesis and oxygen for respiration. They also need to let the waste products of these processes escape.

Plant leaves are therefore adapted for efficient diffusion.

Leaves are adapted to increase the rate of diffusion of gases by having:

> usually a large surface area

> specialist openings called stomata

> spaced out stomata

> gaps between the spongy mesophyll cells.

Diffusion and cells

Many molecules can easily pass into and out of cells through:

> the outer cell wall which is completely permeable to many molecules

> the cell membrane which is partially-permeable.

A **partially-permeable membrane**, such as a cell membrane, allows the passage of small molecules such as water, carbon dioxide and oxygen but not large complex molecules such as starch.

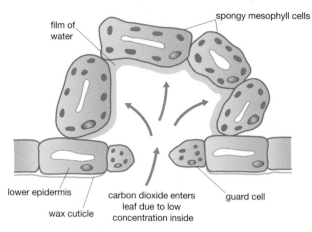

film of water
spongy mesophyll cells
lower epidermis
carbon dioxide enters leaf due to low concentration inside
guard cell
wax cuticle

FIGURE 2: Why are the stomata spaced out?

Questions

4 Which processes in a plant need:
a oxygen and
b carbon dioxide?

5 Suggest what would happen if the surfaces of all a plant's leaves were sealed with Vaseline.

6 Suggest why it is important for plant leaves to be adapted to increase the rate of diffusion.

Changing the rate of diffusion

The rate of diffusion can be increased by:

> having a shorter distance to travel

> having a greater difference in concentration (concentration gradient) by using up the particles in one area such as in photosynthesis in leaf cells

> having a greater surface area for particles to enter or escape from.

Questions

7 Suggest one problem to the plant of having stomata.

8 Suggest which external conditions will increase the rate of diffusion.

Did you know?

A "nose" is someone who develops new perfumes.

There are about 50 famous "noses" in the world. They can identify thousands of different scents.

Supporting cells

The presence of water in a plant cell gives it shape and support. This is similar to the way air gives shape and support to a bicycle tyre. The cell wall of a plant cell also provides support. It is made of cellulose, a strong substance.

Cells and osmosis

When potato chips are placed in a bowl of water they absorb water. This causes them to swell and become firm.

This is because water can enter and leave cells by passing through the partially-permeable membrane. The process is called **osmosis**. Water will also enter and leave an animal cell through the cell membrane.

FIGURE 3: This geranium plant has wilted due to a lack of water. Its cells do not have enough water. What could be done to make it recover?

FIGURE 4: These potatoes have swollen in the water. What has caused this?

Did you know?

The largest known living thing is the General Sherman tree in California.

This redwood tree is over 90 m tall and over 2 000 years old. The strength of its cell walls and the water pressure inside the cells keeps it upright. Plant cell walls are made of cellulose so they are inelastic (do not stretch).

GENERAL SHERMAN

Questions

9 When a potato is boiled it becomes very soft. Suggest why.

10 Why does soaking potato chips in water before cooking produce better chips?

Movement of water in plant cells

If a potato chip is put into pure water it swells up and hardens. Water has entered its cells by osmosis. If the chip is taken out of pure water and put into salt water it turns soft and floppy. Water has left its cells by osmosis.

Plant cells depend on their inelastic cell wall and water inside their cells for support. A living plant cell has a membrane just inside its cell wall. This membrane is partially-permeable so only small molecules such as water can pass through it. Salt molecules outside the cell and sugar molecules inside the cell cannot pass through it. Water molecules will diffuse from a high concentration of water to a low concentration of water across the partially-permeable membrane.

Osmosis is therefore a type of diffusion. The partially-permeable membrane restricts the diffusion of large molecules such as sugars but allows the passage of water molecules.

Osmosis can be described as:

> the movement of water molecules from an area of high water concentration to an area of low water concentration, across a partially-permeable membrane.

A dilute sugar solution will have a high concentration of water, a concentrated sugar solution will have a low concentration of water.

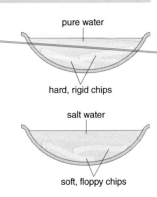

pure water

hard, rigid chips

salt water

soft, floppy chips

FIGURE 5: Potato chips in pure water and in salt water.

Q osmosis in cells   plasmolysis

Turgor pressure

When water enters a cell it swells up and there is an increase in water pressure against the cell wall. This **turgor pressure** causes the cell to become **turgid** (hard and rigid).

When cells lose too much water, the low water pressure causes the cells to collapse and the whole plant will wilt.

Osmosis in animal cells

An animal cell does not have a cell wall. It will easily:

> swell and burst when too much water enters

> shrink when too much water leaves.

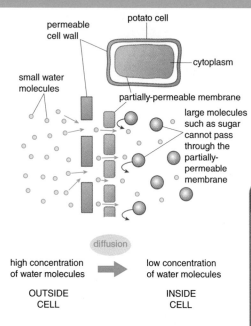

FIGURE 6: Osmosis in a plant cell. Which part of the cell is partially-permeable?

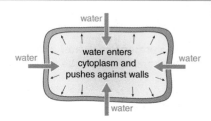

FIGURE 7: Turgor in a plant cell.

Questions

11 What happens to potato chips in salt water?

12 What happens to the water inside these chips?

13 Why do sugar molecules not diffuse through a plant cell membrane?

More on osmosis

When a potato chip is put into salt water, the potato cells become **flaccid** (soft and floppy). This is because there is a higher concentration of water molecules inside the potato cells than outside. Water molecules in the cells and solution will move randomly, some molecules will move into the cells and some will move out. However, there is a net movement of water molecules out of the cells due to osmosis. More water leaves the cell than enters, the cell contents shrink and there is less water pressure against the cell wall. The cell contents become **plasmolysed** and the cell collapses. Therefore a whole plant, such as a geranium, wilts.

If more water enters the cell, it becomes turgid. This is important in supporting plants.

Osmosis in animal cells

If an animal cell, such as a blood cell is placed in salt water or pure water it loses or gains water by osmosis. Since animal cells have no cell walls for support they:

> lose water and shrink (becomes **crenate**) in salt water

> gain water and burst (**lysis**) in pure water.

The results are more extreme than in plants cells since animal cells are not surrounded and supported by a cell wall.

Questions

14 Predict the movement of water when a potato chip is placed in a weak salt solution.

15 If a potato chip is placed in boiling water osmosis does not take place. Suggest why.

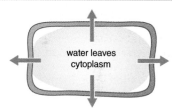

FIGURE 8: Plasmolysis in a plant cell. As water leaves the cytoplasm pulls away from the cell wall.

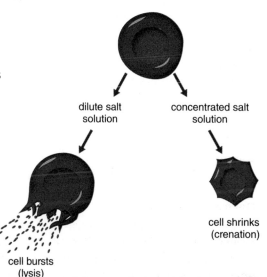

FIGURE 9: Why is the blood concentration important?

Preparing for assessment: Applying your knowledge

To achieve a good grade in science, you not only have to know and understand scientific ideas, but you need to be able to apply them to other situations and investigations. These tasks will support you in developing these skills. **Material on pages 72 and 79 will help with these tasks.**

✺ Roses are red, violets are blue...

Many people know that *Alice in Wonderland* was written by Lewis Carroll. In fact his real name was Charles Dodgson and he was a Professor of Mathematics. Many of the ideas in his two books relate to problems of logic, but Dodgson would also have been aware of scientific experiments that were popular at the time. Changing the colour of flowers was one of these. It is very likely that he made the playing-card figures in *Alice in Wonderland*, shown in this scene, paint the roses a different colour (particularly since they started out white) as a reference to this.

You can do the experiment at home. You need some white flowers, such as carnations, and some food colouring.

Add several drops of food colouring to a vase full of water. Place the flowers in the vase and then leave them for a few hours. The colour will appear in the flower petals.

Now remove the flowers and make a transverse (horizontal) cut through the stems (be very careful using a knife) and you will see tiny coloured dots on the cut face of the stem.

If you choose a kind of plant with multiple white flowers on one stem it is even possible to get different flowers on the same stem to be different colours! Slit the stem vertically part of the way up from its base so that the end is in two or three parts, and place each part in a small beaker. Support the stem. Put a different coloured food colouring in each beaker and leave for a few hours.

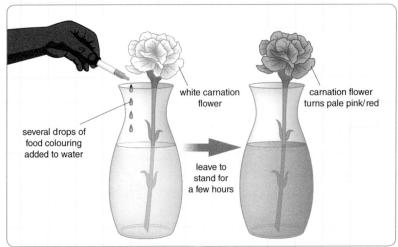

several drops of food colouring added to water

white carnation flower

leave to stand for a few hours

carnation flower turns pale pink/red

Turning a white flower a different colour using food colouring.

Task 1

The solution from the beakers or vase moves up the flower stem. What are the vessels called that the solution travels through?

 Task 2

Cutting through the stem of a flower that has been in food colouring for several hours shows a number of dark dots. What are these?

Task 3

What makes the solution travel up the stem against gravity?

Task 4

Water travels up the stem into the flowers. Why do the flowers not become bloated with water?

Task 5

What do you think would happen if the experiment was repeated with a whole plant using black ink instead of food colouring? Remember that black ink is not a pure substance – it is made from a number of different coloured inks.

Maximise your grade

Answer includes showing that you can...
Explain that the solution is pulled up the stem by transpiration pull (evaporation of water from the leaves).
Understand that the ink-coloured dots show which parts of the stem carry water and the food colouring.
Understand that the ink-coloured dots show the cut ends of groups of xylem cells.
Know that in young stems and leaves the groups of xylem cells are called vascular bundles.
Explain that the solution is pulled up the stem by trans-piration pull caused by evapora-tion of water from leaves.
Explain that the water is used in photosynthesis or lost in transpiration from leaves.
Combine the ideas that black ink contains different chemicals and plant roots are able to selectively absorb minerals to explain why different colours appear in the petals.
Explain how the carrier system in the membrane of root hair cells uses energy and absorbs different colours, therefore different colours will appear.
As above, but with particular clarity and detail.

Transport in plants

You will find out:
> about xylem and phloem cells
> about how water travels through a plant
> about water loss from leaves

A party trick!

A stick of celery is placed in water. A few drops of red dye are added to the water.

After a few hours, parts of the celery are red. The water has travelled through the celery.

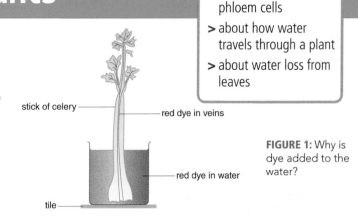

stick of celery

red dye in veins

red dye in water

tile

FIGURE 1: Why is dye added to the water?

How does the party trick work?

In a complete plant, root hairs absorb water from the soil. The water then passes into the root and moves up into the stem. The stem then carries the water up the stem to the leaves.

Evaporation takes place from the surface of plant leaves. **Evaporation** is the process of a liquid turning into a gas. The evaporation of water from leaves is called **transpiration**.

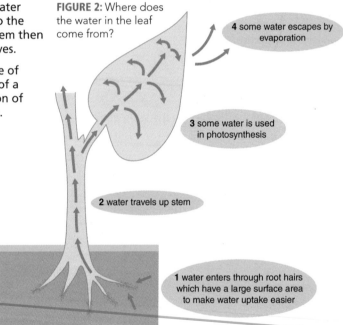

FIGURE 2: Where does the water in the leaf come from?

4 some water escapes by evaporation

3 some water is used in photosynthesis

2 water travels up stem

1 water enters through root hairs which have a large surface area to make water uptake easier

Questions

1 Which part of a plant absorbs water?

2 Suggest why plant leaves need water.

3 What would happen to a plant if its roots were destroyed?

4 What materials does a plant stem carry?

Transport system in a plant

Xylem and **phloem** are specialised cells in a plant. They form continuous **vascular bundles** from the roots to the stems and leaves. This forms the transport system for the plant.

> Xylem cells carry water and minerals from the roots to the leaves. Water is needed by leaves for photosynthesis. Some water evaporates and escapes by diffusion from the leaves. This process of transpiration helps to pull up the water from the roots.

> Phloem cells carry dissolved food such as sugars from the leaves to other parts of the plant. This is called translocation. The sugars can be used for

growth or stored as starch. Many root vegetables such as turnips and carrots are eaten because they contain a lot of food.

More on movement of water in plants

A root hair is a long thin cell. Its large surface area absorbs a lot of water from the soil by osmosis.

Evaporation of water from a leaf creates a 'suction' effect. This pulls the water up through the xylem cells in the stem. This movement of water is useful because:

> evaporation of water cools a plant (just as sweating cools the body)

> photosynthesis needs a supply of water

> a cell full of water gives support

> the water carries useful dissolved minerals.

Spongy mesophyll cells in a leaf are covered with a film of water. Carbon dioxide dissolves in the water and is used in photosynthesis.

Transpiration involves the evaporation and diffusion of water from inside leaves. If too much water evaporates and diffuses out of the leaf via the stomata the plant wilts and may die.

To prevent too much water loss a leaf has:

> a waxy covering called the cuticle

> stomata, which can open or close, mainly on its shaded underside.

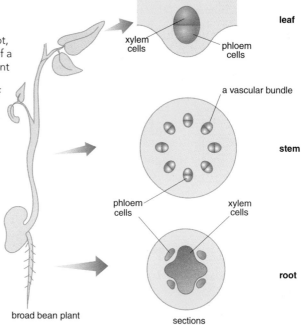

FIGURE 3: Horizontal cross-sections through the root, stem and leaf of a broad bean plant to show the arrangement of xylem and phloem.

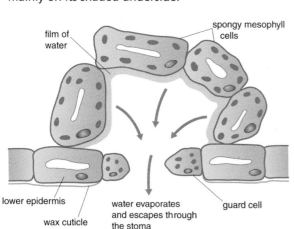

FIGURE 4: Movement of water from spongy mesophyll cells.

Remember!

It's easy to get confused between xylem and phloem cells. Remember phloem carries food.

Questions

5 What are the specialised cells called in a vascular bundle?

6 What type of foods do phloem cells carry?

7 Apart from their taste, why are turnips and carrots eaten?

More on xylem and phloem

The cellulose cell wall of a xylem cell has extra layers of a chemical called **lignin**. This makes it very strong. The cells then die, making long thin tubes with a hollow lumen. These cells are called vessels and are ideal for carrying water. Xylem cells form the wood in a tree.

Wood is so strong it is used to make furniture and even houses.

Phloem cells are also long and thin, forming columns, but unlike xylem cells, they stay alive to be able to pass chemicals such as sugars from cell to cell.

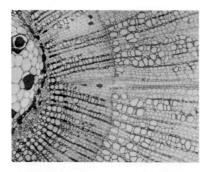

FIGURE 5: Xylem cells showing extra thickening. What is the chemical that gives these vessels their strength?

Questions

8 Explain how xylem cells support a tree.

9 Why is a tree more likely to die if its phloem cells rather than its xylem cells are damaged?

10 In a tree there are more xylem cells than phloem cells. Suggest why.

vascular bundles in plants

How does a plant lose water?

You will find out:

> more about transpiration

> what conditions affect transpiration

> how conditions affect transpiration

Transpiration is water loss from a plant leaf.
The water evaporates from the surface of the leaf.

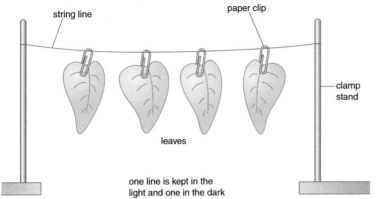

FIGURE 6: Experiment to show the effect of light on transpiration rate. Why are some leaves kept in the light and some in the dark?

Transpiration is affected by:

> temperature
> wind
> amount of light
> humidity (the amount of water in the air).

The effect of light on transpiration rate

Leaves from the same type of plant that are about the same size are weighed. They are then attached to string lines. One of the lines is kept in the light and the other line is kept in the dark for one day. The leaves are then re-weighed. The average loss in weight of leaves kept in the dark is compared to the average loss in weight of leaves kept in the light. Loss of water in transpiration causes leaves to lose weight.

Balancing water loss and water uptake

A plant needs a constant flow of water. Too little water will cause cells to collapse and the plant will be unable to photosynthesise and cool itself.

Questions

11 Choose one of the following words to describe the process of clothes drying:
evaporation transpiration photosynthesis

12 What type of weather is good for drying clothes?

13 a What is water loss from a plant leaf called?

b Suggest how: i) windy ii) warm iii) humid conditions could be set up in the laboratory to investigate their effects on transpiration rate.

Transpiration rate

A high rate of transpiration happens when:

> light intensity increases

> temperature increases

> air movement (wind) increases

> humidity (amount of water in the atmosphere) falls.

Measuring the rate of transpiration

A potometer can be used to measure the rate of water uptake. The uptake of water is shown by the movement of the water in the capillary tube.

It is assumed to be the same as the amount of water lost by transpiration.

FIGURE 7: Cacti plants live in dry hot deserts. Can you suggest why they have very small leaves?

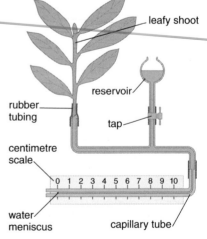

FIGURE 8: A potometer.

lignin

Which combinations of results produce the fastest and the slowest rate of transpiration? (See table below.)

conditions	rate of water uptake cm/min
light, warm, calm	6.3
dark, warm, calm	3.1
light, warm, windy	9.6
light, cold, calm	4.2
dark, cold, calm	0.4

Questions

14 Suggest why the rate of water uptake is actually not the same as the rate of water loss.

15 Explain why cacti must maintain a low rate of transpiration.

16 When plants are protected from wind they grow better. Suggest why.

Controlling water loss

A leaf is adapted for efficient photosynthesis.

Spongy mesophyll cells in a leaf are covered with a film of water. Carbon dioxide dissolves in the water and is used in photosynthesis. However, the water around the spongy mesophyll cells easily escapes. By being efficient in photosynthesis, the plant has a problem with transpiration.

If too much water is lost in transpiration the plant wilts and dies. The plant's cellular structure is adapted to reduce water loss by:

> having the ability to close their stomata. The changes in light intensity and water availability will affect the rate of photosynthesis in guard cells. This changes the amount of sugar in the guard cells and affects osmosis. When more water leaves the guard cells they lose turgor and the stomatal apertures become smaller. This reduces water loss.

> having few, small stomata on the lower leaf surface.

How is transpiration rate affected?

> When light intensity increases, the stomata open. This allows more water to escape.

> As the temperature increases, the random movement of water molecules increases and more water escapes.

> Wind causes more water molecules near stomata to be removed. This increases evaporation and diffusion of water from inside the leaf.

> In dry conditions there is a very low concentration of water molecules outside the leaf. This causes more diffusion of water from inside the leaf to the outside.

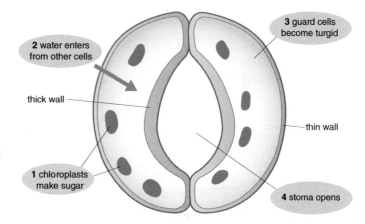

3 guard cells become turgid

2 water enters from other cells

thick wall

thin wall

1 chloroplasts make sugar

4 stoma opens

FIGURE 9: How a stoma opens. How does this process affect the plant?

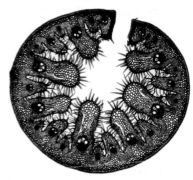

FIGURE 10: a Marram grass grows on exposed sand dunes. **b** To reduce transpiration it has leaves with buried stomata surrounded by leaf hairs. The leaves are curled so the stomata are surrounded by high humidity and are not exposed to wind.

Questions

17 Explain why you should not transplant young plants on a hot windy day.

18 How is Marram grass adapted to live on sand dunes?

19 The stomata of Marram grass may remain open when the plant is short of water and wilting. Suggest why.

Q controlling transpiration

Plants need minerals

You will find out:
> why plants need fertilisers
> which minerals are in fertilisers
> what the minerals are used for

Slash and burn

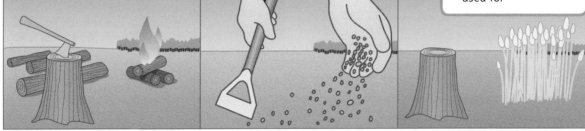

FIGURE 1: The ancient slash and burn method of farming. Why is it not used today in the UK?

trees cut down and burned

wheat seeds planted in the warm ash and covered with soil

crops harvested after six weeks

Thousands of years ago people struggled to grow crops for food. They cut down trees and burned them. This was called 'slash and burn'. They planted their crops. After a few years their crops were poor so they had to move on. The plants had used up the chemicals in the soil. This method is still being used in clearing forests in the Amazon.

Fertilisers

Modern farmers put **fertilisers** on their crop plants. They can then grow the same crop in the same field for several years.

A fertiliser contains **minerals** such as:

> nitrates, containing nitrogen (chemical symbol, N)

> phosphates, containing phosphorus (P)

> potassium (K)

> magnesium (Mg).

Plants use these minerals to grow.

Plants must be watered to keep them alive. Gardeners add fertiliser to the water every week.

Fertiliser contains minerals that dissolve in water. The minerals are in a solution. Plants need minerals to thrive. If a plant does not get one or more of these minerals its growth is poor.

The soil around a plant is watered (not the leaves) because it is the root hairs that absorb the water and the minerals.

FIGURE 2: Spreading slurry (liquid manure) on a field. Why do farmers fertilise their fields?

FIGURE 3: Why is the soil around a plant watered and not its leaves?

Questions

1 Why did early farmers have poor crops?

2 Which chemical elements have the symbols N, P and K?

3 Why do plants need minerals?

Q Slash and burn agriculture   artificial fertilisers

What minerals do plants need?

Nitrates

> A plant uses nitrates to make proteins.

> Proteins are needed for cell growth.

Phosphates

> A plant uses phosphates in respiration (releasing energy).

> Phosphates are also needed for growth, especially in roots.

Potassium compounds

> Plants use these in respiration and in **photosynthesis**.

Magnesium compounds

> A plant needs these in **photosynthesis**.

Did you know?

The longest running experiment in the world is at Rothamsted Research. The Park Grass Experiment started in 1856 to investigate the effects of fertilisers and manures on hay fields.

 Questions

4 Which minerals are needed for photosynthesis?

5 Which minerals are needed for respiration?

6 Which minerals are needed for growth?

7 Suggest why a gardener gives his plants fertiliser every week in the summer months.

What happens to these minerals?

Nitrates

> Nitrogen in the nitrates is used to make amino acids.

> Amino acids are joined together to make different proteins such as enzymes.

Phosphates

> Phosphorus in phosphates is used to make cell membranes and **DNA**.

> DNA carries genetic information.

Potassium compounds

> Potassium is used to help make some enzymes. Enzymes speed up chemical reactions.

> Enzymes are needed in photosynthesis and respiration.

Magnesium compounds

> Magnesium is used to make **chlorophyll** molecules.

Did you know?

The first manufactured fertiliser was called 'superphosphate'. It was made in 1842 by adding sulfuric acid to bones.

 Questions

8 Which elements are used to make enzymes?

9 Why are enzymes important to a plant?

10 Write down one element which helps to make a plant green?

11 Which element is needed to make DNA?

🔍 Rothamsted research

More on fertilisers

Mineral deficiencies

Plants can be grown, without soil, in special solutions to see the effect of the lack of individual minerals. One plant is grown with all the required minerals being present.

Other plants are grown in different culture solutions each lacking only one mineral.

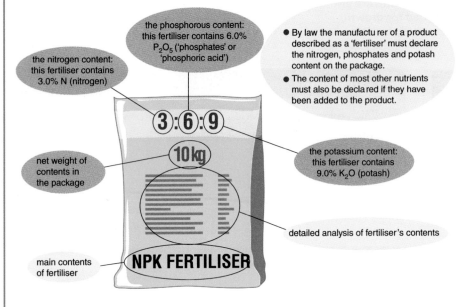

the phosphorous content: this fertiliser contains 6.0% P_2O_5 ('phosphates' or 'phosphoric acid')

the nitrogen content: this fertiliser contains 3.0% N (nitrogen)

● By law the manufacturer of a product described as a 'fertiliser' must declare the nitrogen, phosphates and potash content on the package.
● The content of most other nutrients must also be declared if they have been added to the product.

3:6:9

10 kg

net weight of contents in the package

the potassium content: this fertiliser contains 9.0% K_2O (potash)

detailed analysis of fertiliser's contents

main contents of fertiliser

NPK FERTILISER

FIGURE 4: Suggest why it is important to read the information on a sack of fertiliser.

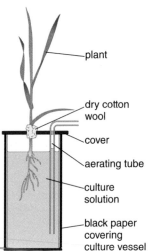

plant

dry cotton wool

cover

aerating tube

culture solution

black paper covering culture vessel

FIGURE 5: Investigating mineral deficiencies. Why are the roots aerated?

Questions

12 Why do gardeners add fertiliser to the water before watering their plants?

13 Name three minerals a fertiliser should contain.

14 The fertiliser sack has a mass of 10 kg. What mass of **a** nitrogen fertiliser, **b** phosphorus fertiliser and **c** potassium fertiliser does it contain ?.

15 In culture solutions investigating mineral deficiencies, the equipment must have no impurities. Suggest why.

Mineral deficiencies

Each mineral is used by a plant for different things. A mineral deficiency in a plant is easy to detect and correct.

Soil contains only small amounts of minerals. Fortunately, a plant needs minerals in very small amounts.

lacking → • poor growth
nitrates   • yellow leaves

lacking → • poor root growth
phosphate • discoloured leaves

lacking → • poor fruits and
potassium   flowers
         • discoloured leaves

lacking → • yellow leaves
magnesium

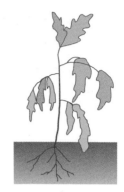

FIGURE 6: Symptoms of different mineral deficiencies in a plant.

Questions

16 What causes poor root growth in plants?

17 What causes discoloured leaves?

18 Why does a sack of fertiliser last a long time?

19 Why is it easy to know when a plant is not getting enough of a mineral?

Active transport

Diffusion and osmosis explain how gases and water enter living things. Minerals are taken up into root hairs by another method.

In 1938 scientists discovered that an increase in the uptake of minerals by a plant is matched by an increase in its respiration rate. This shows that energy is necessary for the uptake of minerals. This is because the minerals are absorbed against a concentration gradient.

Experiments also showed that a plant absorbs different minerals in different amounts. It selects the minerals it needs.

Special carrier molecules take the mineral ions across plant cell membranes. This process is called **active transport**. Different carriers take different minerals.

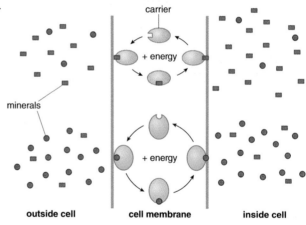

FIGURE 7: Active transport in a plant. Why does this process require energy?

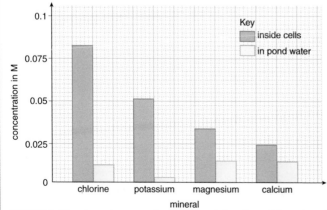

FIGURE 8: Graph to show uptake of minerals by algae. The minerals are selectively absorbed against a concentration gradient.

Questions

20 Scientists have found that some minerals can stop the uptake of a different mineral. How does the theory of active transport explain this?

21 Explain why it is important for a plant to select the minerals it needs.

Decay

Fungus

These 'mushrooms' are really the reproductive structures of a fungus.

The main part of the fungus is underground. Many small tubes called hyphae digest rotting material in the soil.

You will find out:

> what causes decay
> why decay is useful
> what conditions speed up decay
> about detritivores and saprophytes

FIGURE 1: What part of a fungus are these 'mushrooms'?

The process of decay

When plants and animals die their bodies break down. This is called **decay**.

When plant and animal bodies decay they break down into simpler chemicals such as minerals. These minerals can be recycled and used by living plants for growth.

Materials such as plastics and metals do not decay.

Microorganisms such as bacteria and fungi cause decay. They are called **decomposers**.

A microorganism needs oxygen, water and a suitable temperature to survive.

Microorganisms break down sewage (human waste), dead plants and plant waste such as dead leaves into compost.

An experiment to show how decomposers cause decay can be set up. Two samples of soil are collected. One sample is heated to a high temperature (but not burned). Both samples are weighed and then put in sealed flasks containing limewater. After two days the soil samples are re-weighed. Only the fresh soil sample loses mass. The limewater in the flask containing the fresh soil turns from clear to milky. This shows that carbon dioxide is produced. The living organisms in the soil were respiring and breaking down material in the soil.

Questions

1 When does decay happen?

2 Name two decomposers.

3 What do decomposers need to survive?

4 Name one way in which decomposers are useful.

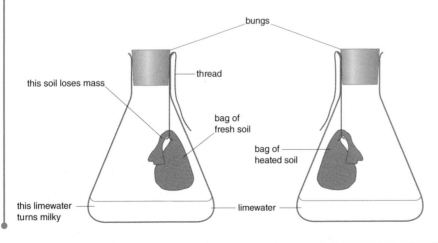

bungs

thread

this soil loses mass

bag of fresh soil

bag of heated soil

this limewater turns milky

limewater

FIGURE 2: Why did the fresh soil sample lose weight? What types of organisms produce carbon dioxide?

Q decay of plants animals

Decay food chains

The remains of dead and decaying plants and animals are called **detritus**.

Animals such as earthworms, maggots and woodlice depend on detritus for their food and are called **detritivores**.

A maggot is the young stage of an insect. As a maggot eats it breaks detritus down into small pieces. This increase in surface area makes digestion easier and also enables other organisms to feed, thereby increasing the speed of decay.

Detritivores are important in food chains as they recycle chemicals from dead plants and animals.

Ideal conditions for making compost

It takes many months for kitchen and garden waste to decay into **compost**.

Conditions that speed the composting process up are:

> warmth, such as placing a compost bin in a sunny area

> moisture, but not too much

> good aeration, such as regular mixing of the contents to allow oxygen in.

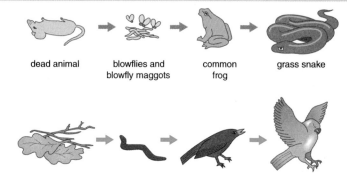

dead animal → blowflies and blowfly maggots → common frog → grass snake

decaying leaves → earthworms → blackbird → sparrowhawk

FIGURE 3: Two food chains to show how dead animals and decaying plant material are recycled. Which organisms in the food chains are detritivores?

Questions

5 What is detritus?

6 How do detritivores recycle chemicals?

7 Write down three reasons to explain why unmixed plant material kept in a sealed bin in the shade is slow to decay.

8 Suggest why gardeners make compost.

More on speeding up compost production

An increase in temperature and a good supply of water and oxygen speed up decay. Processes such as digestion, respiration, growth and reproduction in microorganisms depend on enzyme action, so conditions that speed up enzyme action also speed up decay.

The **optimum temperature** is approximately 37 °C for bacteria and 25 °C for fungi. This is the temperature at which their enzymes work best.

As microorganisms living in a compost bin respire, heat is produced which warms up the compost, which in turn speeds up decay. A compost heap often appears to be steaming. In some cases so much heat is generated that the compost catches alight.

Aerobic bacteria

Much of the decomposition is carried out by aerobic bacteria. They need oxygen for respiration, to release energy for growth and reproduction. The more oxygen there is, the more decomposition will take place.

What is a saprophyte?

Organisms such as fungi that feed off dead and decaying material are called **saprophytes**. They usually live in or on the decaying matter. Its digestive enzymes are released onto the food and break it down into simple, soluble substances. The digested food is absorbed. This type of digestion is called extracellular since it takes place outside cells.

FIGURE 4: Why does a compost heap give off 'steam'?

Questions

9 Explain why a good compost heap becomes hot.

10 Why is it necessary to aerate a compost heap?

11 Suggest why saprophytes need damp conditions to feed.

Preserving foods

You will find out:
> about food preservation methods
> how food preservation works

A person who eats decaying food may suffer from **food poisoning**.

Fungi and bacteria grow on decaying food and it is these organisms that produce the chemicals that cause food poisoning.

Preserving food stops it from decaying.

Ways of preserving food:

> adding sugar or salt
> canning
> cooling
> cooking
> freezing
> drying
> adding vinegar.

adding sugar or salt

canning

cooking

freezing

drying

adding vinegar

FIGURE 5: Why does a fridge keep food for longer?

FIGURE 6: Ways of preserving food

Questions

12 Why should slightly mouldy strawberries not be eaten?

13 Suggest two ways in which fresh strawberries can be preserved.

14 Write down the name of one food preserved by drying.

15 Write down the names of two foods preserved by canning.

Q food preservation

More on preserving foods

Each year there are thousands of cases of food poisoning in the UK. Some people die from it.

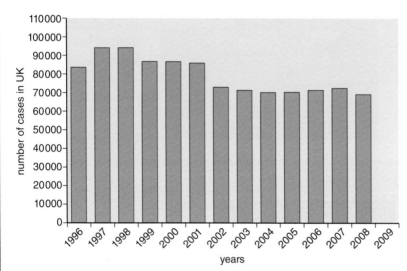

FIGURE 7: Bar chart to show the incidence of food poisoning in England and Wales.

If decomposers such as bacteria and fungi on food are killed or prevented from reproducing it does not decay.

Canning

In modern canning, food is heated to kill bacteria. The food is then put into cans and sealed while it is still hot. This forms a vacuum and prevents the entry of oxygen and bacteria. The steel used in cans has a tin covering so it does not rust. Canned food keeps for many years.

Cooling

A refrigerator cools food down to about 5 °C, which is slightly above freezing. This slows down respiration and reproduction of decomposers.

Freezing

Freezing kills some bacteria and slows down the growth and reproduction of others. Some frozen food is best eaten within a few months and some can last several years. Defrosted food should not be refrozen since the bacteria will have started to reproduce.

Drying

Many foods such as fruits can be dried. Without water, bacteria and fungi cannot feed and grow.

Adding vinegar

Vinegar is an acid. Very few bacteria can grow in acid conditions. Foods such as pickled eggs and chutney are preserved in this way.

Adding sugar or salt

A high concentration of sugar or salt solution kills some bacteria and fungi and stops the growth of others; this is an example of osmosis. Sugar is used in jams and salting is used on meat.

Questions

16 Barbecuing food results in many cases of food poisoning. Suggest why.

17 How many people in the UK suffered from food poisoning in 2004?

18 In 2008 the number of food poisoning cases had slightly fallen. Suggest why.

19 Suggest why food should not be eaten from a damaged can.

food poisoning

Farming

Modern farming

In 1970, people in the UK spent 25 per cent of their income on food. Today only about 15 per cent of a person's income is spent on food. Modern farming methods have enabled farmers to produce cheaper food.

You will find out:
> about pesticides and herbicides
> why intensive farming methods are efficient
> about the advantages and disadvantages of intensive farming
> about growing plants without soil

FIGURE 1: What foods can be made from harvested grain?

Intensive farming

Modern farming methods use machines and chemicals to produce as much food as possible, quickly and cheaply, from the land, plants and animals that are available. This type of farming is called **intensive farming**.

An intensive farmer uses **pesticides** to kill pests that damage crops.

Examples of pesticides are:
> **insecticides** that kill insects
> **fungicides** that kill fungi
> **herbicides** to kill unwanted plants (weeds).

These chemicals can cause harm to the environment and damage human and animal health.

Some farmers use special intensive farming methods to increase their production of food such as:
> glasshouses
> **hydroponics**
> fish farms
> battery farming.

Did you know?

The most famous insecticide is DDT (dichlorodiphenyltrichloroethane).

It was introduced in 1939 to control malaria and typhus among army troops. Later it was extensively used in agriculture with a huge increase in crop yields.

In 1962, its effects on wildlife and human health were realised. Its use is now banned in many countries.

FIGURE 2: Growing plants without soil is called hydroponics. What are the advantages of intensive farming methods?

Questions

1 What type of chemical does a farmer use to kill unwanted insects?

2 Write down one reason why a farmer keeps battery hens.

3 Write down one disadvantage of intensive farming.

4 Name one type of fish that is reared in fish farms.

A farmer's dilemma

Intensive farmer

I use intensive farming methods to produce as much food as possible from my land as quickly as possible. I am an efficient farmer.

His neighbour

Keeping many hens in a small space is cruel and disease spreads easily. I don't like being close to where chemicals are being sprayed. Pesticides build up in food chains and reach toxic levels. Many are persistent – remaining in the soil for many years. Insecticides kill useful insects such as bees as well as the harmful insects. Bees pollinate flowers which then produce seeds.

Q pesticides   intensive farming methods

Build-up of pesticides in food chains

Concentration of DDT in parts per million (ppm) in the food chain

  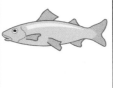 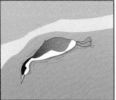

| lake, 0.02 | microscopic life, 5 | fish, 2000 | grebes get a lethal dose |

FIGURE 3: How does the pesticide DDT build-up in a food chain?

The use of DDT is now restricted all over the world since its effects were first proved in Clear Lake, California. It was used to kill gnats (a biting insect) on the lake. The DDT remained in the organisms low down the food chain and predators higher up accumulated a high and eventually lethal dose. People were advised not to eat fish from the lake. Can you suggest why?

Hydroponics

A hydroponics system can be used to grow glasshouse tomato plants. The system does not use soil, so there is less chance of disease or pests affecting the plants.

The plant roots are in specially treated water that contains the required amounts of fertiliser and oxygen. The system is also useful in countries with poor soil or little water to irrigate fields.

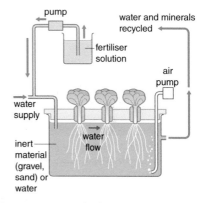

FIGURE 4: A hydroponics system used to grow lettuce plants. Why is an air pump needed?

Questions

5 Explain why killing bees using insecticide causes a problem.

6 Why do some people object to intensive farming?

7 Suggest why DDT in Clear Lake did not kill microscopic life.

8 Suggest why plant roots in a hydroponics system need oxygen.

Intensive farming

Intensive farming is very efficient. More energy is usefully transferred because:

> there are fewer plant pests (weeds) in crops competing for water, minerals and light

> there are fewer animal pests to attack and eat crops or cause disease in livestock

> less heat is lost from animals kept in sheds, such as in battery farming, and their movement is restricted, so energy from food is used for growth rather than for movement.

Hydroponics

A hydroponics system has many advantages:

> mineral supply is controlled and unused minerals are recycled, so costs are lower and there are no pollution problems

> it is set up under cover so there is better control of conditions and disease.

There are some disadvantages:

> manufactured fertilisers must be bought

> tall plants such as tomatoes need support.

FIGURE 5: Intensive pig farming. Do you think this is a good way of farming?

Questions

9 Explain why it is cheaper to keep animals in sheds over winter.

10 Scientists are planning to use hydroponics to feed astronauts on long journeys into Space. Suggest what types of plants could be used.

Organic farming

You will find out:

> about organic farming
> advantages and disadvantages of organic farming
> how biological controls work
> advantages and disadvantages of biological controls

A farmer who does not use manufactured chemicals is called an organic farmer.

An organic farmer does not use:

> artificial fertilisers

> artificial pesticides.

Organic farmers and gardeners use other methods to control pests. One method is to use other animals such as predators. This is called **biological control**.

> Ladybirds are used because they eat small flies called aphids. Aphids damage plants by sucking sap from them.

> A special wasp is used in glasshouses. The wasp is a predator on aphids. Aphids damage plants such as tomato plants.

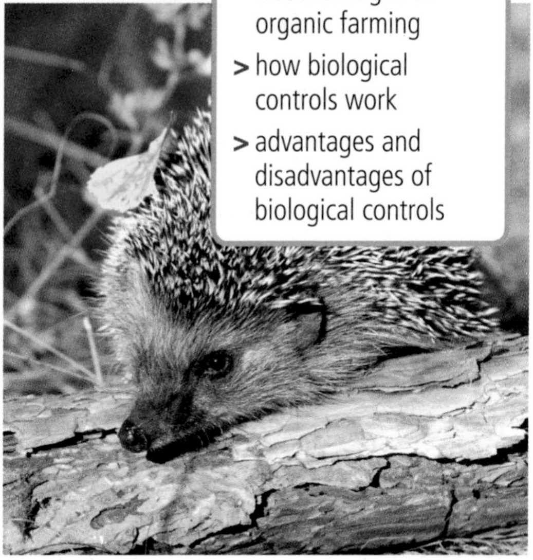

FIGURE 6: Organic gardeners encourage hedgehogs to visit their gardens because the hedgehogs eat slugs and snails that damage plants.

Using pesticides

Many countries in the world cannot afford expensive pesticides.

Food crops show reduced yields due to weeds, disease and insect pests.

	Worldwide harvest in 1 000 tonnes		Percentage losses (%)		
Crop	Potential harvest	Actual harvest	weeds	disease	insects
rice	715 000	378 000	10	9	27
maize	563 000	362 000	13	9	13
wheat	578 000	437 000	9	9	5
sugar cane	1 603 000	737 000	15	19	19

These data show that rice crop is reduced by 47%. Use your calculator to work out the reduction in the sugar cane crop.

Questions

11 What does an organic farmer not use?

12 Why are hedgehogs useful to an organic gardener?

13 How are ladybirds useful to an organic rose grower?

14 What is the method of controlling pests using other animals called?

15 Look at the data in the table above. Which crop would most benefit by using more insecticide?

Other organic methods

> Weeds are removed by hand to reduce competition for light and minerals.

> Seeds are sown at different times. In this way more plants survive animal pests and 'gluts' (plants all ready at the same time) are avoided.

> Peas and beans are grown on a rotating basis because bacteria in their roots 'fix' nitrogen from the air. Their roots are left in the soil to provide a source of nitrogen fertiliser.

Organic farming methods

> Animal manure is used in place of artificial fertilisers. Kitchen and garden wastes are composted.

> **Crop rotation** is used so the same crops do not grow in the same place each year. Different crops use different minerals.

Cost and suitability

Farmers can produce enough food to feed the world's population.

> Very efficient farmers in some countries that have a good climate produce excess food.

> Other countries that have a poor climate, few resources and little money cannot produce enough food or afford to buy food from other countries. If natural disasters such as drought occur the countries depend on other countries for aid. The type of farming practised is a balance between cost and suitability.

> In many developed countries it is fashionable to want **organic food** since it is seen to be healthier and people can afford to pay more.

> In less well developed countries the farmers try to grow enough food, by any means possible, to survive.

Advantages and disadvantages of organic methods

Advantages are:

> expensive chemicals do not have to be bought

> there is no chemical pollution or build-up in food chains

> some people think the products taste better.

Disadvantages are:

> biological control methods are often slow and do not kill all the pests

> crop yields are reduced and the cost of products is higher.

Biological control does not always work

Biological control can replace the use of expensive and potentially harmful chemical pesticides. Repeated treatments are also not needed. However, by introducing a new predator or removing an existing one, food chains or webs can be disrupted. Some animal populations increase because few are eaten, or decrease because predators are short of food and have to eat something else.

FIGURE 7: Flood-affected people queue up to wait for their turn to get food relief in Karachi August 2010.

Questions

16 How do pea and bean plants increase the amount of nitrogen fertiliser in the soil?

17 Suggest why organic foods tend to be expensive.

18 Why do organic gardeners compost their kitchen waste?

19 Suggest why cane toads have become a pest in Australia.

20 If you had the choice between eating organic and non-organic food, which would you choose? Explain your answer.

21 In shops, organic foods usually cost more than intensively produced foods. Explain why.

FIGURE 8: In 1935 large cane toads were introduced into Queensland, Australia to control insect pests on sugar cane. The toads had little impact on the insect population. The insects are now controlled by insecticides. The toads have spread throughout Australia and feed on the smaller native toad and are a pest. By introducing an organism the food web was disrupted.

biological control

Preparing for assessment: Analysis and evaluation

To achieve a good grade in science, you not only have to know and understand scientific ideas, but you need to be able to apply them to other situations and investigations. These tasks will support you in developing these skills.

✸ Task

- Analyse and evaluate an investigation to compare the rate of transpiration from the upper and lower surfaces of different types of leaves.

- Formulate and test hypotheses on the basis of your analysis and evaluation.

✸ Context

Plant leaves lose water by evaporation and diffusion of water vapour. This is called transpiration.

The water vapour escapes from small pores called stomata. Plant leaves show a wide variation in the distribution of stomata.

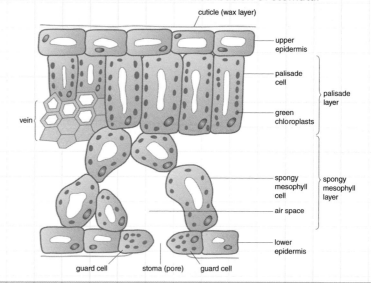

✸ How to do this experiment

General points:

1. Use geranium, laurel and pampas grass leaves.
Geranium is a dicotyledon plant with veins arranged in a network.
Laurel is also a dicotyledon plant but it has a very thick cuticle.
Pampas grass is a monocotyledon plant with parallel veins in its upright leaves.

2. Use cobalt chloride paper. This indicator paper is blue in the absence of water vapour but pink/white in the presence of water vapour.

3. Use thin plastic gloves and/or forceps when handling the cobalt chloride paper and leaves to avoid sweat affecting the results.

4. Run a trial experiment first to gain experience of handling Sellotape® and cobalt chloride paper.

5. The Sellotape® seals off the cobalt chloride paper from the surrounding air so any colour change is only due to water vapour produced by the leaf.

6. Different class groups will investigate one type of leaf.

7. The strips of cobalt chloride should be the same size.

8. The leaves should be kept in the same conditions, i.e. not near a window or Bunsen burner.

Sequence of steps:

1. Warm three short strips of cobalt chloride paper until they turns blue.

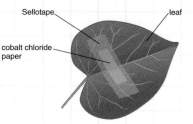

2. Using forceps, place one strip on the upper surface of a leaf and cover it with a larger strip of Sellotape® and start a stopwatch.

3. Repeat for the lower surface.

4. Leave one strip exposed to the air.

5. Time (to the nearest half minute) how long it takes for the cobalt chloride papers to turn completely pink/white.

6. Repeat experiment on the upper and lower surface for three similar sized leaves from the same plant.

7. Record results in a table.

✳ Results, analysis and evaluation

These were the results from three groups of students:

test	Time for cobalt chloride paper to turn pink using geranium leaves/ min		test	Time for cobalt chloride paper to turn pink using laurel leaves/min		test	Time for cobalt chloride paper to turn pink using pampas grass leaves/min	
	upper epidermis	lower epidermis		upper epidermis	lower epidermis		upper epidermis	lower epidermis
1	4.0	2.0	1	13.0	6.0	1	9.0	12.5
2	4.5	1.0	2	13.0	7.5	2	11.0	10.5
3	3.0	0.5	3	15.5	8.0	3	15.0	11.5
4	3.5	1.5	4	10.5	6.0	4	9.5	10.5
5	3.5	0.5	5	10.0	6.5	5	10.5	10.0
mean			mean			mean		

> Why do the students take five sets of readings?

> Explain your answer by referring to the mean values.

1. Calculate the mean values.

2. Plot the results as bar charts using the same scales for the axes. Use the y-axis for the time taken (in minutes) and the x-axis for types of leaves.

3. Explain the link between the time taken for the cobalt chloride paper to turn pink and the rate of water loss.

4. Is there a simple pattern for water loss from the upper and lower surface of all leaves?

> Use your knowledge of leaf structure to suggest reasons for any differences.

5. Is there a difference in water loss from the three types of leaves?

6. Do the results match the hypotheses?

7. What possible sources of error or limitations are there in the experimental technique?

> Describe how they could have been avoided.

8. Suggest what further experiments could be done on the leaves to link up water loss with stomatal distribution.

B4 Checklist

To achieve your forecast grade in the exam you'll need to revise

Use this checklist to see what you can do now. It gives you many of the important points you will need to know. Refer back to the relevant pages in this book if you're not sure and to see if there is anything else you need to know. Look across the three columns to see how you can progress.

Remember you'll need to be able to use these ideas in various ways, such as:

> interpreting pictures, diagrams and graphs
> applying ideas to new situations
> explaining ethical implications

> suggesting some benefits and risks to society
> drawing conclusions from evidence you've been given.

Look at pages 278–299 for more information about exams and how you'll be assessed.

To aim for a grade E	To aim for a grade C	To aim for a grade A
describe how to use collecting and counting methods **explain** what affects the distribution of organisms **define** biodiversity as the variety of species in a habitat	**use** data from capture–recapture to calculate population size **describe** how to map the distribution using a transect line **interpret** data from kite diagrams **compare** biodiversity in natural and artificial ecosystems	**understand** that capture–recapture data has limitations **describe** zonation of organisms in a habitat **explain** how a gradual change in abiotic factors affects distribution **explain** reasons for differences in biodiversity in natural and artificial ecosystems
recall and use the word equation for photosynthesis **explain** why plants grow faster in summer rather than in winter **understand** that plants carry out respiration as well as photosynthesis	**recall** and use the balanced symbol equation for photosynthesis **describe** the discoveries of Greek scientists, Van Helmont, Priestley **describe** how the rate of photosynthesis can be increased **explain** why plants carry out respiration at all times	**explain** the use of isotopes to show the splitting of water by light energy **explain** the effects of limiting factors on rate of photosynthesis **explain** the difference in gaseous exchange in plants during day and night
know that chlorophyll pigments in chloroplasts absorb light energy **know** the entry and exit points of materials for photosynthesis **understand** that broad leaves enable a lot of light to be absorbed	**explain** how different pigments absorb light from different parts of the spectrum **explain** how leaves are adapted for efficient photosynthesis **name** and locate the parts of a leaf	**interpret** data on the absorption of light **explain** how the cellular structure is adapted for photosynthesis

To aim for a grade E

know the definition of diffusion

know that water enters and leaves a cell by osmosis

recall that the cell wall provides support

know that water moves in and out of animal cells through a membrane

---

describe how water travels through a plant

describe experiments showing how transpiration is affected by external conditions

---

know that fertilisers contain minerals

describe experiments to show the effects of mineral deficiencies

describe how minerals are absorbed by plants

---

know the main factors in decay and that it is important for plant growth

describe an experiment to show decay depends on decomposers such as bacteria and fungi

know that food preservation depends on reducing the rate of decay and give examples

---

analyse data to show that more food is produced using pesticides but their use can harm the environment

describe how intensive farming methods increase productivity

describe organic farming methods and biological control

To aim for a grade C

explain diffusion using the idea of random movement of particles

explain osmosis using the idea of a partially-permeable membrane

explain how plants are supported by turgor pressure in its cells

describe the effects of uptake and loss of water from animal cells

---

describe the arrangement of xylem and phloem in a plant

describe how transpiration causes water movement in stems

interpret data from experiments on transpiration

---

explain why nitrates, phosphates, potassium and magnesium compounds are needed by plants

use symptoms to identify mineral deficiencies

know that minerals are present in the soil in only low amounts

---

describe how temperature, oxygen and water affect decay

know that detritivores feed on dead and decaying material and also increase the surface area of the material

explain how food preservation methods reduce the rate of decay

---

explain the advantages and disadvantages of using pesticides

describe hydroponics and possible uses

explain the advantages and disadvantages of organic farming and biological control

To aim for a grade A

explain how the rate of diffusion can be increased

predict water movements during osmosis

explain the words flaccid, plasmolysed, turgid

understand and use the words crenation and lysis

---

describe the structure of xylem and phloem

explain how transpiration is a result of leaf adaptations for photosynthesis

explain how the leaf's cellular structure is adapted to reduce transpiration

---

describe what is made from the minerals

explain how minerals are taken in by active transport

---

explain why temperature, oxygen and water affect decay

explain what is meant by a saprophyte and how they carry out extracellular digestion

---

explain the advantages and disadvantages of using hydroponics

explain how intensive food production improves the efficiency of energy transfer

B4 Exam-style questions

Foundation Tier

1 Look at the kite diagram. What does it show?

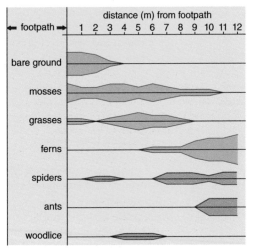

AO1 **(a)** Select two correct descriptions. [2]

A the distribution of organisms in a habitat

B the population size of an organism

C the size of animals in a quadrat

D an artificial ecosystem

E results from a transect line

AO2 **(b)** Suggest why few animals are found near the path. [2]

[Total: 4]

2 Look at the experiment on plant leaves.

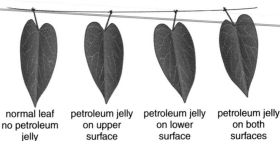

normal leaf no petroleum jelly | petroleum jelly on upper surface | petroleum jelly on lower surface | petroleum jelly on both surfaces

AO1 **(a)** The leaves will lose weight because they lose water. What is this loss of water called? [1]

AO1 **(b)** Describe how this water loss takes place. [3]

AO2 **(c)** Which leaf will lose the most water? Explain your answer. [2]

[Total: 6]

AO1 **3**
AO2 Look at the diagram of a fish and a green plant in an aquarium exposed to a bright light. Describe and explain the differences in gas levels in the aquarium during the day and night. [6]

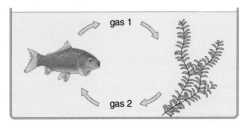

4 A fertiliser sack shows information on three important minerals

AO1 **(a)** Name these three minerals shown by the symbols N, P, K. [1]

AO2 **(b)** The fertiliser sack contains 1.66 kg of the fertiliser with the symbol N. Calculate the amounts of the other two minerals. [2]

[Total: 3]

3 : 6 : 9
10kg
NPK FERTILISER

5 Some scientists have investigated using different amounts of fertiliser on different species of rice.

amount of fertiliser in kg/hectare	mass of rice harvest in tonnes/hectare		
	variety A	variety B	variety C
0	3.9	4.4	5.2
20	4.4	5.6	5.5
40	5.1	6.8	4.9
60	4.4	8.0	4.0

AO3 **(a)** Farmers in poor, underdeveloped countries cannot afford to buy fertilisers. Which rice variety should they grow? Explain your answer. [2]

AO1 **(b)** Organic farmers do not use artificial pesticides. How do they control pests? [2]

[Total: 4]

AO1 recall the science   AO2 apply your knowledge   AO3 evaluate and analyse the evidence

✳ Worked Example – Foundation Tier

Jackie looks at cells of two different microscopic plants called algae. Each cell contains some green chloroplasts.

(a) Which alga will produce the most food? [3]

alga A   alga B

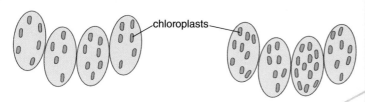

chloroplasts

Explain your answer.

Alga B because it has the most chloroplasts.

(b) Jackie then tested the algae for the presence of glucose. She found only a small amount. Suggest what could have happened to the glucose. [3]

The algae have used the glucose for energy. Any excess would be stored.

(c) Jackie grows the algae in containers inside a greenhouse.

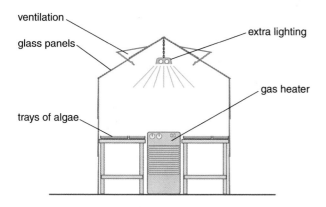

ventilation

glass panels

extra lighting

gas heater

trays of algae

Describe how the conditions in the greenhouse affect the rate of photosynthesis. [3]

The extra lighting will increase the rate of photosynthesis. The extra heat and carbon dioxide from the gas heater will also affect the rate of photosynthesis.

How to raise your grade!

Take note of these comments – they will help you to raise your grade.

The student has correctly stated alga B but since there are only two choices, examiners do not award a mark. All marks are for explanations. The student should have linked up more chloroplasts using more light with more photosynthesis.   1/3

The student correctly stated that glucose is used for energy but it is not stored. A more detailed answer points out that glucose is converted into starch (for storage), cellulose (to make cell walls), proteins and fats.
1/3

The student has correctly identified extra lighting and that it will increase the rate of photosynthesis.

The student has shown a common error. By using the word "affect" the student has not described whether the rate will increase or decrease.   1/3

This student has scored 3 marks out of a possible 9. This is below the standard of Grade C. With more care the student could have achieved a Grade C.

Higher Tier

1 Look at the kite diagram. It shows zonation of organisms.

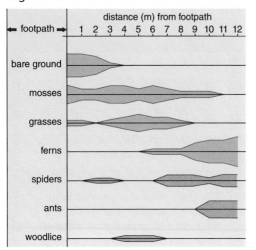

AO1 **(a)** What causes this zonation of organisms? [2]

AO2 **(b)** If the investigation was done on a path through a forestry plantation instead of a native woodland, the results would be different. Explain how and why they would be different [3]

[Total: 5]

2 Diagrams A and B show two different models of photosynthesis.

A   B

light energy   light energy

water   carbon dioxide   carbon dioxide   water

oxygen   hydrogen   oxygen   carbon
gas   ions   gas   monoxide

glucose and   glucose and
water   water

AO2 **(a)** Describe the important difference between the two models. [2]

AO1 **(b)** Explain how experiments proved A to be
AO2 correct. [4]

AO1 **(c)** Write down a balanced symbol equation for photosynthesis. [2]

[Total: 8]

AO2 **3** Potato cores were cut to the same length. Each
AO3 core was placed in a different concentration of sugar solution. After 30 minutes they were measured.

Describe and explain the results using the terms flaccid, plasmolysed and turgid. [6]

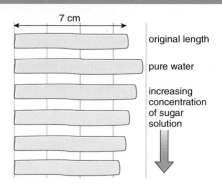

7 cm

original length
pure water
increasing concentration of sugar solution

AO1 **4 (a)** Phosphorus and potassium are two elements important to plants. Describe their importance to plants. [3]

AO1 **(b)** The two graphs show potassium and
AO2 bromine ion uptake in dandelion roots. Use these results to explain ion uptake in plants such as dandelions. [4]

[Total: 7]

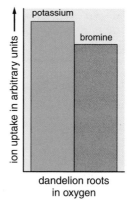
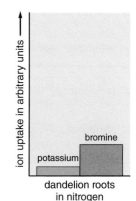

potassium / bromine

ion uptake in arbitrary units

dandelion roots in oxygen

bromine / potassium

ion uptake in arbitrary units

dandelion roots in nitrogen

5 Look at the diagram of a leaf.

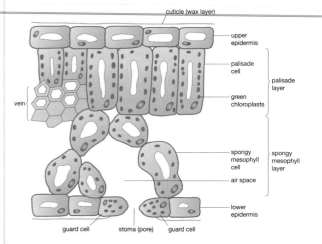

cuticle (wax layer)
upper epidermis
palisade cell
palisade layer
green chloroplasts
vein
spongy mesophyll cell
spongy mesophyll layer
air space
lower epidermis
guard cell   stoma (pore)   guard cell

Discuss this statement. [4]

"The leaf of a green plant is adapted for efficient photosynthesis rather than efficient transpiration".

AO1 recall the science   AO2 apply your knowledge   AO3 evaluate and analyse the evidence

✳ Worked Example – Higher Tier

(a) A potometer can be used in experiments to measure the rate of transpiration in plants.

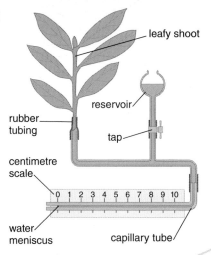

leafy shoot
reservoir
rubber tubing
tap
centimetre scale
0 1 2 3 4 5 6 7 8 9 10
water meniscus
capillary tube

How to raise your grade!
Take note of these comments – they will help you to raise your grade.

(i) What does a potometer actually measure? [1]

It measures water uptake by the stem.

> The student is correct. 1/1

(ii) Explain why this may not be the same as water loss. [1]

Some is used in respiration.

> The student is incorrect. Some water is used in photosynthesis. 0/1

(iii) Results of the experiment showed that the rate of transpiration was high in windy conditions. Explain why. [4]

Transpiration depends on diffusion of water. The wind will blow the water molecules away from the leaves.

> The student is correct in referring to diffusion and the removal of water molecules. However there is no information on evaporation of water or the effects of the wind removing water molecules from near the leaf surface so maintaining a concentration gradient for diffusion. 2/4

(b) The graph shows the rate of transpiration for two different plants, A and B, in the same conditions. Describe and explain the difference in results for plants A and B [4]

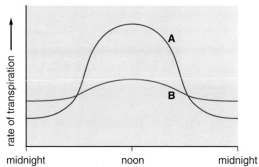

rate of transpiration
A
B
midnight   noon   midnight

For plant A, the rate of transpiration increases during the day reaching a maximum around noon. However the rate of transpiration for plant B remains fairly constant during day and night.

Plant A must have larger leaves than plant B

> The first part of the answer is correct.

> This is a poor explanation since there are no references to stomata. Apart from external conditions, the rate of transpiration will depend on the number, distribution, position and size of stomata. Plant A must have many, large exposed stomata. 2/4

> This student has scored 5 marks out of a possible 10. This is below the standard of Grade A. With more care the student could have achieved a Grade A.

C3 Chemical economics

Ideas you've met before

Particle theory

The three states of matter are solid, liquid and gas.

Changing state involves energy (in or out).

Every element is made of the same kind of atom.

Diagrams can be used to show elements, compounds and mixtures.

 What is the name ending of a two-element compound?

Chemical techniques

Separation involves using dissolving, filtering and crystallisation and chromatography.

Mass is conserved during chemical reactions.

Word equations are used to represent reactions.

 How do you know when a chemical reaction is taking place?

Energy in reactions

Fossil fuels can be burnt to provide energy.

Many other reactions also give out heat.

Heat is energy measured in joules.

Temperature is measured in °C and shows how hot an object is.

Change of state is linked to melting and boiling points.

 Can you name three fossil fuels?

The chemical industry

Metals can be extracted from ores.

Reduction is removing oxygen.

Rocks containing useful metals are called minerals.

The thermit reaction can be used to weld railway lines.

The reactivity series can be used to decide if a reaction will take place.

 Which metals occur naturally?

In C3 you will find out about...

> reaction rates and how they can be changed

> comparing reaction rates using graphs and gradients

> collision theory

> factors affecting reaction rates

> relative atomic mass and relative formula mass

> calculating the mass of reactants needed and products formed

> percentage yield

> atom economy

> the UK chemical industry

> sustainable use of resources

> energy changes in reactions – exothermic and endothermic

> calculating energy transfer

> testing for purity

> batch and continual processes

> factors affecting the cost of chemical manufacture

> allotropes of carbon and their uses – diamond, graphite and buckminsterfullerene

> nanochemistry and its uses

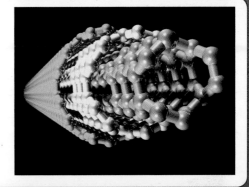

Rates of reaction (1)

You will find out:
> about reaction rates
> how reaction rates can be changed

Explosions

Some of the most impressive chemical reactions that you can see are explosions:

> An explosion is a very fast reaction that makes large volumes of gas.

> The gas molecules moving away from the centre of an explosion produce the 'explosive force'.

Some explosions are very useful, such as those used in mining produced by, for example, dynamite, TNT (trinitrotoluene) or nitroglycerine.

The explosive forces split large sections of rock so that the metal ores can be extracted.

Other uses of explosives can cause devastation and destruction.

FIGURE 1: Why are explosives used in a quarry?

Speed of reaction

In a chemical reaction reactants are made into products.

reactants → products

The **rate of reaction** measures how much product is made each second. Some reactions are very fast and others are very slow:

> Rusting is a very slow reaction.

> Burning and explosions are very fast reactions.

The **reaction time** is the time taken for the reaction to finish. The shorter the reaction time, the faster the reaction.

In factories it is important to be able to control the speed of a reaction:

> If a reaction is too fast there could be a dangerous explosion.

> If a reaction is too slow then the factory may not be able to make a material efficiently and will lose money.

Questions

1 Look at the word equation to show how ammonia can be made.

   hydrogen + nitrogen → ammonia

   List the reactants needed to make ammonia.

2 Name one reaction which is

a very fast

b very slow.

Reaction rates

The rate of reaction shows how much product is formed in a fixed time (or how fast a reactant disappears). The rate changes during a reaction.

Reactions are usually faster at the start, and then they slow down as the reactants are used up.

An example is the reaction between magnesium ribbon and hydrochloric acid.

$Mg + 2HCl \rightarrow MgCl_2 + H_2$

Hydrogen is a gas so it is easy to measure how much is produced.

If the reaction is slow, the rate of reaction might be measured in cm^3/min. If it is faster, it might be measured in cm^3/s.

Q chemical explosions   reactant

Sometimes it is easier to measure the mass of a product formed. The rate of reaction is then measured in g/s or g/min.

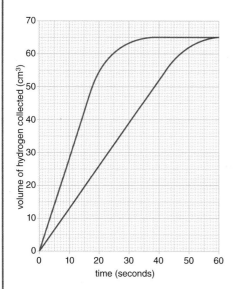

FIGURE 2: This graph shows what happens when the quantity of reactants changes. At the start, the gradient is steeper on the blue line. What does this mean?

Questions

3 Brian does two experiments on reacting magnesium ribbon with hydrochloric acid. He measures the volume of hydrogen produced every minute. Here are his results.

Time (minutes)	0	1	2	3	4	5
Experiment 1. Hydrogen produced (cm³)	0	30	33	38	40	40
Experiment 2. Hydrogen produced (cm³)	0	60	67	76	79	80

Use the data in the table to:

a plot one graph that shows both sets of results.

b label the graph to show where the reaction is fastest.

c work out the time when each reaction has finished.

d Find out what Brian changed in the experiment.

Remember!

The steeper the **gradient** (slope) on the graph, the faster the reaction.
As the gradient decreases, the reaction slows down.
When the line is horizontal the reaction has finished, so no more product is being made. This is because one of the reactants has been used up.

Calculating the rate of reaction

The rate of reaction can be worked out from the gradient of a graph.

The gradient is found by drawing construction lines. Choose a part of the graph where there is a straight line (not a curve). Measure the value of y and x and remember to use the scale of the graph carefully. Then divide y by x. Determine the units from those given on the axes.

gradient = y/x = 51/25 = 20.5 cm³/s

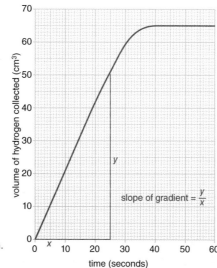

FIGURE 3: Calculating the gradient of a graph. Why is this useful?

slope of gradient = $\frac{y}{x}$

Questions

4 Magnesium ribbon and hydrochloric acid are put into a beaker. Hydrogen escapes from the reaction mixture. Describe an experiment to find out the total mass of hydrogen given off every 10 seconds.

5 If 30 cm³ are made in 20 seconds, what is the reaction rate?

Measuring the rate of reaction

It is difficult to measure the rate of very fast or very slow reactions.

However, the rate of reaction between magnesium and dilute hydrochloric acid can be measured easily in the laboratory.

During the reaction, magnesium ribbon fizzes in dilute hydrochloric acid and colourless bubbles of hydrogen are given off. The volume of hydrogen collected in the gas syringe can be measured every 10 seconds.

magnesium + hydrochloric acid → magnesium chloride + hydrogen

The results of the experiment can be plotted on a graph, as shown in Figure 5.

The **gradient** (slope) of the line shows how fast the reaction is.

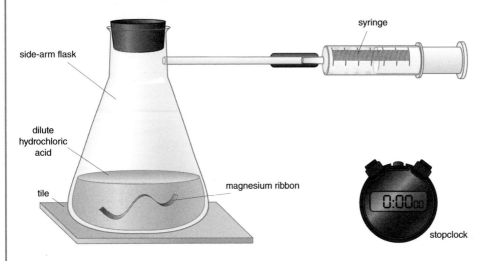

FIGURE 4: Measuring the rate of reaction between magnesium and dilute hydrochloric acid. What will you see in the flask that shows the reaction is taking place?

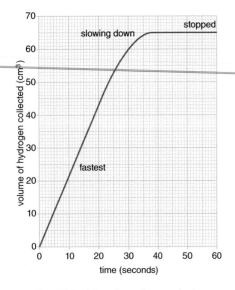

FIGURE 5: What does the graph show about the rate of reaction? Why does the line flatten out?

Questions

6 Look at the line on the graph in Figure 5.

a What is the volume of gas given off after 10 seconds?

b How can you tell from the graph when the reaction has finished?

c Why does the reaction finish?

d What is the total volume of hydrogen made when the reaction has finished?

e When the reaction is fastest, what is the gradient like?

Q science line graphs   word equations

Limiting reactants

$Mg + 2HCl \rightarrow MgCl_2 + H_2$

In this reaction, both Mg and HCl are the **reactants**. To find the rate of reaction we can only change the amount of one of these.

To make sure the one we change fully reacts, the other reactant needs to be in excess (use more than needed).

The one *not* in excess is called the **limiting reactant**, and this is the one that is used up by the end of the reaction.

How much product?

The amount of **product** is *directly proportional* to the amount of limiting reactant used.

So, if the amount of this reactant doubles, the amount of product doubles.

The more reactant that is used, the more **product** it forms.

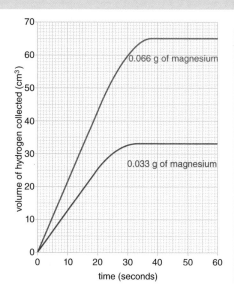

FIGURE 6: The graph shows the volume of hydrogen produced over time when different amounts of magnesium ribbon are added to dilute hydrochloric acid. The red line shows that 0.033 g of magnesium produces 33 cm³ of hydrogen. How much hydrogen is formed if the amount of magnesium is doubled to 0.066 g?

Questions

7 Look at the graph in Figure 6.

a How much hydrogen would be produced if 0.0165 g of magnesium is used?

b Why are the amounts of hydrogen produced proportional to the amount magnesium used?

c Which reactant is used up by the end of a reaction?

More on the limiting reactant

Reactions occur when particles collide together with sufficient energy to react.

The amount of product formed depends on the amount of reactant used.

If the number of particles of the limiting reactant is doubled, the number of successful collisions is also doubled. This should double the number of product particles made.

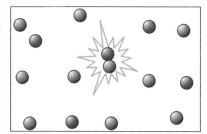

FIGURE 7: Which colour represents the limiting reactant?

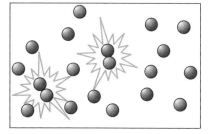

FIGURE 8: What is the effect on the product of doubling the amount of limiting reactant in Figure 7?

Questions

8 Explain what is meant by 'limiting reactant'.

9 What effect would tripling the number of reactant particles have on the amount of product produced?

Rates of reaction (2)

You will find out:
> about collision theory
> some factors that affect reaction rate

Fireworks

Reactions in fireworks need to be controlled.

Firework rockets need to get high before colourful explosions take place.

The time between explosions also needs controlling for maximum effect.

FIGURE 1: Why do reaction rates need to be controlled in fireworks?

Collision theory

Chemical reactions take place when reactant particles hit or collide with each other.

To make the reaction faster, we need to increase the number of collisions.

To make the reaction slower, we need to reduce the number of collisions.

We can make the reaction faster by:

> increasing the **concentration**

> increasing the temperature

> increasing the pressure (but only if the reactants are gases).

The concentration or molarity of solutions is measured in mol/dm^3.

The higher the mol/dm^3, the more concentrated the acid is.

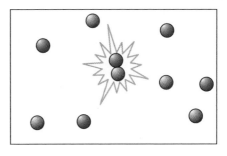

FIGURE 2: The blue and red spheres represent reacting particles. How many particles have reacted? How many more could react?

Remember!
Concentration is the amount of chemical dissolved in a certain volume. For example, 20 g dissolved in 50 cm^3 is a higher concentration than 10 g dissolved in 50 cm^3.

Questions

1 Name three ways to change reaction rate.

2 How could you slow down a reaction?

3 For what type of reactants can pressure be used to change the reaction rate?

🔍 chemical concentration   pressure in gases

More on collision theory

Scientists use models to explain ideas like collision theory.

Concentration and collision theory

As the concentration increases the particles become more crowded.

For example, at the low concentration in Figure 3 there are four particles of A. At the higher concentration there are ten particles of A in the same volume.

The particles are more crowded *in the same volume* so there are more collisions and the rate of reaction increases.

Temperature and collision theory

Particles move faster as the temperature increases. The reacting particles have more **kinetic energy** and so the number of collisions increases and the rate of reaction increases.

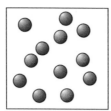

low concentration — high concentration

reacting particle of substance **A**
reacting particle of substance **B**

FIGURE 3: Why does the rate of reaction increase when the concentration of reactants increases?

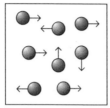
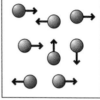

low temperature — high temperature

reacting particle of substance **A**
reacting particle of substance **B**

FIGURE 4: The thickness of the arrows indicates the amount of kinetic energy that the particles have. Why does an increase in temperature makes a reaction go faster?

Questions

4 a Draw diagrams to show the particles in a gas at low pressure and the particles in a gas at high pressure.

b Use your diagram to explain why a reaction between two gases happens faster at high pressure than at low pressure.

Even more on collision theory

It is not just the number of collisions that determines the rate of a reaction, it is the **collision frequency**.

Collision frequency describes the number of successful collisions between reactant particles that happen each second. The more successful collisions per second, the faster the reaction.

For a successful collision to occur each particle must have sufficient energy to react.

As the concentration increases the number of collisions per second increases and so the rate of reaction increases.

As the temperature increases the reactant particles have more kinetic energy and so the collisions are more energetic and successful and, consequently, the rate of reaction increases.

As the pressure on reacting gases increases, particles are squeezed closer together, and the number of successful collisions per second increases and so the rate of reaction increases.

Remember!
In a Higher Tier paper always refer to 'more collisions per second', not just 'more collisions'.

Questions

5 Explain why reactions between gases happen faster at high temperature than at low temperature.

6 Suggest why increasing the pressure has little effect on the rate of a reaction between two solids.

You will find out:

> how to analyse experimental data on rates of reaction

Interpreting data

To measure the rate of reaction you can change one of these:

> the temperature

> the concentration

> the pressure.

You could obtain results by collecting gas in a syringe or in a measuring cylinder.

The results make a graph shape that looks like the one in Figure 5.

You could obtain results by measuring the mass loss.

The results make a graph shape like the one in Figure 6.

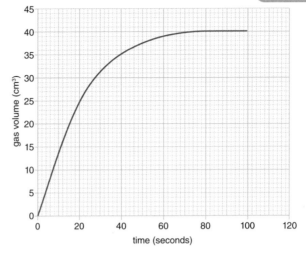

FIGURE 5: Is the gas made quickly or slowly at first? What happens to the rate of gas made between 40 and 60 seconds?

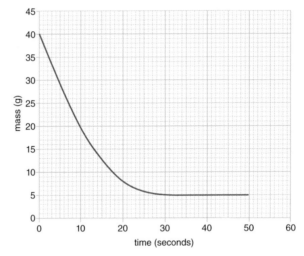

FIGURE 6: How do you know when the reaction has finished?

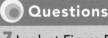

Questions

7 Look at Figure 5.

a How much gas is made after 20 seconds?

b What volume is gas is made in total?

8 Look at Figure 6.

a Which part of the graph shows the fastest reaction?

b Which part of the graph shows where the reaction is complete.

c Why does the graph slope downwards?

Rate of reaction and concentration

Figure 7 shows the results from two different experiments. In both experiments a 3 cm length of magnesium ribbon and 50 cm³ of hydrochloric acid were used. The blue line shows the results of the experiment with concentrated hydrochloric acid. The red line shows the results of the experiment with dilute hydrochloric acid.

The gradient of the blue line is greater than that of the red line. This shows that the rate of reaction is faster when concentrated acid is used.

The total volume of hydrogen produced during both experiments is the same. This is because excess acid and the same mass of magnesium are used.

Rate of reaction and temperature

Figure 8 shows the results from two different experiments. In both experiments a 3 cm length of magnesium and 50 cm³ of hydrochloric acid were used.

The blue line shows the results with the acid at a temperature of 30 °C. Its gradient is greater than that for the red line, which describes the experiment in which the acid was at a temperature of 20 °C.

The total volume of hydrogen produced during both experiments is the same. This is because excess acid and the same mass of magnesium are used.

Q create a graph   reading science graphs

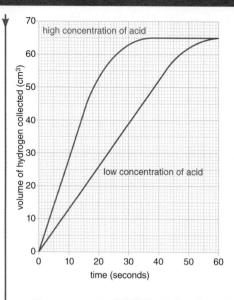

FIGURE 7: Graph to show how rate of reaction changes with a change in concentration of the reactants. How can you tell that one of these reactions proceeded faster than the other?

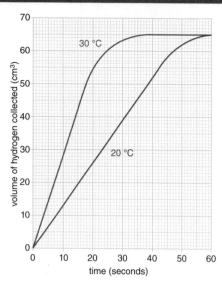

FIGURE 8: Graph to show how the rate of reaction changes with a change in temperature. After what time has one of the reactants been used up?

Questions

9 Explain why the reaction between magnesium and dilute acid is slower than the reaction with concentrated acid.

10 Explain why the reaction between magnesium and dilute hydrochloric acid becomes faster as the temperature of the acid increases.

11 Sketch a graph to show the effect of changing pressure on:

a the rate of reaction

b the amount of product formed.

Extrapolation and interpolation

Extrapolation – making an estimate beyond the range of results.

Interpolation – making an estimate between results in a range.

It is useful to be able to do both these for results in tables and graphs.

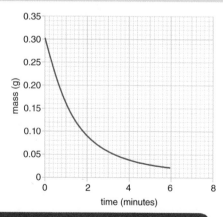

FIGURE 9: Graph of reaction measurements.

Questions

12 a Interpolate the reading at 3 minutes.

b Extrapolate the reading to 8 minutes.

c What pattern does the graph show?

d How can you find the rate of reaction after 2 minutes?

Did you know?

As a very rough guide, in many reactions starting at room temperature, a rise of 10 °C doubles the rate of reaction.

Q reaction rate graphs   interpolation and extrapolation

Rates of reaction (3)

You will find out:
> about explosions
> how changing the surface area affects the rate of reaction
> what catalysts are and what they do

The Hindenburg disaster

In 1937, the German airship LZ 129 Hindenburg was about to dock at Lakehurst Naval Air Station in America. In less than a minute, 200 000 m³ of hydrogen had exploded and burned. Modern airships are filled with non-combustible helium.

FIGURE 1: The Hindenburg airship disaster

Explosions

An **explosion** is a very fast chemical reaction that makes and releases large volumes of gases. The gases move outwards from the reaction at great speed, which causes the explosive effect.

Useful commercial explosives are TNT (trinitrotoluene) and dynamite.

Remember the 'squeaky pop' test for hydrogen gas? This is a very quick reaction when a mixture of hydrogen and oxygen gas is ignited.

Fine powders in the air, like flour can also explode if they are ignited.

Powder or lump?

Figure 2 shows two beakers in which reactions are taking place:

> Each beaker contains 1.0 g of calcium carbonate reacting with 40 cm³ of dilute hydrochloric acid.

> The temperature of the acid is the same in both beakers.

> The only difference is that one beaker contains powdered calcium carbonate and the other contains a lump of calcium carbonate.

Powdered reactants always react faster than lumps. Breaking up a substance into smaller bits makes the reaction faster.

FIGURE 2: Experiment to see if a reaction is faster if the solid reactant is a powder. How can you tell that the reaction on the left is faster?

Did you know?

A factory making custard powder was once destroyed in an explosion.

Fine custard powder in the air was ignited by an electrical spark.

Questions

1 What is an explosion?

2 How is the reaction speed changed when a chemical is broken up into smaller bits?

More on explosions

Fine powders, such as flour, custard, sulfur and coal dust, are all flammable if they spread out into the air and catch fire. This can be a potential danger in some factories. The reason for this is to do with the increased surface area in contact with oxygen.

Surface area and collision theory

The surface area of a powdered reactant is much larger than the surface area of a block of reactant with the same mass. As a block is broken into smaller bits, more surfaces are created.

In the left-hand diagram in Figure 3 the blue particles are in a block. The red particles can only react when they collide with the surface of the block, so the reaction is slow. In the right-hand diagram, the block is broken up, so the surface area is increased and there are more collisions between the particles. This means the rate of reaction increases.

small surface area   large surface area

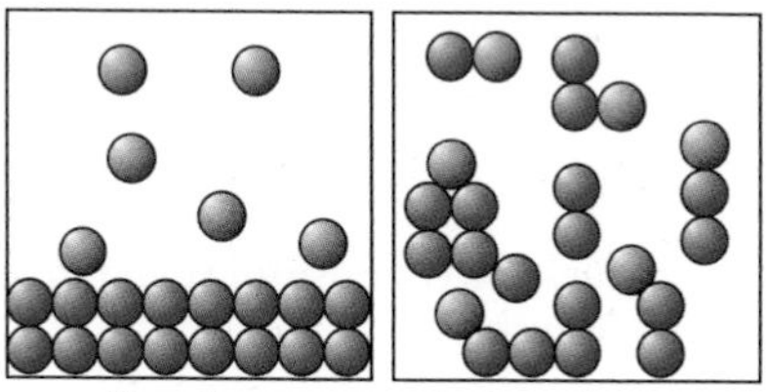

FIGURE 3: Explain why when a solid is made into a powder it reacts faster. What do the red and blue dots represent?

Questions

3 Explain why flour dust in the air can explode?

4 Explain why it is quicker to fry 100 g of thin chips than 100 g of thick chips.

5 A factory uses combustible powdered sulfur. One of the safety rules in the factory is that workers should report all spillages of powdered sulfur. Suggest, with reasons, some other safety rules.

More on surface area and collision theory

It is the collision frequency between reactant particles that is important in determining how fast a reaction takes place. The more successful collisions there are each second, the faster a reaction.

When the surface area of a solid reactant is increased there are more collisions each second. This means the rate of reaction increases.

Questions

6 Explain why it is difficult to find the rate of reaction when changing the surface area of marble chips (calcium carbonate).

7 If only the surface area is changed, explain why the reaction rate changes, but not the total volume of product made.

🔍 surface area reaction rate   factory explosions

Catalysts

A **catalyst**:

> is a substance added to a chemical reaction to make the reaction go faster

> does not change how much of the product is made.

A catalyst is very useful when making chemicals. When a catalyst is added to a reaction the same amount of product is produced but in a much shorter time.

Analysis of experimental results

Paul and Meghali investigate the reaction between zinc and sulfuric acid. Zinc sulfate solution and hydrogen gas are formed.

They want to see if they can find a catalyst for this reaction.

In each experiment they use 1.5 g of zinc powder and 50 cm³ of dilute sulfuric acid. They also add 0.1 g of another substance.

Paul and Meghali measure the time it takes to collect 100 cm³ of hydrogen in the gas syringe. Their results are shown in the table.

Experiment number	Substance	Time to collect 100 cm³ of gas (seconds)	Colour at the start	Colour at the end
1	no substance	150	—	—
2	magnesium chloride	150	white	white
3	copper chloride	15	green	pink
4	copper powder	25	pink	pink
5	iron(II) sulfate	20	green	grey

Did you know?

The *time taken* for a reaction to finish can be longer or shorter, but not faster or slower.

Questions

8 Use the information, and the table to answer the following.

a What is the name of the gas collected in the gas syringe?

b Which substance does not speed up the reaction?

c Copper powder made the reaction faster but was not used up. What type of substance is copper powder in this reaction?

Catalysts and rate of reaction

A catalyst increases the rate of a reaction. Some other properties of a catalyst are:

> it is unchanged at the end of a reaction

> only a small mass of catalyst is needed to catalyse a large mass of reactants

> specific to one single reaction.

Most catalysts only make a specific reaction faster. They do not make all reactions faster. Although copper catalyses the reaction between zinc and dilute sulfuric acid it will not catalyse other reactions. This means that scientists have had to discover many different catalysts. Examples are:

> zeolites or aluminium oxide in the cracking of long-chain hydrocarbons

> rhodium-based catalysts in a catalytic converter.

everyday uses of catalysts   chemical reactions

A catalyst does not increase the number of collisions per second. Instead, it works by making the collisions that take place more successful.

> It helps reacting particles collide with the correct orientation.

> It allows collisions between particles with less kinetic energy than normal to be successful.

Scientists are still searching for new catalysts that can make a specific reaction faster, more energy efficient and more cost effective.

Analysis of experimental results

Look at Paul and Meghali's results on the facing page. It is possible to conclude that *only* copper powder is a catalyst for the reaction between zinc and sulfuric acid.

The results show that when copper powder is added, the reaction is faster than when zinc and sulfuric acid alone are used. There was no change in the appearance of the copper by the end of the reaction.

In another experiment, Paul and Meghali use 0.27 g of zinc powder, 0.1 g of copper powder and 50 cm³ of sulfuric acid. The acid is in excess.

This time they measure the total volume of gas collected in the gas syringe every
10 seconds.

The graph in Figure 4 shows that the reaction stops when the volume of hydrogen remains constant. This occurs after 20 seconds.

If a 0.27 g lump of zinc is used instead of zinc powder, the graph obtained is different. The initial gradient of the line is less, but at the end of the reaction the same volume of gas (100 cm³) is produced.

If a 0.135 g lump of zinc is used, the final volume of gas produced is 50 cm³.

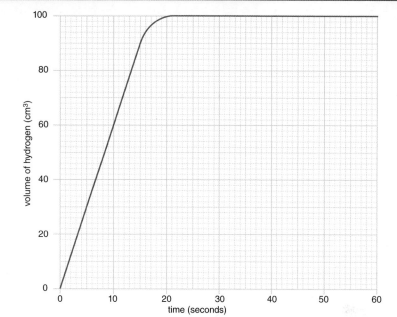

FIGURE 4: Graph to show the volume of gas collected when zinc powder and sulfuric acid react in the presence of copper powder. What would the graph look like if copper powder was not added?

Questions

9 Using information on this page and from the table on the opposite page explain:

a why copper chloride is not a catalyst for the reaction between zinc and sulfuric acid

b why excess acid is needed.

c what effect would using a block of copper instead of powder have?

10 Paul and Meghali repeat the experiment with the gas syringe. This time they use a 0.27 g lump of zinc and 100 cm³ of sulfuric acid.

Describe and explain how the rate of reaction and the total volume of hydrogen gas formed will change.

11 Write a balanced symbol equation for the reaction between zinc and sulfuric acid.

Reacting masses

You will find out:
> about relative atomic mass
> how to work out relative formula mass

Q. If an atom of hydrogen reacts with an atom of chlorine, why doesn't a gram of hydrogen react with a gram of chlorine?

A. Because hydrogen and chlorine have different atomic masses.

Element	Symbol	Relative atomic mass
hydrogen	H	1
carbon	C	12
nitrogen	N	14
oxygen	O	16
sulfur	S	32
chlorine	Cl	35.5

Relative atomic mass

Atoms of different elements have different masses. We compare their masses using the relative atomic mass scale:

> Hydrogen has a mass of 1 on this scale.

> Other elements are heavier. For example, carbon has a relative atomic mass of 12.

Look at the periodic table on page 318 and find the **relative atomic mass** of some of the other elements.

Relative formula mass

What if the atoms are combined?

Simply add up all the masses in the formula.

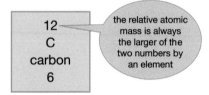

the relative atomic mass is always the larger of the two numbers by an element

an element in the periodic table

FIGURE 1: What is the relative atomic mass of carbon?

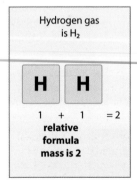

Hydrogen gas is H_2

$1 + 1 = 2$
relative formula mass is 2

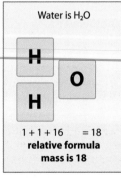

Water is H_2O

$1 + 1 + 16 = 18$
relative formula mass is 18

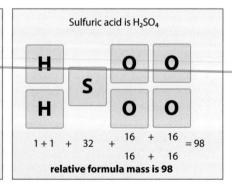

Sulfuric acid is H_2SO_4

$1 + 1 + 32 + \begin{matrix}16 + 16 \\ 16 + 16\end{matrix} = 98$
relative formula mass is 98

FIGURE 2: How to work out relative formula masses.

Questions

You will need to use the periodic table on page 318 to help you to answer the following questions.

1 Lead, Pb, is a very dense metal. Find its relative atomic mass.

2 Uranium, U, is even more dense. Find its relative atomic mass.

3 Work out the relative formula mass of carbon dioxide, CO_2.

4 Work out the relative formula mass of propane, C_3H_8.

Did you know?

A relative mass is just a number. It doesn't have 'grams' after it.

More on relative formula masses

Relative formula masses need more thought to work out when the atoms are within brackets.

Find the relative formula mass of ammonium carbonate, $(NH_4)_2CO_3$ (N=14, H=1, C=12, O=16).

1. Work out the inside of the bracket.	$14 + (1 \times 4)$	= 18
2. Multiply the total by the number outside the bracket.	18×2	= 36
3. Work out the remaining part.	$12 + (16 \times 3)$	= 60
4. Add the first part, $(NH_4)_2$, to the second part, CO_3		96

Remember!

The shorthand for relative atomic mass is A_r and for relative molecular mass is M_r

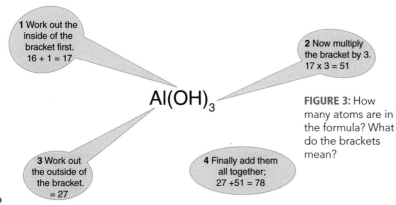

1 Work out the inside of the bracket first.
$16 + 1 = 17$

$Al(OH)_3$

2 Now multiply the bracket by 3.
$17 \times 3 = 51$

3 Work out the outside of the bracket.
$= 27$

4 Finally add them all together;
$27 + 51 = 78$

FIGURE 3: How many atoms are in the formula? What do the brackets mean?

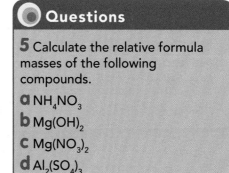

Questions

5 Calculate the relative formula masses of the following compounds.

a NH_4NO_3

b $Mg(OH)_2$

c $Mg(NO_3)_2$

d $Al_2(SO_4)_3$

The history of relative atomic mass

A relative mass is, in fact, a ratio, so there are no units.

It is a way of comparing the same number of atoms in different elements.

In 1803 John Dalton arranged six known elements in sequence by comparing their mass to the lightest element, hydrogen. He defined that the relative atomic mass of an atom was the number of times it was heavier than hydrogen. Later, in 1869, Dmitri Mendeleev used this idea to create the periodic table we still use today.

However, as most elements have isotopes (atoms with the same atomic number, but different atomic masses) this put some elements in the wrong places. Oxygen was tried next as a comparison, but eventually in 1961 carbon-12 was chosen as it was easy to measure accurately. Everything is now compared to one-twelfth of an atom of carbon (which is effectively the same as comparing everything to hydrogen).

To eliminate the problem of isotopes altogether, the modern periodic table is arranged in order of atomic number.

Questions

6 Look at the periodic table on page 318. It is arranged by atomic number. How different would it be if the first 36 elements were arranged by relative atomic mass?

Q relative formula mass practice   atomic mass history

Conservation of mass

The total mass in a reaction never changes.

An example is heating copper carbonate. It breaks down into copper oxide and carbon dioxide.

Figure 4 shows that:

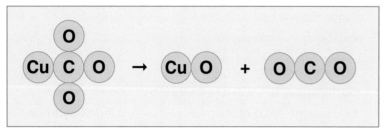

FIGURE 4: In this reaction, what can you say about the total mass on each side?

> the same number (five) of atoms are on the left as on the right side

> the number of each element on each side is also the same

> all the atoms in the reactant (left side) are rearranged into the products (right side).

In a reaction the total mass of the products is exactly the same as the total mass of the reactants. If the mass of the reactants increases, the mass of the products increases.

Simple calculations

You can work out how much product is made in a reaction without knowing any chemical equations.

Using conservation of mass

When calcium carbonate is heated it turns into calcium oxide and carbon dioxide.

10 g of calcium carbonate produces 5.6 g of calcium oxide.

How much carbon dioxide is made?

If you start with 10 g of reactant, you must end up with 10 g of product.

5.6 g of the product is calcium oxide, so 10 – 5.6 = 4.4 g of carbon dioxide.

Questions

7 In the reaction shown in Figure 4, when 12.4 g of copper carbonate are heated, 8 g of copper oxide are made. How much carbon dioxide will be made?

8 2.4 g magnesium makes 4.0 g of magnesium oxide.

a How much magnesium oxide would 4.8 g of magnesium make?

b How much oxygen would react with 2.4 g of magnesium to make 4.0 g of magnesium oxide?

Calculating reacting masses

Copper carbonate decomposes (breaks down) when it is heated into copper oxide and carbon dioxide.

The word equation for the reaction is:

copper carbonate → copper oxide + carbon dioxide

The symbol equation is:

$CuCO_3 \rightarrow CuO + CO_2$

The relative atomic masses are:

$64 + 12 + (16 \times 3) \rightarrow (64 + 16) + (12 + (16 \times 2))$

The relative formula masses are:

$124 \rightarrow 80 + 44 = 124$

So the total mass of the reactants = the total mass of the products:

124 g → 80g + 44 g = 124 g

This is conservation of mass. When chemicals react, the atoms inside the reactants swap places to make new compounds – the products.

These products are made from exactly the same atoms as before. There are the same number and type of atom on each side of the equation so the overall mass stays the same.

Atoms are not created or destroyed during a chemical reaction.

Limiting factors

When magnesium ribbon is put into acid, it reacts giving off hydrogen gas.

To make sure all the magnesium is dissolved, excess acid is used.

In excess acid, the amount of gas produced is directly proportional to the amount of magnesium used.

Remember!

Check the relative formula mass calculations by checking both sides are equal.
If a question asks for tonnes rather than grams, this is not a problem – any units can be assigned once the relative formula mass is calculated.

 Questions

9 In the copper carbonate reaction on page 112, what would happen if you:

a Doubled the amount of copper carbonate?

b Halved the amount of copper carbonate?

10 hydrogen + oxygen → water

32 g of oxygen reacted with hydrogen to make 36 g of water.

How much hydrogen was used?

11 32 g of copper made 40 g of copper oxide.

How much copper is needed to make 2 g of copper oxide?

12 In excess acid, 2.4 g of magnesium makes 0.2 g of hydrogen.

a How much magnesium would be needed to make 1g of hydrogen?

b Describe why adding more acid would not make more hydrogen.

Predicting the yield of a reaction

Predictions can be made by looking at the equation for a reaction.

Example 1

$CaCO_3 \rightarrow CaO + CO_2$
How much CO_2 is made when 500 g of $CaCO_3$ decomposes?
Find the relative formula mass of $CaCO_3$:
$40 + 12 + 16 + 16 + 16 = 100$
Now find the relative formula mass of CO_2:
$12 + 16 + 16 = 44$
So 100 g of $CaCO_3$ gives 44 g of CO_2.
So 500 g of $CaCO_3$ gives $5 \times 44 = 220$ g of CO_2.

Example 2

How much iron is produced from 480 tonnes of iron (III) oxide?
$Fe_2O_3 + 3CO \rightarrow 2Fe + 3CO_2$
Find the relative formula mass of Fe_2O_3:
$(56 \times 2) + (3 \times 16) = 160$
Find 2× the relative formula mass of Fe:
$(56 \times 2) = 112$
So 160 tonnes of Fe_2O_3 make 112 tonnes of iron.
The question states that 480 tonnes of iron (III) oxide are used, which is three times 160, so the amount of iron made is $112 \times 3 = 336$ tonnes.

 Questions

You will need to use the relative atomic mass numbers on page 318.

13 $H_2 + Cl_2 \rightarrow 2HCl$

How much HCl is made from 2 g of hydrogen?

14 The equation for burning propane is:

$C_3H_8 + 5O_2 \rightarrow 3CO_2 + 4H_2O$

a How much water is produced by burning 44 g of propane?

b How much water is produced by burning 22 g of propane?

c How much carbon dioxide is produced by burning 22 g of propane?

An alternative method is to find how much 1 tonne of iron oxide makes, then multiply by the amount asked for:

1 tonne of iron (III) oxide makes:

$\frac{112}{160} = 0.7$ tonnes

0.7 tonnes $\times 480 = 336$ tonnes

Preparing for assessment: Applying your knowledge

To achieve a good grade in science, you not only have to know and understand scientific ideas, but you need to be able to apply them to other situations and investigations. These tasks will support you in developing these skills.

✳ Fighting a war with chemicals

Fritz Haber's name is associated with the Haber process used to make ammonia. Haber was born in 1868 in Prussia – a part of Europe that was to become Germany.

Ammonia was important to Germany (and to many other countries) for two reasons – it is used to make fertilisers and explosives.

When Haber started looking at the supply of ammonia, the industry was already well established but relied on a natural source – bird droppings.

Haber, together with another German scientist, Carl Bosch, developed a way of manufacturing ammonia. Ammonia has the formula NH_3 and they came up with a way of making it from nitrogen and hydrogen, at conditions of high temperature and pressure and with an iron catalyst.

At the outbreak of World War I, Haber placed himself and his team of scientists at the service of the German government. His development of ammonia production meant that Germany could produce large quantities of explosives for the war effort and undoubtedly helped to increase its effectiveness as a fighting force. Haber also developed methods of chemical warfare and he organised and directed the use of chlorine gas at Ypres in 1915, which killed thousands of Allied troops and hundreds of German troops as well.

Troops in the trenches in World War I, wearing gas masks to protect themselves against chemical attack.

Germany lost the war in 1918, but in that year Haber was awarded the Nobel Prize for his work on the synthesis of ammonia. There was criticism of him though in the scientific community because of his willingness to use his skills in developing weapons of war.

In Chile, South America, there was one of the world's largest deposits of bird droppings. It was about 220 miles long and over 1.5 metres thick. The deposits are called guano.

Guano deposits produce a natural source of ammonia.

Maximise your grade

	Answer includes showing that you can...
F	State that for nitrogen and hydrogen to react their particles must collide.
	Explain what happens to the rate of reaction when gases react and the pressure is increased.
	Explain the term 'catalyst' and explain why a catalyst (iron) is needed in the reaction to make ammonia.
	Calculate the formula mass of ammonia (NH_3) from the atomic masses of nitrogen (14) and hydrogen (1).
C	Explain why the formula of ammonia (NH_3) contains two elements and four atoms and that the equation for the formation of ammonia is balanced. $N_2 + 3H_2 \rightleftharpoons 2NH_3$
	Explain why batch processes are often used for the production of pharmaceutical drugs but continuous processes are used to produce chemicals such as ammonia.
	Explain what type of energy transfer takes place when nitrogen or hydrogen bonds are broken and what type of energy transfer takes place when nitrogen bonds with hydrogen to make ammonia.
A	Calculate the atom economy of the reaction to make ammonia using the equation and atomic masses given above and explain why this reaction has such a beneficial atom economy.
	Explain why the reaction of nitrogen and hydrogen to make ammonia is exothermic overall. Use ideas about bond making and bond breaking.
	Explain, using the reacting particle model, why changes in pressure change the rate of reaction in terms of successful collisions between particles.
	As above, but with particular clarity and detail.

 Task 1

In the period following World War I, Haber received recognition at the highest level for his work on ammonia synthesis, but he was criticised by many other scientists in the international community. Explain why opinions about him varied.

 Task 2

Why is it true that ammonia has the ability to sustain life and to destroy it?

 Task 3

Ammonia is made by a continuous process, not a batch process. Explain why chemicals like ammonia are made using a continuous process whereas pharmaceutical drugs are made by batch processes.

 Task 4

Iron has the symbol Fe. If iron is needed in the synthesis of ammonia, why does iron not appear in the formula for ammonia, NH_3? Use ideas about catalysts.

 Task 5

Do some research on the synthesis of ammonia from nitrogen and explain why it is not surprising that the reaction needs a very high temperature and a high pressure to keep a high rate of reaction.

Percentage yield and atom economy

You will find out:
> about the chemical industry
> about percentage yield

Making chemicals

Making chemicals is one of the biggest industries in the UK. Over two million scientists are employed directly by the industry.

The companies make millions of different compounds for use as drugs, health-care products, food additives, agricultural products, cleaning materials, fuels, packaging, etc. Products are sold worldwide.

About 10% of the profits are put back into research.

It can take up to 10 years to test and trial a new medical drug to make sure it is safe to use. Companies need to get a licence to manufacture and sell a new drug.

FIGURE 1: Chemicals are made in the laboratory as well as on an industrial scale.

Yield and percentage yield

The **percentage yield** of a reaction is a way of comparing the amount of product made (the actual yield) to the amount we expect to make (the predicted yield).

The more reactant is used, the more product is made.

Sometimes a reaction does not seem to produce as much chemical as it should.

In real experiments, 100% of all of the chemicals do not react and some always get left behind.

The amount of product collected is usually less than the amount expected:

> 100% yield means that no product has been lost.
> 0% yield means that no product has been collected.

If all the lost bits of chemical could be collected, the total mass of product would be the same as the total mass of reactant.

How do bits of chemical get lost? Look at this example.

Ben adds zinc to acid. When no more reacts, he filters off the unreacted zinc. Next, Ben evaporates the solution to make crystals. The following are possible reasons why some of the product could be lost:

> the zinc and acid might not have finished reacting
> some of the solution would stick to the beaker when it was poured
> some solution would soak into the filter paper
> some product might spit out during evaporation.

Questions

1 What does predicted yield mean?

2 What does actual yield mean?

3 Joy makes a chemical by pouring two solutions into a beaker and then filtering off the solid.

Suggest two reasons why the yield is less than 100%.

4 A reaction produces less than half the expected product. Choose the correct percentage yield from the list.

30%   50%   70%   100%

5 State three possible reasons for not getting 100% yield.

🔍 folding filter paper   soluble

Calculating percentage yield

To calculate percentage yield the following two things must be known:

> the amount of product made, the 'actual yield'

> the amount of product that should have been made, the 'predicted yield'.

$$\text{percentage yield} = \frac{\text{actual yield} \times 100}{\text{predicted yield}}$$

Example

A company making sulfuric acids gets an actual yield of 74 tonnes. They predicted a yield of 85 tonnes.

What is the percentage yield?

$$\text{Percentage yield} = \frac{74}{85} \times 100 = 87\%$$

 Questions

6 a Fred reacted potassium iodide with lead nitrate. He expected to make 20 g but only made 18 g.

What was the percentage yield?

b Suggest two reasons why some product was lost.

7 Mohammed makes some copper oxide. The reaction produces 48 g.

The predicted yield was 64 g.

What is the percentage yield of his reaction?

Making a profit

Industry needs to reduce costs and maximise profits.

To do this, yields need to be as high as possible.

To get the yield high all the reactants need to be converted into the product.

To prevent reactants being wasted, chemists need to know the exact amount to use to make a certain amount of product. Too much reactant is wasteful and costly, and too little reduces the amount of product.

While the exact amount of reactants to use can be calculated, reactions are never 100% efficient. Some product can be lost during stages like separating, purification and packaging.

Questions

8 Which two of the following actions will increase profits?
a reuse reactants that have not reacted **b** make sure excess reactants are used to make the product **c** package products in recyclable materials **d** thermally insulate the equipment.

9 Why is it important for companies to cut costs?

Atom economy

Atom economy is a way of measuring the amount of atoms that are wasted or lost when a chemical is made.

100% atom economy means that all the atoms in a reactant have changed into the desired product.

Calcium hydroxide (limewater) can be made like this:

calcium oxide + water → calcium hydroxide solution (limewater)

All the reactants are used to make a single product.

This means the atom economy is high, so the process is 'greener', meaning it is more environmentally friendly.

If less than 100% of the atoms are changed, some of the reactant molecules have been wasted.

An example is making magnesium sulfate for use in bath salts.

The bath salts can be made like this:

magnesium hydroxide + sulfuric acid → magnesium sulfate + water

As two products are made, not all the reactants are used to make the magnesium sulfate. Some are used to make the water. This means there is a lower atom economy, so the reaction is more wasteful, or less 'green'.

You will find out:

> about atom economy as a way of measuring how efficient a reaction is

> how to determine atom economy

> about linking atom economy to more sustainable 'greener' reactions

FIGURE 2: Which gas does limewater test for and what would you see?

FIGURE 3: Bath salts. The active ingredient in bath salts is magnesium sulfate. Why do people use bath salts?

Questions

10 Look at these two word equations:

magnesium + oxygen → magnesium oxide

magnesium + hydrochloric acid → magnesium chloride + hydrogen

a Which reaction has the best atom economy?

b Which reaction is more sustainable or 'greener'?

11 Why is atom economy important?

Atom economy formula

Atom economy can be found using a formula.

$$\text{atom economy} = \frac{\text{relative formula mass of the desired product}}{\text{relative formula mass of all the products}} \times 100$$

M_r is usually used as an abbreviation for relative formula mass (A_r stands for relative atomic mass).

Look at this example:

A company needs to make magnesium sulfate solution for use as an indigestion remedy, or at a higher concentration as a laxative. They need to find the best method.

Remember!

You need to be able to recall and use this formula:

$$\text{atom economy} = \frac{M_r \text{ of the desired product}}{\text{sum of } M_r \text{ of all the products}} \times 100$$

Q making bath salts   calcium sulfate uses

Method 1:

$$MgO + H_2SO_4 \rightarrow MgSO_4 + H_2O$$

A_r $\quad$ $24 + 16$ $\quad$ $\underline{2 + 32 + 64}$ $\rightarrow$ $24 + 32 + 64$ $\quad$ $2 + 16$

M_r $\quad\quad$ 40 $\quad\quad\quad$ 98 $\quad\quad\rightarrow\quad\quad$ 120 $\quad\quad\quad$ 18

$$\text{atom economy} = \frac{M_r \text{ of desired product}}{\text{sum of } M_r \text{ of all the products}} \times 100$$

$$= \frac{120}{138} \times 100 = 86.96\%$$

Method 2:

$$MgCO_3 + H_2SO_4 \rightarrow MgSO_4 + H_2O + CO_2$$

M_r $\quad$ 84 $\quad\quad$ 98 $\quad\rightarrow\quad$ 120 $\quad\quad$ 18 $\quad\quad$ 44

$$\text{atom economy} = \frac{M_r \text{ of desired product}}{\text{sum of } M_r \text{ of all the products}} \times 100$$

$$= \frac{120}{182} \times 100 = 65.93\%$$

Which method is best? The higher the atom economy, the fewer atoms are wasted, so the first method is a 'greener' process.

Questions

12 A company has two methods for making a drug.

The drug has a M_r of 568.

Method 1 makes a second non-useful product with a M_r of 68.

Method 2 makes three non-useful products with a total M_r of 56.

a Calculate the atom economy for each method.

b Which method is best to use?

c Explain why.

More on atom economy

Every industrial process wants as high an atom economy as possible to reduce the formation of unwanted products. This makes the process more sustainable as it is more efficient and it also makes better use of the reactants, conserving raw materials.

Some reactions can give a very low atom economy.

An example is making hydrogen as a fuel source for vehicles.

Hydrogen can be made from water:

$$2H_2O \rightarrow O_2 + 2H_2$$

A_r H = 1, O = 16

Using the atom economy formula (see Standard Demand section), the atom economy for making hydrogen is 12.5% and for making oxygen is 87.5%.

As both products are useful, the overall atom economy can be considered to be 100%.

FIGURE 4: Hydrogen fuel dispenser. Why is the hydrogen under pressure? What do the hazard symbols represent?

Questions

13 a Put these reactions in order of their atom economy by looking at their equations only (the desired product is in **bold**):

i $Zn + 2HCl \rightarrow$ **$ZnCl_2$** $+ H_2$

ii $Zn + 2HCl \rightarrow ZnCl_2 +$ **H_2**

iii $2Mg + O_2 \rightarrow$ **$2MgO$**

b Check you answer by calculating the atom economy for each reaction (Zn = 65, H = 1, Cl = 35.5, Mg = 24, 0 = 16]

14 How can a reaction have an atom economy of 100%?

Q atom economy $\quad$ hydrogen fuel

Energy

Energy in cooking

Energy is taken in by an egg as it changes during cooking. This change is a chemical reaction.

The burning gas gives out the energy the egg needs to cook. This is also a chemical reaction.

Almost all chemical reactions give off or take in energy.

FIGURE 1: Frying eggs needs energy.

Exothermic and endothermic reactions

Chemical reactions happen when reactants change into products.

Chemical reactions can either give out or take in energy.

Exothermic reactions give out energy to the surroundings (release energy).

Endothermic reactions take in energy from the surroundings (absorb energy).

Measuring temperature changes shows the type of reaction:

> if the temperature goes up, it is an **exothermic reaction**

> if the temperature drops, it is an **endothermic reaction**.

FIGURE 2: What happens during an exothermic reaction?

Remember!

Energy is not a product and it is not a chemical.
Energy is measured in joules (J) or kilojoules (kJ).
Temperature is measured in degrees Celsius (°C).

Questions

1 What are reactants turned into in a chemical reaction?

2 How can an energy change be recognised?

3 What unit is temperature measured in?

4 When gas burns, it gives out energy. Which type of reaction is this?

Q chemical change   temperature

Making and breaking bonds

During chemical reactions, chemical bonds are broken and made. Bond making is an exothermic process. Bond breaking is an endothermic process.

 Questions

5 In a reaction, during which process is energy taken in?

6 In a reaction, during which process is energy given out?

FIGURE 3: What happens during an endothermic reaction?

 ## 'Make or break'

> Photosynthesis is an endothermic reaction.

> Energy is taken in from the Sun.

The word equation for photosynthesis is:

carbon dioxide + water → glucose + oxygen

The energy taken in is used to break the bonds in the carbon dioxide and water molecules to form separate atoms. Bond breaking is an endothermic process as it requires energy.

When the separate atoms reform into glucose and oxygen, new bonds are made. Bond making is an exothermic process, as it releases energy.

To decide if a chemical reaction is exothermic or endothermic, the amount of energy needed for bond breaking must be compared with the amount of energy released when new bonds are formed.

Chemical reactions can be thought of as happening in two stages:

> In stage 1, energy is needed to break the bonds in the reactants into separate atoms. This energy is called **activation energy**.

> In stage 2, the separated atoms combine to form the products, and energy is released.

> Bond breaking requires energy.

> Bond making releases energy.

> In an **endo**thermic reaction, more energy is required than released.

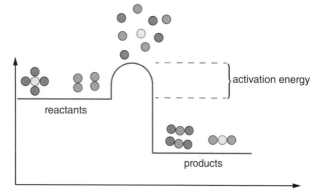

activation energy

reactants

products

FIGURE 4: Exothermic reaction pathway. How would you define an exothermic reaction?

 Questions

7 What happens when new bonds are formed?

8 Why is bond breaking an endothermic process?

9 What do you need to know to decide if a reaction is exothermic?

10 What type of reaction occurs when more energy is given out when the products are formed than taken in to break the bonds of the reactant?

11 Sketch a graph to show the energy changes in an endothermic reaction.

Combustion

Combustion is the scientific word for burning, and it is an exothermic reaction.

Fuels need oxygen to burn.

Ethanol burns in oxygen to make carbon dioxide and water:

ethanol + oxygen → carbon dioxide + water

In this word equation the reactants are ethanol and oxygen and the products are carbon dioxide and water.

Comparing fuels

The apparatus in Figure 5 can be used to measure the energy released by a fuel.

The fuel needs to be burnt in a bottled gas burner or spirit burner.

Water is heated in a copper calorimeter.

To compare different fuels fairly:

> use the same size and type of calorimeter

> measure out the same volume of water

> use water that starts at the same temperature

> adjust the flame so it is the same size

> ensure the flame is the same distance from the calorimeter

> burn the fuels for the same amount of time.

To compare fuels, find out how much the temperature of the water goes up.

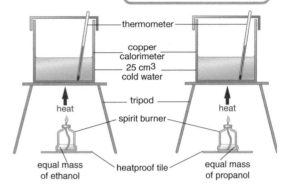

FIGURE 5: Why is the same mass of water used in the experiments?

Questions

12 What does ethanol make when it burns in oxygen?

13 The table shows some results from burning the same amounts of different fuels for 30 seconds.

Fuel	Temperature at the start (°C)	Temperature at the end (°C)
methanol	25	35
ethanol	26	46
propanol	24	53

a Work out the temperature rises.

b Which fuel released the most energy?

Comparing the energy from 1 g of fuel

It is useful to compare how much energy is given out by 1 g of different fuels.

For solid fuels, 1 g can be weighed out and then burnt.

For liquid fuels, one way to do this is to use the method shown in Figure 5, but also place the burner on a balance. Note the mass of the fuel plus burner, then heat water in the calorimeter until the mass of fuel has reduced by 1 g.

To make sure results are precise:

> repeat the experiment three times

> exclude as many draughts as possible.

This formula needs to be used to find the energy:

energy transferred (in J) = $m \times c \times \Delta T$

where m is the mass of water in grams, c is the specific heat capacity of water (= 4.2 J/g °C) and ΔT = temperature change of water in °C

Q burning fuels   carbon monoxide   naming alcohols

During the experiment, the following need to be measured:

> the mass of water placed in the calorimeter (remember 1 g of water has a volume of 1 cm³, so 50 cm³ of water has a mass of 50 g)

> the temperature at the start and at the end, to work out the change in temperature.

Example

Calculate the energy transferred if 100 g of water is heated from 20 °C to 70 °C:

energy   = 100 × 4.2 × (70 − 20)

               = 420 × 50

               = 21 000 J

               = 21 kJ

 Questions

14 How does repeating an experiment three times check the precision?

15 1 g of propanol is used to heat 50 g of water. The temperature of the water starts at 24 °C, and rises to 56 °C. Calculate the energy transferred.

16 Explain what needs to be done to ensure a fair comparison of the energy in different fuels.

Comparing fuels using calculations

This formula is used mainly to calculate the energy in a fuel:

energy transferred (in J) = $m \times c \times \Delta T$

It can also be rearranged so you could calculate:

> m (the mass of water heated)

> ΔT (the change in temperature).

Example

A model immersion heater contains 200 g of water.

Water has a specific heat capacity of 4.2 J/g °C.

It is heated by a fuel that gives out 24 600 J of energy when it is burnt. Calculate the temperature rise to the nearest degree.

energy transferred (in J) = $m \times c \times \Delta T$

Rearranging gives:

$$\Delta T = \frac{\text{energy transfer}}{m \times c}$$

$$\Delta T = \frac{24\ 600}{200 \times 4.2} = 29.3\ °C$$

The energy output of a fuel is can be calculated using the formula:

$$\text{energy per gram} = \frac{\text{energy supplied}}{\text{mass of fuel burnt}}$$

The unit is joules per gram (J/g).

Example

If this water is heated by 3.0 g of fuel then the energy output is:

$$\text{energy per gram} = \frac{24\ 600}{3.0} = 8200\ J/g$$

Remember!
Always set your working out clearly and show all the steps.

 Questions

17 Calculate the energy per gram released by 5.0 g of fuel that raises the temperature of 100 g of water from 18 °C to 78 °C.

18 A fuel produces 14 670 J of energy and raises the temperature of water by 45 °C.

a Calculate the mass of water heated.

b If 6 g of the fuel was burnt, calculate the energy produced by 1 g of fuel.

Batch or continuous?

You will find out:
> about the differences between batch and continuous processes

How are chemicals made?

> A sulfuric acid plant makes thousands of tonnes of sulfuric acid every day.

> A drugs company may make less than a tonne of a medicine in a whole year.

These amounts are so different that the chemicals have to be made in very different ways.

FIGURE 1: Why are drugs made in small batches?

Batch and continuous processes

Continuous process

If something is needed in large amounts it is usually made by a **continuous process**. Production takes place night and day throughout the year, non-stop.

Two examples of chemicals made by a continuous process are:

> ammonia, for fertilizers

> sulfuric acid, which is used to make thousands of different compounds.

Batch process

If something is needed in small amounts they are made using a **batch process**.

Examples are **pharmaceuticals** (medical drugs), which are only made when they are needed.

Did you know?

Some bread is made in bulk. It is cooked by a continuous process. The bread sits on a slow-moving conveyor belt through a very long oven.

Questions

1 Name one substance that is made by a batch process.

2 Name two substances that are made by a continuous process.

Comparing the processes

Batch process

Drugs companies make most medicines in small amounts – batches – and store the medicines in a warehouse. New batches are then made whenever the stored medicine runs low.

If a lot of one medicine is needed, several batches can be made at the same time.

Small batches are manufactured to meet the expected demand. Making very large batches of most pharmaceuticals is not done, as drugs have a 'best before' date and so cannot be stored for long.

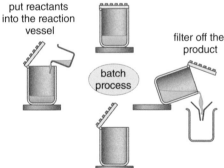

carry out the reaction

put reactants into the reaction vessel

filter off the product

batch process

clean out the reaction vessel and find out what needs to be made for the next batch

FIGURE 2: A batch process is used to make small amounts of a product. Medicines are often made in this way. Can you suggest why?

Continuous process

A continuous process plant is effective because it works at full capacity all the time to manufacture chemicals which have a very high demand. It costs an enormous amount to build, but once running it makes a large amount of product. Providing the raw materials are available, the plant can be kept running at the optimum (best) conditions. Once this is achieved, it takes less energy to maintain than it would to stop and start the process.

Questions

3 The local baker makes her bread in a batch process. Suggest the main stages in making a batch of bread.

4 The local baker makes batches of different 'product' every day. Suggest what might be made in the different batches.

5 When would a continuous process be used?

Evaluating the processes

Aspirin is the most widely used over-the-counter drug in the world. World production is estimated at 45 000 tonnes a year. Aspirin is used to relieve pain, reduce inflammation, and lower fever.

Aspirin tablets of the same dosage amount are manufactured in batches. After careful weighing, ingredients are mixed and compressed into units of mixture called slugs. The slugs are then filtered to remove air and lumps, and are compressed again to form individual tablets. The number of tablets made depends on the size of the batch, the dosage amount, and the type of tablet machine used. Documentation for every batch is kept throughout the manufacturing process, and finished tablets undergo several tests before they are bottled and packaged for distribution.

Batch processes are flexible. It is easy to change from making one compound to another. When enough batches of aspirin have been made, production can be switched to another pharmaceutical. Each batch has to be supervised, so labour costs are higher. Also, the time spent filling and emptying reaction vessels is time when the vessels are not producing chemicals, so the vessels are not used as efficiently as in a continuous process. In contrast, world production of sulfuric acid is around 165 million tonnes a year. Both sulfuric acid and ammonia are made in a continuous process.

Sulfuric acid and ammonia plants are built to make large but fixed amounts. Production plants are built to meet advanced orders of regular amounts. It is difficult to change this without increasing costs. At full capacity, the plant will be at its most efficient in terms of energy requirements balanced against yield. If production needs to be slowed, efficiency decreases making the product more expensive.

Continuous processes employ very few people, and so the cost per tonne is very small. A disadvantage is that the reaction vessels and pipes are designed only to work well at one level of output. What they make or how much cannot be changed easily.

Questions

6 Suggest why running a continuous process plant at a slower rate increases the running costs.

7 Suggest one advantage and one disadvantage of making:

a aspirin by batch processing

b chemicals such as sulfuric acid and ammonia by continuous processing.

Factors that affect the costs of medicines

The chemicals in medical drugs need to be as pure as possible. Impurities might give side effects. All new drugs need to be tested to make sure they work and to check they do not have serious side effects.

Three tests can be used to find if a chemical compound is pure.

A pure chemical will:

> melt at a fixed temperature

> boil at a fixed temperature

> give the same result when tested using thin layer **chromatography**.

The raw materials for medicines can be chemicals that are man-made (synthetic), or chemicals extracted from plants.

Other medicines are made totally in the laboratory; they are synthetic.

Whichever way the medicines are made, there are costs.

FIGURE 3: The foxglove plant gives the heart medicine called digitalis. Why should we be worried that many wild plants are endangered?

FIGURE 4: Six factors that affect the cost of making a medicine. This is why medicines cost so much.

Remember!

Make sure you learn the six factors that affect the cost of making a medicine.

Questions

8 List six factors that affect the cost of making a medicine.

9 Why are some flowering plants important to drugs companies?

10 Name **three** ways to test the purity of a chemical.

Why are medicines so expensive?

> It can take years to develop and safety test a new medicine. Each country has strict safety laws to guide drugs companies. The price for a new medicine is set so that the investment costs of its research and development can be recouped.

> The raw materials needed may be rare and costly. The medicines based on compounds extracted from plants are difficult to find and are often only found in inhospitable places. Raw materials are sometimes difficult to extract. A flowering plant may contain thousands of similar chemicals, so separating the desired chemical is time consuming and expensive.

> Medicines are made by batch processes, so less automation can be used and their manufacture is labour intensive.

🔍 making medicines   plants used for medicines

Extracting chemicals from plants

Chemical compounds in plants are held in cells. Plants are crushed to break the strong cell walls and boiled so the chemicals can dissolve. Different solvents are used to separate compounds using chromatography.

Thin layer chromatography is a method for finding the purity of a compound. It works by comparing the speed of movement against a sample known to be pure.

Pure compounds also have definite melting and boiling points. If a compound is impure it will melt and boil over a range of temperatures; usually raising the boiling point and lowering the melting point.

FIGURE 5: Thin layer chromatography. How is this method used to separate compounds and check purity?

Questions

11 List the main processes in extracting a chemical from a plant.

12 What is meant when it is said that chromatography can 'purify' a chemical compound?

13 A compound melts at 159 °C and boils at 211 °C. During chromatography the sample moves 2 cm in 5 minutes.

How would the results be different if the compound was impure?

Drug development

It is difficult and costly to develop and license new pharmaceutical drugs.

Some reasons are:

> thousands of compounds often need to be tested before an effective one is found

> likely compounds need to be manufactured and tested on living tissue to unsure safety

> long term **trials** are needed to identify possible side effects

> lots of similar compounds can be developed to try to reduce side effects

> drug companies need to prove the drug is actually effective at a certain dose

> research needs to be evaluated before a patent is granted

> drug companies need to recoup development costs prior to the patent expiring, when other companies can make their own generic version.

Questions

14 Explain why generic drugs have no initial development costs.

15 Suggest why a drugs company has to submit a new drug for approval before it is sold.

16 Suggest why it takes so long to test the safety of a drug.

Q developing new medicines   cost of medicines

Allotropes of carbon and nanochemistry

You will find out:

> about the properties of diamonds
> how to recognise the structures of diamond, graphite and buckminsterfullerene
> about allotropes

At the cutting edge

Soot and diamonds are formed from carbon.

Pencil leads are also a form of carbon, called graphite.

So is buckminsterfullerene. (Have you heard of it?)

They are all the same element, carbon.

The reasons for the differences between them are at the 'nano' level – the level where atoms fit together.

FIGURE 1: Why are the diamonds used in jewellery cut and polished?

How are diamond and graphite different?

Diamond and graphite are both made from **carbon**. The table shows their physical properties.

Diamond	Graphite
lustrous (shiny), transparent and colourless	lustrous, opaque and black
very hard	soft and slippery
very high melting point	very high melting point
insoluble in water	insoluble in water
does not conduct electricity	conducts electricity

Industrial diamonds are nowhere near as pretty. They are often opaque and dark brown or black. But they have all the other properties, especially hardness!

You need to recognise the structure of diamond and graphite (shown in Figure 3).

FIGURE 2: Industrial diamonds. What is the difference between jewellery diamonds and industrial diamonds?

Questions

1 Name the element that diamonds are made from.

2 List four differences between diamond and graphite.

Q properties of diamonds   uses of diamonds

Why diamond and graphite are useful

Both diamond and graphite are giant structures of carbon atoms.

Diamond forms when extreme heat and pressure forces carbon deposits to crystallise deep inside the Earth.

Graphite is made when carbon deposits are changed when metamorphic rock forms.

The different conditions affect the way the covalent bonding takes place.

Diamonds are the hardest natural substance known. Diamonds also have a very high melting point. When diamonds form, slight imperfections occur forming cleaving planes that allowing them to be shaped. These properties make them ideal for industrial cutting tools, such as rock saws and grinding wheels.

Pencils contain graphite, which is black and slippery. When a pencil is used, some of the graphite slides off and sticks to the paper, making black marks. The slipperiness also makes graphite a good lubricant, even though it is a solid. Powdered graphite is often used to lubricate door locks.

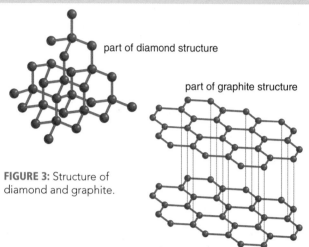

part of diamond structure

part of graphite structure

FIGURE 3: Structure of diamond and graphite.

Questions

3 Suggest why a high melting point is important for the blades of industrial cutting tools.

4 Which properties of graphite make it suitable for use in pencils?

5 Suggest advantages of using graphite as a lubricant rather than an oil-based lubricant.

Explaining the properties of diamond and graphite

Carbon atoms have four electrons in the outer electron shell.

In diamond every carbon atom is joined to four others in a three-dimensional tetrahedral lattice. The atoms join together by strong covalent bonds that involve electron sharing. This arrangement gives strength in all directions, and requires a large amount of energy to break the bonds, giving a very high melting point of 3350 °C.

No free electrons exist in the structure, so diamond does not conduct electricity.

Graphite has a similar melting point to that of diamond. It has a layered arrangement in which each carbon atom is covalently bonded to three others, forming single layers of regular hexagons.

This formation means each carbon atom has an unshared electron in its outer shell, free to move anywhere along the layer. So graphite is an electrical conductor, often used as electrodes in the electrolysis of molten electrolytes.

The distance between the layers is greater than between covalently bonded atoms, so the layers in graphite are only weakly attracted to each other. This means that when a force is applied, the layers can slide over each other, which makes graphite a high-temperature lubricant.

In pencils, different amounts of clay are mixed with graphite to make harder grades.

Questions

6 Explain, in terms of structure and bonding, why diamond is the hardest substance known to man.

7 Describe why graphite is an electrical conductor.

8 Explain why both diamond and graphite have high melting points.

Nanotubes

Carbon can also be used to make very small structures called **nanotubes**.

Nanotubes have special properties and are useful as:

> semiconductors in electrical circuits

> industrial catalysts

> for reinforcing graphite in tennis rackets.

FIGURE 4: Carbon nanotubes. How many carbon atoms are in each ring?

Did you know?

Nanotubes are 20 times as strong as steel!

Did you know?

Graphite pencils were first made in 1565. All the graphite came from a mine in Cumbria.

Questions

9 Name three different carbon structures.

10 Give two uses of carbon nanotubes.

Allotropes

Diamond, graphite and **buckminsterfullerene** are all **allotropes** of carbon. Allotropes are different structures of the same element. They each have different physical properties but similar chemical properties.

Carbon atoms can also be joined into structures called **fullerenes**.

Fullerene uses

Ball-shaped fullerenes – buckyballs – can act as hollow cages to trap other molecules. Researchers are finding ways to use them to:

> carry drug molecules around the body and deliver them to where they are needed

> trap dangerous substances in the body and remove them.

There are 60 carbon atoms in buckminsterfullerene. It is written as C_{60}. These molecules are so small that they are measured in **nanometres**. A nanometer is one thousand millionths of a metre (10^{-9} metres).

FIGURE 5: As well as their arrangements in diamond and graphite, carbon atoms can arrange as a sphere.

Q making pencils   using graphite

Questions

11 Describe why allotropes have different properties.

12 Explain why fullerenes can be used in drug-delivery systems.

13 Buckminsterfullerene has lower melting and boiling points than diamond. Use your knowledge of their structures to explain why.

More on giant molecular structures

Carbon is not the only element to form **giant molecular structures**. Giant structures can contain more than one element. Examples are quartz (silicon dioxide) and polymers. Giant molecular structures have many **covalent bonds**.

Covalent bonding usually means that all the available electrons are shared, so they do not conduct electricity. Graphite and carbon nanotubes are exceptions.

Generally, melting point depends on the size of a giant covalent structure because of the energy required to break all the bonds.

Large structures like quartz are strong because of a regular three-dimensional pattern in the bonding of their atoms.

Quartz has a high melting point due to the high number of strong covalent bonds that need to be broken, which requires a large amount of energy.

FIGURE 6: The structure of quartz.

Nanotubes as catalysts

Nanotubes can be used in catalyst systems. Atoms of the catalyst can be attached to the nanotubes. The nanotube has a large surface area, so there is more chance that the reactants collide with the catalyst.

Questions

14 Describe properties that giant covalent structures have.

15 Scientists plan to attach atoms of precious-metal catalysts to nanotubes instead of using small lumps of solid metal as a catalyst. Suggest and explain one advantage of this method.

Preparing for assessment:
Planning and collecting primary data

To achieve a good grade in science, you not only have to know and understand scientific ideas, but you need to be able to apply them to other situations and investigations. These tasks will support you in developing these skills.

✳ Tasks

> Suggest a hypothesis that relates the **mass of fuel** burnt to the **temperature rise** of water in a calorimeter.

> Plan an investigation to test your hypothesis. It is unlikely you will have the exclusive use of a digital balance, so you could burn the fuel for different lengths of time, weighing the fuel before and after.

> Write a risk assessment for you plan.

> Once your plan has been approved, perform the investigation, record the results and write a simple conclusion.

✳ Context

Fuels release energy when they burn.

This energy can be used directly such as in vehicles or for heating our homes, or indirectly to make electricity.

The energy released can be found by measuring the mass of fuel used and the temperature rise.

✳ Planning your investigation

You can measure the mass of fuel before and after, and measure the temperature rise.

These are the things you will need to consider when planning your investigation. (You can develop your plan in groups of two or three.)

1. What will you need to measure to obtain the results?

2. How will you change the amount of fuel burnt?

3. What do you need to keep the same to make it a fair test?

4. How many different results will you need before you can identify a trend?

5. Will you need to repeat your readings? If so, how many times?

6. You should carry out a risk assessment before you start the investigation. What precautions should you take?

7. Write the plan for the investigation.

> Try to write the plan in a logical order and ask yourself if someone can perform the investigation following just your plan.

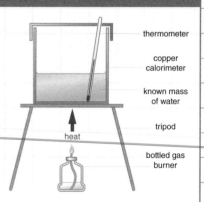

thermometer

copper calorimeter

known mass of water

tripod

heat

bottled gas burner

✸ Performing the investigation

Once your plan has been approved you can perform the investigation.

1. Find the mass of fuel used each time.

2. If you repeated any readings, all of these will need to be recorded as well as the average result.

3. Record your results in a table like this. You may need to add extra rows.

mass of fuel at the start (g)	time burnt (s)	mass of fuel at the end (g)	temperature at the start (°C)	temperature at the end (°C)

4. If you were to complete this as a GCSE controlled assessment, you would go on to plot a graph and evaluate the investigation.

5. Is there any way in which you could have improved on how you performed the investigation?

6. What have you found out about how the mass of fuel burnt affects the temperature rise?

How many sets of readings will you need to take to identify a trend?

What graph would you draw?

What would the labels be on the axes?

How would you use the graph to decide on the answer to the task?

Think about accuracy and precision. How are they different?

C3 Checklist

To achieve your forecast grade in the exam you'll need to revise

Use this checklist to see what you can do now. It gives you many of the important points you will need to know. Refer back to the relevant pages in this book if you're not sure and to see if there is anything else you need to know. Look across the three columns to see how you can progress.

Remember you'll need to be able to use these ideas in various ways, such as:

> interpreting pictures, diagrams and graphs
> applying ideas to new situations
> explaining ethical implications
> suggesting some benefits and risks to society
> drawing conclusions from evidence you've been given.

Look at pages 278–299 for more information about exams and how you'll be assessed.

To aim for a grade E	To aim for a grade C	To aim for a grade A
recognise that rusting is a slow reaction, but burning is a fast reaction **label** the apparatus needed to measure how fast a gas is made **plot** a graph of how fast a gas is made, or mass is lost **read** data from graphs and know steeper lines mean a faster reaction **explain** why a reaction stops	**understand** that reaction rate tells you how fast a product is being made **know** the common units for measuring reaction rate (g/s for mass loss and cm3 /s for gas made) **interpret** data from graphs and tables to compare reaction rates **recognise** the amount of product made is directly proportional to the amount of the limiting reactant used **recall** the limiting reactant is the one not in excess	**calculate** the rate of reaction from the slope of a graph using the gradient **explain** using reacting particles why the amount of product formed is directly proportional to the amount of limiting reactant used

recognise chemical reactions involve collisions
describe the effect of changing temperature, concentration and pressure on the rate of reaction
interpret data from tables and graphs to:
  • **read** off values
  • **compare** rates by comparing gradients
  • **compare** rates using reaction times
	understand that reaction rate depends on collisions
	explain in terms of reacting particles why temperature, concentration and pressure affect reaction rate
	interpret data to:
	  • **decide** when a reaction has finished
	  • **compare** rates during a reaction
		understand that rate depends on collision frequency, and successful collisions depend on energy transfer
		explain reaction rates for changing temperature, concentration and pressure in terms of collision theory
		interpret data by:
		  • calculation from the slope of a graph
		  • extrapolating readings
		  • interpolating readings

recall catalysts increase reaction rate **recall** powdered reactants react faster than solid lumps **describe** explosions as very fast reactions releasing large volumes of gas **interpret** data to compare rates using reaction times	**describe** a catalyst as a chemical used in small amounts and specific to one reaction that changes reaction rates without being used up **explain** how changing surface areas affects reaction rate **explain** why airborne powders can be dangerous in factories **interpret** data to sketch graphs to show rate and amount of product formed	**explain** in terms of collisions how changing surface area affects reaction rate **interpret** data to sketch graphs to show rate and amount of product formed

To aim for a grade E

calculate the relative formula mass when given the data

understand that mass is conserved during a chemical reaction

calculate the mass of reactants, product or gas formed, when given the other values

---

understand percentage yield compares the amount actually made to the expected amount

recognise possible reasons for not obtaining 100% yield

understand atom economy measures the amount of atoms wasted during manufacture

interpret simple percentage yield and atom economy data

---

recall that exothermic reactions release heat to the surroundings

recall that endothermic reactions take in heat from the surroundings

describe the apparatus used to burn a fuel so that the energy can be compared

---

describe the difference between batch and continuous processes

list factors that affect the cost of manufacture

explain why new drugs need to be tested

recall raw materials for drugs can be synthetic or extracted from plants

describe how to test for purity – melting point, boiling point and chromatography

---

explain why diamond, graphite and buckminsterfullerene are all forms of carbon

know the differences between diamond and graphite

recall nanotubes are used to make materials stronger and are used as semiconductors

To aim for a grade C

calculate relative formula mass for formulae that include brackets

use relative formula mass to show and explain why mass is conserved during a reaction

recognise and use the ideas that the mass of produce is directly proportional to the limiting reactant used

---

recall percentage yield = × 100

recall that atom economy = × 100

interpret complex percentage yield and atom economy data

---

recall bond making is an exothermic process and bond breaking endothermic

describe how the energy can be calculated from one gram of fuel

---

explain why batch processing is used to make drugs, but **continuous** processing for ammonia

explain why drug s are needed, and why development is expensive

describe how chemicals can be extracted from plants

interpret data about purity of a substance

---

explain why diamond, graphite and fullerenes are allotropes

explain in terms of properties the uses of diamond and graphite

explain why diamond and graphite have a giant covalent structure

explain why fullerenes can be used in new drug delivery systems

To aim for a grade A

use relative formula mass and symbol equations to show mass is conserved in a reaction

interpret chemical equations quantitatively

calculate the mass of products or reactants from balanced symbol equations using relative formula masses

---

explain why industry wants a high percentage yield to:
- reduce reactants wasted
- reduce costs

calculate atom economy using balanced symbol equations and relative formula masses

explain why industry wants a high atom economy to:
- reduce unwanted products
- increase sustainability

---

explain why a reaction is exothermic or endothermic using energy changes during bond making and breaking/

rearrange the formula energy transfer (in J) = m × c × Δt to calculate m or t

calculate the energy in J/g using energy per gram =

---

evaluate the advantages and disadvantages of batch and continuous processes

explain why is it difficult to test and develop new pharmaceutical drugs that are safe to use

---

explain the properties of diamond and graphite in terms of structure and bonding

predict the properties of substances with giant molecular structures

explain why nanotubes enable them to be used as catalysts

Foundation Tier

AO1 **1 (a)** Name **three** ways to speed up a chemical reaction. [3]

(b) The graph shows how much gas is made during a chemical reaction.

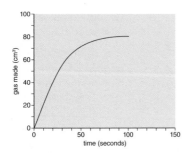

AO2 **(i)** What volume of gas was made after 50 seconds? [1]

AO2 **(ii)** What was the total volume of gas produced? [1]

AO2 **(iii)** According to the graph, was the reaction fastest:

between 0 and 20 seconds ☐

between 40 and 60 seconds ☐

between 60 and 80 seconds ☐ [1]

AO1 **(iv)** Explain why the reaction stops. [1]

AO3 **(v)** Sam thinks the graph shows the amount of gas made increases by the same amount each minute. Is Sam right? Explain your answer. [1]

[Total: 8]

2 Copper metal reacts with oxygen to make copper oxide.

AO2 **(a)** The formula of copper oxide is CuO. Work out its relative formula mass. (Cu 64, O 16) [1]

AO2 **(b)** If 3.2 g of copper make 4.0 g of copper oxide, how much oxygen reacts? [1]

[Total: 2]

3 Kasia measured out enough sulfuric acid to make 2.0 g of magnesium sulfate. She reacted the acid with excess magnesium oxide, filtered the mixture and left the magnesium sulfate solution to crystallise.

AO2 **(a)** Which chemical was the limiting reactant? [1]

AO2 **(b)** The actual yield was 1.5 g of magnesium sulfate. Calculate the percentage yield. [1]

AO2 **(c)** Suggest two possible reasons why the percentage yield was less than 100%. [2]

[Total: 4]

4 Reactions A and B both make useful salts, but carbon dioxide and water are unwanted waste products.

A ammonia + sulfuric acid → ammonium sulfate

B magnesium carbonate + sulfuric acid → magnesium sulfate + water + carbon dioxide

AO2 **(a)** Which reaction has the highest atom economy? [1]

AO1 **(b)** Which reaction shows the 'greenest' process? [1]

[Total: 2]

AO1 **5 (a)** Describe or draw a diagram to show the apparatus needed to find out how much energy is made when a potato crisp burns. [4]

AO3 **5 (b)** This table shows the results for different crisps.

Crisp	A	B	C	D
Energy(J)	115	118	118	117

Jai thinks all the crisps give out the same amount of energy. Is Jai right? Explain your answer. [4]

[Total: 8]

AO1 **6 (a)** List five factors that add to the cost of developing a new drug. [5]

AO1 **(b)** Describe how drugs are tested for purity. [3]

[Total: 8]

AO1 **7** Draw a table to show differences between graphite and diamond. [4]

AO1 **8** Give two uses of carbon nanotubes. [2]

AO1 recall the science   AO2 apply your knowledge   AO3 evaluate and analyse the evidence

✳ Worked Example – Foundation Tier

A student placed 50 cm³ of hydrochloric acid in an open flask on a top-pan balance. She added excess marble chips and measured the mass of the flask once a minute.

Time (min)	1st mass (g)	2nd mass (g)
0	70	69
1	69	68
2	67.5	67
3	66.5	66
4	66	65.5
5	66	65.5

How to raise your grade!
Take note of these comments – they will help you to raise your grade.

(a) Why did the mass of the flask go down? [1]

Reactants were used up.

> The statement is true but does not explain why the mass is dropping. It was important to state that a gas is given off. **0/1**

(b) How many minutes did it take for all the acid to react? [1]

4 minutes.

> This is correct because the mass stops changing at 4 minutes. **1/1**

(c) Which is the limiting reactant? [1]

Hydrochloric acid.

> This is correct. The limiting reactant is the one that is not in excess. **1/1**

(d) In another experiment the temperature of the acid is increased, but everything else is kept the same.

Predict the final mass of the flask and give a reason for your answer. [2]

66 grams.

> The answer is incomplete because no reason is given. The correct explanation is that the amount of acid, which is the limiting reactant, is unchanged in the second reaction. **1/2**

(e) Use the reacting particles model to explain why increasing the temperature makes the reaction finish faster. [3]

Particles collide quicker.

> This answer is too simplistic. Key points are that the acid particles have more energy at a higher temperature, so they move faster, and collide with the marble chips more frequently. **0/3**

> This student has scored 3 marks out of a possible 8.
>
> This is below the standard of Grade C. With a little more care the student could have achieved a Grade C.

Higher Tier

1 In an experiment, a student reacted magnesium with acid at different temperatures. The hydrogen produced was collected in a gas syringe. The table shows the results.

Temperature (°C)	Volume of gas collected in 10 seconds (cm³)		
20	30	35	37
25	40	45	36
30	54	55	52
35	69	68	68

AO2 **(a)** Which variable is being: (i) investigated; (ii) measured? [2]

AO2 **(b)** Which result is probably an outlier? [1]

AO3 **(c)** The student found a website which claimed that reaction rates double for every 10 °C rise in temperature. Explain how well the data in the table supports this claim. [3]

[Total: 6]

AO2 **2 (a)** Calculate the relative formula mass of these compounds
(Fe 56, Na 23, Mg 24, O 16, S 32):
(i) magnesium sulfate, $MgSO_4$ [1]
(ii) iron(III) sulfate, $Fe_2(SO_4)_3$ [1]

AO2 **(b)** Iron(III) sulfate reacts with sodium to form sodium sulfate and iron. Construct a balanced symbol equation and show that mass is conserved during the reaction. [2]

[Total: 4]

AO2 **3 (a)** A company making nitric acid expects to make 40 tonnes but only manages 38 tonnes. Calculate the percentage yield. [2]

AO2 **(b)** When calcium carbonate ($CaCO_3$) is heated strongly it decomposes into calcium oxide (CaO) and carbon dioxide (CO_2). The carbon dioxide is a waste product. Calculate the atom economy for the production of calcium oxide (C 12, Ca 40, O 16). [3]

AO1 **(c)** Explain why industrial processes need as high an atom economy as possible. [2]

[Total: 7]

AO1 **4 (a)** Describe the main steps to find the energy in one gram of solid fuel. [6]

The quality of written communication ✍ will be assessed in your answer to this question.

AO2 **(b)** The temperature of 50 cm³ of water increased by 36 °C. Water has a specific heat capacity of 4.2 J / °C / g. Calculate the energy transferred using this formula:
energy transferred (in J) = $m \times c \times \Delta T$ [2]

In another experiment, 25 704 J was produced when 3 g of a liquid fuel heated 60 cm³ of water. Calculate the energy 1 g of fuel would make and the mass of water heated. [4]

[Total: 12]

AO1 **5 (a)** Explain why drugs are usually manufactured using batch processing. [4]

AO1 **(b)** Explain why it is expensive and difficult to develop a new drug. [6]

[Total: 10]

AO1 **6 (a)** Describe why graphite is used in pencils. [2]

AO1 **(b)** Explain in terms of its structure and bonding why graphite is an electrical conductor. [2]

[Total: 4]

AO1 recall the science   AO2 apply your knowledge   AO3 evaluate and analyse the evidence

✳ Worked Example – Higher Tier

Ammonia can be manufactured from nitrogen and hydrogen according to this equation.

$$N_2 + 3H_2 \rightarrow 2NH_3$$

(a) The relative formula mass of N_2 is 28, H_2 is 2 and NH_3 is 17. If 14 kg of nitrogen are reacted with an excess of hydrogen in a small-scale process, what is the predicted yield of ammonia that could be formed? [2]

N_2	+	$3H_2$	$\rightleftharpoons$	$2NH_3$
$2 \times 14 = 28$				$2(14 + 3) = 34$
so 28 kg			$\rightarrow$	34 kg
and 14 kg			$\rightarrow$	17 kg

(b) In an industrial-scale process only 12 tonnes of ammonia were made when the predicted possible yield of 48 tonnes was possible. Calculate the percentage yield this represents. [2]

$$\% \text{ yield} = \frac{actual\ yield}{predicted\ yield} \times 100\% = \frac{12}{48} \times 100\% = 25\%$$

(c) Explain why a continuous process is more economic than a batch process for ammonia manufacture. [3]

Continuous processes are better for bulk chemicals like ammonia that are needed continuously. They are expensive to set up because they need more equipment. However they achieve a high rate of production and rarely need to be shut down. They are also easy to automate so only a small workforce is needed, which keeps labour costs down.

(d) Ammonia (NH_3) is reacted with sulfuric acid to make ammonium sulfate, $(NH_4)_2SO_4$, to use as fertiliser.

Write a balanced symbol equation for the reaction. [2]

$$NH_3 + H_2SO_4 \rightarrow (NH_4)_2SO_4$$

(e) State the atom economy of the reaction between ammonia and sulfuric acid. [1]

(f) Explain why it is an advantage for industrial reactions to have a high atom economy. [2]

It cuts the amount of waste.

How to raise your grade!

Take note of these comments – they will help you to raise your grade.

Very clearly set out – a good answer. **2/2**

An excellent answer. The student has stated the equation to be used and worked methodically through the calculation. **2/2**

The answer shows excellent understanding and the information is clear, relevant and well structured. **3/3**

The formulas are correct but the equation is clearly not balanced. Two ammonia molecules are needed to provide the two ammonium ions in the product. **1/2**

There is only one product so none of the atoms in the reactants are wasted. No calculation is needed. The atom economy must be 100%. **0/1**

As this is a two mark answer, more detail is needed. A good answer would add that this makes the process more sustainable. **1/2**

This student has scored 9 marks out of a possible 12. This is below the standard of Grade A. With a little more care the student could have achieved a Grade A.

C4 The periodic table

Ideas you've met before

Atoms and molecules

The Ancient Greeks thought that matter was made up from atoms.

Oxygen, nitrogen and hydrogen are elements that are gases.

Iron, copper and gold are elements that are metals.

A new substance is made by a chemical change, which is not reversible.

 Why doesn't gold lose its shine?

Metals and non-metals

Metals are shiny, conduct electricity, can be hammered flat and conduct heat.

When magnesium and oxygen are heated they make a new chemical compound.

Sulfur and iodine are non-metals and do not conduct electricity.

Bridges and car bodies are made from metals since metals are strong.

 Which metals are best to make saucepans?

Periodic table

There are about 100 elements, which are written in order in the periodic table.

Metals are found on the left-hand side of the table.

Elements have a symbol of one capital letter and sometimes one small letter.

Elements join together chemically to make compounds.

 What is made when sodium is burned in chlorine?

Water purification

When some chemical compounds are heated they are broken down.

When copper carbonate is heated it breaks down and gives off carbon dioxide.

Carbon dioxide turns limewater milky.

Water is conserved in reservoirs and underground in aquifers.

 How is water purified?

In C4 you will find out about...

> how an atom is a nucleus surrounded by electrons

> how isotopes have different mass numbers

> how to recognise an ion from its formula

> how sodium chloride is a giant ionic lattice and so has a high melting point

> how elements in the same group have similar properties

> how elements in group 7 have seven outer electrons

> why potassium reacts in a similar way to sodium

> how to predict the properties of group 1 elements

> how chlorine is used to sterilise water

> how to predict some reactions of group 7 elements

> why compounds of transition metals are often coloured

> how to identify ions of copper using sodium hydroxide

> how to interpret data on the physical properties of metals

> metals that conduct electricity by delocalised electrons

> where important water resources are found

> the important stages in the water purification process

Atomic structure

You will find out:
> the structure of atoms
> charges and masses of particles
> why atoms are neutral

Atoms as models

Atoms are the building blocks of all matter, both living and non-living. They can join together in millions of different ways to make all the materials around us. They even join together to make us!

We are made of complex materials. We can explain how simple materials are made using ideas and models of **atoms**. These ideas have changed over the years, but now scientists believe atoms are very small, have a very small mass and are made of three important particles – **protons**, **electrons** and **neutrons**.

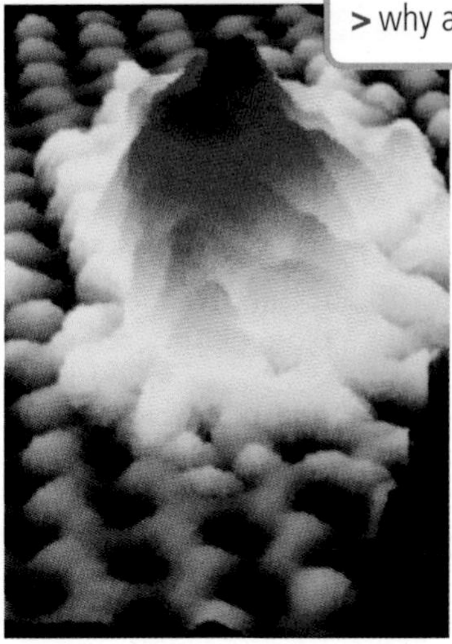

FIGURE 1: Coloured image of gold atoms on a layer of graphite atoms.

Atoms

Individual atoms are very small and have very little mass. There are about ten million million atoms in this full stop.

An atom is made up of a **nucleus** that is surrounded by electrons.

> The nucleus carries a **positive charge**.

> Electrons that surround the nucleus each carry a **negative charge**.

> Overall, the atom has no charge.

Each atom has an atomic number. This number is written next to the symbol of an element in the **periodic table** (see page 164).

If the atomic number of an element is known, it can be identified by looking on the periodic table. Its **symbol** and name are both written there.

The nucleus of an atom is made up of protons and neutrons.

The **atomic number** is the number of protons in an atom.

The atomic number for helium is 2 because it has two protons.

Neutrons have no charge. There are two neutrons in a helium nucleus.

The **mass number** of an atom is the total number of protons and neutrons in an atom.

electrons, each carries a negative charge

nucleus, carries a positive charge

FIGURE 2: The structure of an atom. What charge does an electron carry?

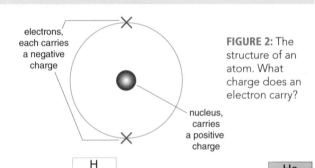

H hydrogen 1							He helium 2
Li lithium 3	Be beryllium 4	B boron 5	C carbon 6	N nitrogen 7	O oxygen 8	F fluorine 9	Ne neon 10
Na sodium 11	Mg magnesium 12	Al aluminium 13	Si silicon 14	P phosphorus 15	S sulfur 16	Cl chlorine 17	Ar argon 18
K potassium 19	Ca calcium 20						

the atomic number of Lithium is 3

FIGURE 3: Part of the periodic table. What is the atomic number of magnesium?

Questions

1 What is the charge on a nucleus?

2 What is the charge on an electron?

3 What is the charge on an atom?

4 Use part of the periodic table above to identify the name of the element with an atomic number of 16.

5 What is the number of protons in an atom of fluorine, F?

electron   element   neutron   nucleus   proton

More on atoms

The nucleus of an atom is made up of protons and neutrons.

> The relative mass of a proton is 1.
> The relative mass of a neutron is also 1.

The relative masses and charges of electrons, protons and neutrons are shown in the table below.

	Charge	Mass
electron	−1	0.0005 (zero)
proton	+1	1
neutron	0	1

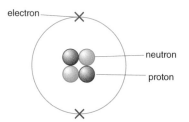

If a particle has an atomic number of 11, a mass number of 23 and a **neutral** charge, it must have:

> 11 protons, because it has an atomic number of 11.
> 11 electrons, as there are 11 protons.
> 12 neutrons, as the mass number is 23 (23 − 11 = 12).

Some examples are given in the table below.

	Atomic number	Mass number	Number of protons	Number of electrons	Number of neutrons
sodium	11	23	11	11	12
$^{19}_{9}F$	9	19	9	9	10
carbon-12	6	12	6	6	6

Isotopes

Elements can exist in more than one form. They have the same atomic number but different mass numbers. These are called **isotopes**.

If an element has the symbol $^{14}_{6}C$, it has no charge so it is an atom. It has six protons and a mass of 14. It must therefore have 8 neutrons (14 − 6).

$^{14}_{6}C$ is sometimes written as carbon-14. It is a radioactive isotope of carbon.

FIGURE 4: Helium has two protons and two neutrons. What is its mass number?

Questions

6 What is the mass of one proton?

7 What is the charge on a proton?

8 What is an isotope?

9 Fill in a row in the table for carbon-14.

Remember!

It is because a helium atom, He, has two protons that it has an atomic number of 2; not the other way around.

Why are atoms neutral?

An atom is neutral because it has an equal number of electrons and protons. The positive charges balance out the negative charges. An atom has a radius of about 1×10^{-10} m and a mass of about 1×10^{-23} g.

What happens if an atom loses electrons?

If a particle has an atomic number of 3, a mass number of 7 and a neutral charge, it must be a lithium atom. If the lithium atom loses one electron it becomes a charged particle with one positive charge.

	Atomic number	Mass number	Charge	Number of protons	Number of electrons	Number of neutrons
lithium atom	3	7	0	3	3	4
Li$^+$	3	7	+1	3	2	4

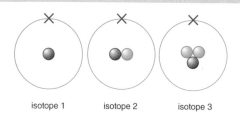

FIGURE 5: Isotopes. What is different in isotopes of the same element?

More on isotopes

Isotopes of an element have different numbers of neutrons in their atoms.

Isotope	Electrons	Protons	Neutrons
$^{1}_{1}H$	1	1	0
$^{2}_{1}H$	1	1	1
$^{3}_{1}H$	1	1	2

Questions

10 What is the number of protons, electrons and neutrons of the charged particle $^{39}_{19}K^+$?

11 What is the isotope that has 17 protons and 18 neutrons?

Compounds

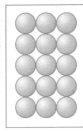

FIGURE 6: Copper and helium are elements. How do you know by looking at their arrangement of atoms?

Elements have the same type of atoms. Their symbol uses one capital letter only or one capital letter and one lower case letter. If two elements join together (chemically combine) a compound is made. A **compound** is a substance that contains at least two different elements chemically combined.

The elements in a compound can be seen from the formula, by using the periodic table (see page 164).

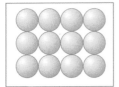
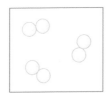
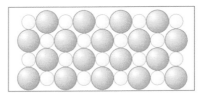

FIGURE 7: Magnesium and oxygen react to produce magnesium oxide, a compound.

Examples

> The compound magnesium oxide, MgO, contains Mg, magnesium, and O, oxygen.
> $CaCO_3$ is a compound made up from three elements, calcium, Ca, carbon, C and oxygen, O.

The arrangement of electrons in atoms

Electrons occupy the space around the protons of the nucleus in 'shells'.

The **electron shell** nearest to the nucleus can take up to two electrons.

The second shell can take up to eight electrons.

Example

Oxygen has an atomic number of 8. It has eight electrons in two shells around the nucleus. The shell nearest to the nucleus takes two electrons. The next six occupy the second shell. This is written as 2.6.

Carbon has an electronic structure of 2.4. What is the number of electrons that carbon has? How many shells do the electrons occupy?

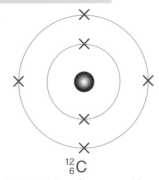

$^{12}_{6}C$

FIGURE 8: A carbon atom. How many shells do the electrons occupy?

Questions

12 Write down whether the following substances are elements or compounds and explain how you can tell. Cu, copper, $CuCl_2$, copper chloride, $CuSO_4$, copper sulfate.

13 Magnesium has an electronic structure of 2.8.2. How many electrons does an atom of magnesium have?

14 Aluminium has an electronic structure of 2.8.3. How many electron shells are occupied in an atom of aluminium?

15 Explain why J. J. Thomson's discovery of the electron was important.

Did you know?

Democritus described atoms in about 400 BC. He had no evidence. It was just an idea. In 1803 John Dalton described atoms in more detail. Again, his was an idea or **model**. As scientists made further discoveries about the behaviour of atoms, the model was changed as new evidence was found.

- In 1897, J. J. Thomson discovered the electron.
- In 1911, Ernest Rutherford said the atom had a nucleus.
- In 1913, Niels Bohr used evidence to explain that electrons orbited the nucleus in orbits.

At each stage the explanations were **provisional** until more convincing **evidence** was found.

Q compound   electron shell   electronic structure

More on the arrangement of electrons in atoms

The elements in the periodic table are arranged in ascending atomic number.

Example

> The atomic number of hydrogen is 1, that of carbon is 6 and that of sodium is 11.

If an atom of an element has the **electronic structure** of 2.8.6 then it is in the third row of the periodic table.

It has three electron shells. It has six electrons in its outer shell. Its atomic number is 16. Looking at the periodic table the atom is of the element sulfur, S.

The theory of electron shells in atoms came much later than the early theories of the atoms. The idea was explored as more **evidence** became available. An early theory was developed by John Dalton. His explanation was **provisional**, but later was **confirmed** by better evidence.

When J. J. Thomson, Rutherford and Bohr found new evidence to add to the understanding of the model of an atom, they were making an explanation that was provisional, because they only had their current evidence. They did not have the evidence that became available in later years. When that happened more observations could be explained and the theory or model developed to give a better explanation.

Questions

16 What is the atomic number of sodium?

17 How many electrons has an atom of sodium?

18 Draw a diagram to show the pattern of electrons in a sodium atom.

Electronic structure

Each element has an electron pattern (electronic structure). The electronic structure of each of the first 20 elements can be worked out using:

> the atomic number of the element

> the maximum number of electrons in each shell.

The third shell takes up to eight electrons before the fourth shell starts to fill.

Example

> The atomic number of lithium, Li, is 3. So the first two electrons of the lithium atom fill the first shell. The third electron goes into the second shell. The electronic structure is 2.1.
>
>
>
> FIGURE 9: How many electrons did it take to complete the inner shell?

> Neon, Ne, has an atomic number of 10 so its electronic structure is 2.8.

> Aluminium, Al, has an atomic number of 13, so the electrons start to occupy the third shell and its electronic structure is 2.8.3.

> Calcium, Ca, has an atomic number of 20, so the electrons start to occupy the fourth shell and its electronic structure is 2.8.8.2.

You can search for information on the experiments that these pioneer scientists did to develop our modern idea of the atom.

Questions

Use the periodic table on page 164 to help you answer the questions.

19 What is the electronic structure of F that has atomic number 9?

20 What is the electronic structure of Mg that has atomic number 12?

21 What is the electronic structure of K?

22 What is the element with electronic structure 2.8.3?

Did you know?

Scientists build on the work of others before them and use models to help explain what they find.

1803: J. Dalton's model was an 'indivisible' atom like a billiard ball.

1897: J. J. Thomson discovered electrons. So his model was now like a 'plum pudding'.

1909: Geiger and Marsden had some unexpected results to an experiment and said an atom has a nucleus.

1911: Rutherford, said an atom had a nucleus at the centre surrounded by a ring of electrons.

In 1913, Bohr, to explain more experimental evidence, said that electrons can only be in orbits and needed a 'quantum of energy' to move to the next orbit.

Each of these scientists needed the work of the one before to build on what was already known.

Q electron pattern   electron shell   Geiger and Marsden experiment

Ionic bonding

You will find out:
> an ion is a charged atom
> how to recognise ions, atoms and molecules from formulae

Bonds

Atoms join together, or bond, in many different ways. They may join by transferring electrons. They may bond by sharing electrons. These different types of bond allow atoms to join in a huge number of ways to make millions of different compounds. One way atoms bond is by electron transfer, which is called ionic bonding.

FIGURE 1: In the sea there are sodium ions that sometimes make salt crystals. What is a sodium ion?

Forming ions and molecules

An **atom** is the smallest particle that can **bond** with another particle.

An atom can be recognised from its symbol. It either has one capital letter or one capital letter and one lower case letter (look at the periodic table on page 164).

The symbol for an atom has no numbers and no charge. The symbol for a magnesium atom is.

Mg   magnesium atom

An **ion** is a charged atom or group of atoms. It has a positive or negative **charge** on it. A calcium ion has two **positive** charges. It is a positive ion.

Ca^{2+} calcium ion

A **molecule** has more than one atom in its formula and no charge.

CO$_2$ carbon dioxide molecule

Ions can have a positive charge, such as the calcium ion, or they can have a negative charge, such as Cl^-. This is a negative ion.

Ions can also be groups of atoms with a charge, NH_4^+ or NO_3^-.

The table shows some examples of atoms, ions and molecules.

Atom	Ion	Molecule
H	H$^+$	H$_2$
O	O^{2-}	H$_2$O
Mg	Mg^{2+}	CO$_2$
Cl	Cl$^-$	Cl$_2$
Na	Na$^+$	Br$_2$
S	SO$_4^{2-}$	F$_2$

Questions

Use the table above to help you to answer these questions.

1 What is the symbol for an atom of oxygen?

2 What is the formula for a magnesium ion?

3 What is the formula for a hydrogen molecule?

4 How many atoms are there in one molecule of carbon dioxide?

Why do atoms form bonds?

Atoms with an outer shell of eight electrons have a **stable electronic structure**. These are normally **unreactive**.

An atom usually has too many electrons or too few electrons to be stable.

> An atom that has too many electrons needs to lose them. These are **metal** atoms.

> An atom that has too few electrons needs to gain them. These are **non-metal** atoms.

Ionic bonds

If an atom loses electrons a positive ion is formed. This is because there will be fewer negatively charged electrons than the number of positively charged protons in the nucleus.

> If an atom loses one electron a (**positive**) 1+ ion is formed, for example, Na → Na⁺.

> If an atom loses two electrons a (**positive**) 2+ ion is formed, for example Mg → Mg²⁺.

If an atom gains electrons a negative ion is formed. This is because there will be more negatively charged electrons than the number of positively charged protons in the nucleus.

> If an atom gains one electron a (**negative**) 1– ion is formed, for example F → F⁻.

> If an atom gains two electrons a (**negative**) 2– ion is formed, for example O → O²⁻.

A metal atom has a small number of electrons in its outer shell. It needs to lose these electrons to be stable. The electrons **transfer** from the metal atom to a non-metal atom to be stable.

A non-metal atom has 'spaces' in its outer shell. It needs to gain electrons to be stable. The electrons transfer to the non-metal atom from the metal atom for the non-metal atom to be stable.

> The metal atom becomes a positive ion.

> The non-metal atom becomes a negative ion.

> The positive ion and the negative ion then attract one another.

> They attract to a number of other ions to make a solid **lattice**.

If magnesium makes a 2+ ion and chlorine makes a 1– ion, then two chlorine ions are needed to bond to one magnesium ion to make $MgCl_2$. There are more examples on page 149.

Questions

5 What is the difference between a metal atom and a non-metal atom? Use ideas about electrons in your answer.

6 Explain how a positive ion is made from a neutral atom.

7 Explain how metals and non-metals combine.

8 Explain why the formula for aluminium chloride is $AlCl_3$, using ideas about transferring electrons.

Describing ionic bonding

We use the **'dot and cross' model** to describe ionic bonding.

Example

Sodium chloride is a solid lattice made up of many pairs of ions.

> Sodium forms a positive sodium ion.

> Chlorine forms a negative chloride ion.

The outer electron of sodium is transferred to the outer shell of the chlorine atom.

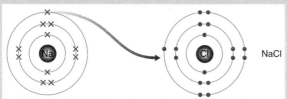

NaCl

FIGURE 2: How many electrons are now in the outer shell of the chlorine model?

The sodium ion and the chloride ion are held together by the attraction of opposite charges. They are represented in this way showing only the outer electrons. No inner shells are shown. The ions are drawn with square brackets and the charge.

FIGURE 3: Why are no electrons represented in the sodium model?

Example

The bonding in magnesium oxide works in exactly the same way as that in sodium chloride, except that the magnesium atom loses two electrons and the oxygen atom gains the two electrons to become an oxide ion.

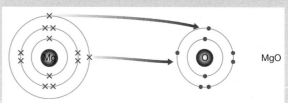

MgO

FIGURE 4: How many electrons are now in the outer shell of the oxygen model?

The 'dot and cross' model of the ions is shown in Figure 5.

FIGURE 5: Why are two of the electrons in the oxygen model represented as crosses (xx)?

Other examples can be found on page 149.

Questions

9 Use dot and cross diagrams to explain the ionic bonding in lithium fluoride from Li⁺ and F⁻ ions.

🔍 'dot and cross' model   electrostatic attraction   lattice   metal

Properties of sodium chloride and magnesium oxide

You will find out:
> properties of sodium chloride and magnesium oxide
> about their giant ionic lattice structure

Melting points

Sodium chloride is the chemical name for the common salt you put on your chips.

> Sodium chloride has a **high melting point** (804 °C).

> Magnesium oxide has a very high melting point (2800 °C).

These melting points are high because the ions form giant lattice structures that need a lot of energy to break down.

FIGURE 6: Sodium chloride can form giant crystals. What do you notice about their shape?

Does sodium chloride conduct electricity?

> Sodium chloride dissolves in water. It **conducts electricity** when it is in solution.

> Sodium chloride conducts electricity when it is melted (molten).

> Sodium chloride does not conduct electricity when it is a solid.

FIGURE 7: Salt extraction at La Palma, Canary Islands. What happens to the water so that this salt is left?

Questions

10 Describe the melting point of sodium chloride.

11 Is sodium chloride soluble or insoluble in water?

12 When does sodium chloride **not** conduct electricity?

13 Describe the melting point of magnesium oxide.

14 When does sodium chloride conduct electricity?

Conducting electricity

Sodium chloride and magnesium oxide do not conduct electricity when they are solids. The electric current cannot pass through because the ions are not able to move.

Solutions

Sodium chloride can dissolve in water though. When it is dissolved in water it makes a **solution**.

Sodium chloride solution can conduct electricity. When it conducts electricity the ions move and can 'split up' from each other. Two electrodes are put into the solution to allow the electricity to pass through.

Molten substances

Sodium chloride and magnesium oxide have high melting points, but they do become molten at much lower temperatures. They become molten substances. If two electrodes are put into the molten liquids both will conduct electricity.

The structure of sodium chloride or magnesium oxide is a **giant ionic lattice**, in which positive ions have strong **electrostatic attraction** to negative ions.

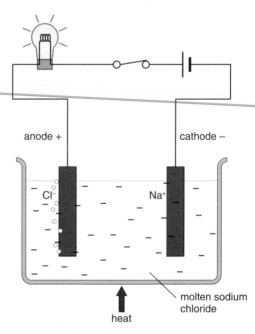

anode +   cathode −

Cl^- Na^+

molten sodium chloride

heat

FIGURE 8: Electrolysis of molten sodium chloride. Which gas is formed at the anode?

molten   dissolve   evaporate   solution   melting point

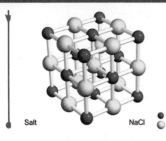

FIGURE 9: There is a strong attraction between positive and negative ions of sodium chloride, forming a giant lattice. What do you notice about the shape of the lattice?

Salt   NaCl

● - Na
○ - Cl

Questions

15 Describe the structure of sodium chloride.

More on ionic bonding

'Dot and cross' models for sodium chloride and magnesium oxide are given on page 147. Other compounds that bond using ionic bonds are sodium oxide and magnesium chloride.

Sodium only has one electron to lose, but oxygen needs to gain two electrons. Two sodium atoms are needed to bond with one oxygen atom.

However, magnesium needs to lose two electrons, but chlorine can only gain one electron. So one magnesium atom needs two chlorine atoms to achieve an ionic bond and make magnesium chloride.

Each atom has either gained or lost the correct number of electrons to achieve a complete outer shell. The outer shell needs eight electrons. It is called a **stable octet**.

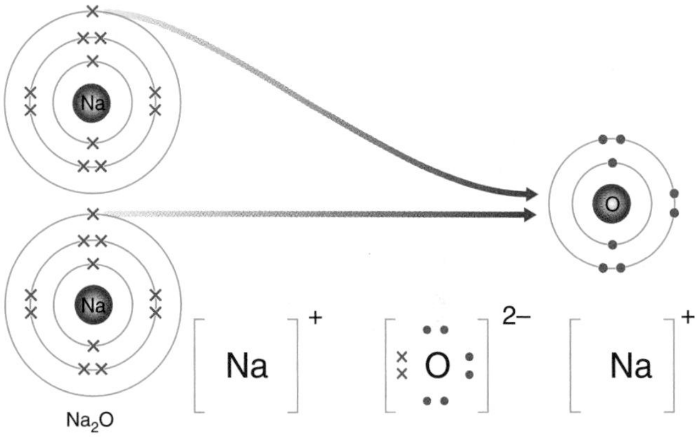

Physical properties

Some of the **physical properties** of sodium chloride and magnesium oxide are explained because:

> there are strong attractions between positive and negative ions, which have to be overcome, so they have high melting points;

> ions cannot move in the solid, so they do not conduct electricity;

> ions can move in solution or in a **molten liquid** so these conduct electricity.

Why is the melting point of magnesium oxide so high?

Both magnesium and oxygen form ions. The ions attract to form a lattice. This gives magnesium oxide a high melting point as a lot of energy is needed to break it apart. This bonding also exists in sodium chloride. Two things make the melting point of magnesium oxide higher.

> Each magnesium atom donates **two** electrons to the oxygen atom, which makes a stronger ionic bond than the bonds made by sodium transferring one electron to chlorine atoms.

> Magnesium ions are very small in radius. So magnesium can get much closer to oxygen, which makes the ionic bond strength higher. More energy is needed to separate these ions than is needed to separate sodium and chloride ions.

Na_2O

FIGURE 10: Why is the formula written as Na_2O?

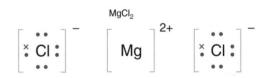

$MgCl_2$

FIGURE 11: Why is the formula written as $MgCl_2$?

Questions

16 What is a stable octet?

17 Draw a diagram to show how an electron is transferred from a lithium atom to a fluorine atom.

18 Draw a 'dot and cross diagram' to show lithium fluoride.

19 Draw a 'dot and cross diagram' to show calcium chloride.

20 Why does magnesium oxide have a high melting point?

21 Why do ionic solids not conduct electricity?

22 Why do molten liquids of ionic compounds conduct electricity?

🔍 giant ionic lattice   stable octet

The periodic table and covalent bonding

You will find out:
> about non-metal atoms bonding together
> how carbon dioxide and water properties relate to covalent bonds
> how to show covalent bonds by 'dot and cross' models

How do atoms join together?

Atoms can join together by **transferring** electrons. However, others may bond by sharing electrons.

Metals and **non-metals** have atoms which join together by transferring electrons. This is called **ionic bonding**.

Non-metals can also **share** electrons between atoms. The compounds have different properties and do not conduct electricity. This type of bonding is known as **covalent bonding**.

These two types of bonding produce different kinds of substances.

> Ionic bonding. Large crystals are bonded by ionic bonds.

FIGURE 1: What do the preservative used in wine making and the gas used as the fuel in the fire have in common?

> Covalent bonding. Molecules are bonded by covalent bonds. Examples are carbon dioxide, which is a gas, and water, which is a liquid. These molecules have low melting points.

 ## Molecules

A molecule forms when two or more non-metal atoms bond together.

If a molecule has the formula O_2 we can see that it has two oxygen atoms in its molecular formula, so the total number of atoms is two. Its displayed formula is shown on the left and a model of it on the right.

$O = O$

If a molecule has the formula CO_2 we can see that it has one carbon atom and two oxygen atoms in its molecular formula, so the total number of atoms is three. Its displayed formula is shown on the left and a model of it on the right

$O = C = O$

The atoms in water, H_2O are also bonded covalently.

Carbon dioxide and water are covalent molecules. They do not conduct electricity.

FIGURE 2: Sulfur is an important industrial chemical. It is covalently bonded. Will it conduct electricity?

FIGURE 3: A molecule forms when two or more atoms bond together. Is this is a model of a diatomic molecule or a triatomic molecule?

Questions

1 How many atoms are there in a molecule of sulfur, S_6?

2 How many different types of atom are there in a molecule of H_2SO_4?

3 How many atoms are there in total in a molecule of H_2SO_4?

4 How many atoms are there in the methane molecule shown here?

 atom   molecule   covalent bonding   methane

Covalent bonding

Non-metals combine together by sharing **electrons**. This is called **covalent** bonding.

Carbon dioxide molecules are held together by **weak intermolecular forces** between the molecules. Water molecules are held together by weak intermolecular forces between the molecules. Carbon dioxide and water are simple molecules.

> ## Questions
>
> **5** Which bonding joins pairs of oxygen atoms?
>
> **6** Draw a model of hydrogen atoms bonding as H_2.
>
> **7** Draw the displayed formula for carbon dioxide.

'Dot and cross' models

The formation of simple molecules that contain single and double covalent bonds can be represented by 'dot and cross' models. The models show the outer electrons only. They do not show the inner electrons. One pair of dot and cross together represents a **single** bond. Two pairs of dot and cross side by side (as in CO_2) represent a **double** bond. They do not use circles to represent shells.

Example

> A molecule of water is made up of three atoms. There are two hydrogen atoms and one oxygen atom in water.
>
> > Oxygen has six electrons in its outer shell. So it needs two more electrons to be complete.
>
> > Hydrogen atoms each have one electron in their only shell. So the oxygen outer shell is shared with each of the hydrogen electrons.
>
> > In this way each of the hydrogen atoms has a share of one more electron so the shell is full.

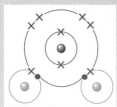

FIGURE 4: A molecule of water. The oxygen atom (red) has a share of eight electrons so has a full outer shell. How many electrons does each hydrogen atom share?

Example

> A molecule of carbon dioxide is also made up of three atoms. There are two oxygen atoms and one carbon atom.
>
> > Carbon has four electrons in its outer shell. So it needs four more electrons to be complete.

> ## Questions
>
> **8** Draw a 'dot and cross' model of the bonding in methane, CH_4.
>
> **9** Draw a 'dot and cross' model of the bonding in NH_3.

> Oxygen atoms each have six electrons in their outer shell. So they each need two more electrons to be complete.

> So each oxygen outer shell is shared with two of the electrons of the carbon outer shell.

> In this way each of the oxygen atoms has a share of two more electrons so the shell is full.

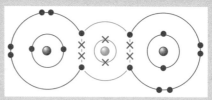

FIGURE 5: A molecule of carbon dioxide. The carbon atom (blue) has a share of eight electrons so has a full outer shell. How many electrons does each oxygen atom share?

Example

'Dot and cross' models for some simple molecules.

H × H
H_2

What does the dot and cross represent in this model of a hydrogen molecule?

× Cl × Cl
Cl_2

How many outer electrons does each chlorine atom have in this model?

O × C × O
CO_2

Why are there two dots and two crosses on each side of the carbon atom?

H × O × H
H_2O

How many shared pairs of electrons are there in this water molecule?

Groups and periods in the periodic table

You will find out:

> about ionic and covalent bonding
> about group numbers
> how a period relates to electronic structure

Elements that are in the same **group** (family) in the periodic table (see page 164 for the whole periodic table) are written in the same vertical **column** on the table.

A group of elements is all the elements in a vertical column of the periodic table. These elements have similar **chemical properties**.

this column has the elements of group 1

| | | | | | | | H
hydrogen
1 | | | | | | | He
helium
2 |

Li lithium 3	Be beryllium 4		B boron 5	C carbon 6	N nitrogen 7	O oxygen 8	F fluorine 9	Ne neon 10
Na sodium 11	Mg magnesium 12		Al aluminium 13	Si silicon 14	P phosphorus 15	S sulfur 16	Cl chlorine 17	Ar argon 18
K potassium 19	Ca calcium 20							

FIGURE 6: Group 1 elements in the periodic table. Name the elements in group 1.

Lithium, sodium and potassium are all in group 1.

A **period** of elements is all the elements in a horizontal **row** of the periodic table.

| | | | | | | | H
hydrogen
1 | | | | | | | He
helium
2 |

Li lithium 3	Be beryllium 4		B boron 5	C carbon 6	N nitrogen 7	O oxygen 8	F fluorine 9	Ne neon 10
Na sodium 11	Mg magnesium 12		Al aluminium 13	Si silicon 14	P phosphorus 15	S sulfur 16	Cl chlorine 17	Ar argon 18
K potassium 19	Ca calcium 20							

this row has the elements of period 3

FIGURE 7: Elements from period 3 of the periodic table. List the elements that belong to this period.

Did you know?

1829: Döbereiner saw three elements had similar properties. He called these 'triads'. He noticed patterns with sodium, lithium and potassium, with calcium, strontium and barium and chlorine, bromine and iodine. These were called Döbereiner triads.

1865: John Newlands. He noticed a pattern every eight elements.

He used his new evidence to make a new theory – the 'law of octaves'.

1869: D. Mendeleev gathered more evidence and put together the first modern periodic table. He even left gaps in his table for elements not discovered.

Questions

10 Which elements are in group 2?

11 Which elements are in period 2?

Group numbers

The group number is the same as the number of electrons in the outer shell.

Group 1 elements have one electron in the outer shell (Figure 8).

FIGURE 8: Lithium has the electron pattern on the left and sodium that on the right.

Group 7 elements have seven electrons in the outer shell (Figure 9).

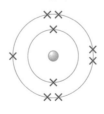
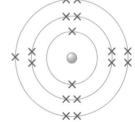

FIGURE 9: Fluorine has the electron pattern on the left and chlorine that on the right.

group of elements   periodic table   period of elements

Group 8 elements have eight electrons in the outer shell (Figure 10).

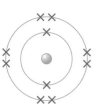

FIGURE 10: Neon has the electron pattern on the left and argon that on the right.

How to tell which period an element belongs to

> If an element has electrons in only one shell it will be found in the first period.

> If an element has electrons in two shells it will be found in the second period.

> If an element has electrons in three shells it will be found in the third period.

Element	Electron pattern	Period
H	1	1
Li	2.1	2
Na	2.8.1	3

Did you know?

1865: John Newlands put the 56 known elements into groups. He noticed that every eighth element behaved similarly. His idea was not accepted until 50 years later when scientists had more evidence.

1869: Mendeleev arranged the elements in order in a table. He noticed a **periodic** change in properties. He predicted that there would be new elements found between zinc and arsenic. What are the elements he did not know?

Questions

12 Sodium has an electron pattern 2.8.1.
To which group does sodium belong?

13 Explain why lithium belongs to group 1.

Predicting physical properties

Carbon dioxide and water are simple molecules with weak intermolecular forces between the molecules.

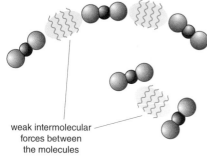

weak intermolecular forces between the molecules

FIGURE 11: Molecules have weak intermolecular forces between them. How does this affect their physical properties?

The physical properties of carbon dioxide and water are related to their structure.

> As they have weak intermolecular forces between the molecules they are easy to separate so the substances have low melting points.

> As there are no free electrons available they do not conduct electricity.

Questions

14 Magnesium has an electron pattern 2.8.2. To which group does magnesium belong?

15 Explain why nitrogen belongs to Period 2.

16 Explain why carbon dioxide has a low melting point.

17 Explain why water does not conduct electricity.

18 The figure shows the structure of phosphorus. In which group is phosphorus and in which period of the periodic table?

Did you know?

1891: Mendeleev refined his table but it still did not contain the noble gases. These were isolated later.

1914: experiments confirmed his idea of periodicity was right if the order was by atomic number, not mass. His predicted missing elements were found. You can find out about pieces of evidence confirming his ideas by searching the scientists' names and following the links.

periodicity   noble gases

The group 1 elements

You will find out:
> group 1 metals are the alkali metals
> Li, Na and K are group 1 metals
> how to predict properties of other alkali metals

The alkali metals

Ever wondered how fireworks are made to have such stunning colours? It is because they have compounds added to them. Sodium compounds, for instance, produce a bright yellow colour against the night sky. Sodium is an element in the periodic table in the first column. When Dmitri Mendeleev first put the elements into an order in his first table, he grouped lithium, sodium and potassium together. We now know these as the group 1 elements.

FIGURE 1: This is lithium. What kind of surface does it have now it has been cut with a knife?

Alkali metals

Group 1 metals are called the **alkali metals** because their reaction with water gives alkaline solutions (not acidic ones).

Alkali metals are stored under oil because they react with air and water.

Alkali metals react vigorously with water to form the metal hydroxide (which is an **alkaline solution**) and hydrogen.

The order of **reactivity** of the alkali metals with water is:

> sodium is more reactive than lithium

> potassium is more reactive than sodium.

The word equations for these reactions are:

sodium + water → sodium hydroxide + hydrogen

potassium + water → potassium hydroxide + hydrogen

lithium, sodium and potassium are metals in group 1

							He helium 2
				H hydrogen 1			
Li lithium 3	Be beryllium 4	B boron 5	C carbon 6	N nitrogen 7	O oxygen 8	F fluorine 9	Ne neon 10
Na sodium 11	Mg magnesium 12	Al aluminium 13	Si silicon 14	P phosphorus 15	S sulfur 16	Cl chlorine 17	Ar argon 18
K potassium 19	Ca calcium 20						

FIGURE 2: Group 1 metals in the periodic table. What is another name for them?

FIGURE 3: The order of reactivity of the alkali metals. Which metal is the most reactive?

Did you know?

The picture shows you what can happen if you get pure sodium hydroxide on your skin. Sodium hydroxide is used in home oven cleaners.

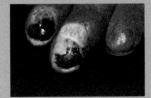

Sodium hydroxide is more dangerous to get into the eyes than acids are. The ions travel to the back of the eyes and can irreversibly damage the retina.

That is why your teacher always tells you to wear safety glasses when handling chemicals, especially alkalis.

Questions

1 Write down the names of three group 1 metals.

2 What do group 1 metals look like when freshly cut?

3 Why are group 1 metals kept under oil?

4 Which metal in group 1 is the least reactive?

5 Lithium reacts with water to form lithium hydroxide and hydrogen. Write down the word equation for this reaction.

alkali   alkali metal   alkali metal density

Properties of alkali metals

When lithium, sodium and potassium react with water they float on the surface. This is because their **density** is less than the density of water.

A gas is given off. This gas is hydrogen. The metal reacts with water to form an **alkali**. This is the **hydroxide** of the metal.

> Sodium forms sodium hydroxide.

> Potassium forms potassium hydroxide.

Is there a pattern in the reactivity of alkali metals with water?

> Lithium reacts quickly and vigorously with water.

> Sodium reacts very quickly and very vigorously with water.

> Potassium reacts extremely vigorously with water and produces a lilac flame.

FIGURE 4: When lithium is added to water it floats and gives off bubbles of a colourless gas. What is the gas?

To construct the **balanced symbol equation** for these reactions you need to start with the word equation.

lithium + water → lithium hydroxide + hydrogen

Next put in the symbols formula of compounds, including lithium hydroxide, LiOH:

$Li + H_2O \rightarrow LiOH + H_2$

These formulae are now correct and cannot be changed, but this is not **balanced**.

Only the number of whole molecules can now change. To increase the number of hydrogen atoms on the left hand side use two molecules of water:

$Li + 2H_2O \rightarrow LiOH + H_2$

This is still not right. You will need 2LiOH:

$Li + 2H_2O \rightarrow 2LiOH + H_2$

The H atoms balance and the O atoms balance. Now the Li atoms need to balance:

$2Li + 2H_2O \rightarrow 2LiOH + H_2$

This is the correct balanced symbol equation.

 Questions

6 Why does sodium float on water?

7 Why are group 1 metals called alkali metals?

8 How does the reactivity of the alkali metals with water change going down the group?

9 Will the reaction of caesium with water be more or less vigorous than sodium with water?

Predicting the properties of alkali metals

You need to construct the balanced symbol equation for the reaction between group 1 elements and water, with no formulae given.

The word equation is written first.

sodium + water → sodium hydroxide + hydrogen

Next represent this in symbols:

$Na + H_2O \rightarrow NaOH + H_2$

Now balance the whole equation:

$2Na + 2H_2O \rightarrow 2NaOH + H_2$

The same can be done for potassium on water, which is even more reactive.

You should be able to predict the way that other alkali metals behave from the patterns so far.

> How does rubidium react with water?

> Does it react more or less vigorously than potassium?

> How do other properties of rubidium and caesium compare to potassium?

Think about the trends you have seen so far.

reactivity increases down the group

	Melting point in °C	Boiling point in °C
$_3$Li	179	1317
$_{11}$Na	98	892
$_{19}$K	64	774

FIGURE 5: Reactivity of the alkali metals with water increases down group 1.

 Questions

10 Will the melting point of rubidium be higher or lower than that of lithium?

11 Write a balanced symbol equation for the reaction of potassium with water.

12 What alkali will be made when caesium reacts with water?

Q Mendeleev   firework colour   chemical reactivity

Flaming metals

You will find out:

> group 1 elements produce coloured flames

> how to carry out a flame test

> how properties are related to electronic structure

FIGURE 6: If you put compounds of lithium in a flame, the flame turns red.

FIGURE 7: If you put compounds of sodium in a flame, the flame turns yellow.

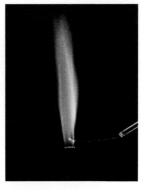

FIGURE 8: If you put compounds of potassium in a flame, the flame turns lilac.

Did you know?

When a **flame test** is used the outer electron is 'excited' out of its shell and when it returns a colour is produced. The colours are different because the outer electron of different atoms are excited by different amounts.

In Figures 6–8, which chloride compound produces the yellow flame?

Whizzbang Firework Company wants to make a new set of 'starburst' fireworks for its anniversary range.

It wants to produce a set of lilac 'starbursts'. The technical department suggests adding potassium chloride to the mix. Look at Figures 6–8. Which colour would this metal compound make against the night sky?

The department is also asked to produce a starburst with a red effect and one with a yellow effect.

memo

WHIZZBANG
technical department

Suggestions for chemicals to be added to 'starbursts' for anniversary range:

Chemicals	Colour
potassium chloride	lilac
lithium chloride	red
sodium chloride	yellow

Questions

13 What colour does a compound of sodium produce in a flame?

14 What colour does a compound of potassium produce in a flame?

15 Which chemical produces a red flame?

FIGURE 9: A student performs a flame test. What metal compound is she testing?

Flame tests

Sancha and Alessia want to test the quality of the chemicals that the Whizzbang firework company is going to add to its fireworks.

There is one test that will ensure the colour is correct.

> Put on safety glasses. Moisten a flame test wire with dilute hydrochloric acid.

> Dip the flame test wire into the sample of solid chemical.

> Hold the flame test wire in a blue Bunsen burner flame.

> Record the colour of the flame in a table.

Alkali metal in the compound	Colour of flame
lithium	red
sodium	yellow
potassium	lilac

These group 1 metals have one **electron** in their **outer shell**.

This is why group 1 metals have similar properties.

🔍 flame test   flame colour   group 1 metal   ionic equation

Questions

16 When performing a flame test what is the flame test wire moistened with?

17 What colour Bunsen burner flame is the test wire and sample held in?

18 What alkali metal compound is indicated by a lilac flame?

19 How many electrons do elements in group 1 have in their outer shell?

Explaining reactivity patterns

Alkali metals have similar properties. This is because when they react their atoms need to lose one electron to form full outer shells. This is then a **stable electronic structure**.

When the atom **loses one electron** it forms an **ion**. The atom becomes charged. It has more positive charges in its nucleus than negative electrons surrounding it. So the charge is positive. It has become a positive ion.

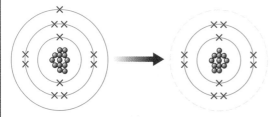

FIGURE 10: How alkali metals achieve a stable electronic structure. 11 protons (+) and 11 electrons (−) make a neutral atom; 11 protons (+) and 10 electrons (−) make a positive ion.

This can be represented by an equation. It shows the formation of an ion from its atom:

Na − e⁻ → Na⁺

Lithium loses its outer electron from its second shell. Sodium loses its outer electron from its third shell. The third shell is further away from the attractive 'pulling force' of the nucleus. This means that the electron from sodium is more easily lost than the electron from lithium. This is why sodium is more reactive than lithium.

Remember!
'OIL': Oxidation Is Loss.

electron in outer shell is further from the nucleus, there is less attractive force, so the electron is more easily lost

electron in outer shell is closer to the nucleus, there is more attractive force, so the electron is less easily lost

FIGURE 11: The more easily a metal atom loses its outer electron the more reactive it is.

The easier it is for an atom of an alkali metal to lose one electron, the more reactive the alkali metal.

The outer electron of potassium is even further away from the nucleus so it is even more easily lost from the atom. So potassium is more reactive than either sodium or lithium.

Oxidation

If electrons are lost, the process is called **oxidation**.

K − e⁻ → K⁺

You can see why this process is oxidation from its **ionic equation**. An atom of potassium loses one electron. It now becomes a potassium ion. The potassium ion is a positive ion. This is an example of oxidation.

Questions

20 Why does an atom in group 1 lose electrons?

21 Why is sodium more reactive than lithium?

22 What is the process of losing electrons called?

23 Write an equation to show the oxidation reaction for lithium.

stable electronic structure   oxidation   outer electron shell

Preparing for assessment: Applying your knowledge

To achieve a good grade in science, you not only have to know and understand scientific ideas, but you need to be able to apply them to other situations and investigations. These tasks will support you in developing these skills.

✳ 'Dancing around and delirious with joy!'

What do you see when you look at the periodic table? You see a summary of hundreds of pieces of information about elements, brought together in one chart. This information has been discovered by many different scientists from all over the world, but a few people have made a major contribution and one of those was Cornishman Humphry Davy.

Around 1800, there were a number of scientists experimenting with using electricity to break down substances into their component elements. Davy was convinced he could do this to what were then known as the alkali earths. At first he was unsuccessful – he tried passing current through solutions of alkali earths, but this simply released hydrogen. He then tried melting the compounds and passing a current through them. This worked much better and produced tiny beads of pure metal. When beads of one of these metals were dropped onto water they 'skimmed about excitedly with a hissing sound, and soon burned with a lovely lavender light' according to Humphry's brother John.

Humphrey 'danced around and was delirious with joy' when he saw this lavender light.

Davy continued to use electricity to extract elements from compounds in a process we now call electrolysis and he managed to discover magnesium, calcium, barium and strontium. Another of his discoveries was nitrous oxide, or 'laughing gas', which he is said to have breathed large quantities of himself. Then used by dentists as an anaesthetic, Davy used it for fun. It is likely that this and the other gases he sampled affected his health and caused him to die prematurely.

Task 1

From Davy's earlier electrolysis experiments, how do you think he could show that the gas released was hydrogen?

✸ Task 2

What was the metal that, when added to water, 'skimmed about excitedly with a hissing sound, and soon burned with a lovely lavender light'?

✸ Task 3

Think about the reaction of the metal added to water.

a If Universal Indicator solution was added to the water after the reaction, what colour would it be?

b What product from the reaction causes this colour change?

c How does this help to explain the general name given to the group 1 metals?

d Why is a safety screen necessary during the reaction?

✸ Task 4

Write down an equation to summarise the reaction.

✸ Task 5

When Davy wanted to extract pure sodium he started with table salt.

a Why was this?

b What is the other element released?

✸ Maximise your grade

Answer includes showing that you can...
State how to test for hydrogen.
Explain why safety precautions are necessary when Group 1 metals react with water and explain how a flame test can identify a group 1 metal.
Write a word equation for the reaction between the metal and water. Suggest the colour change for Universal Indicator as it is added to water after the reaction.
State the chemical name for table salt and identify the two elements in the compound.
State the product of the chemical reaction between the metal and water. Explain, using the terms acid and alkali, why Universal Indicator changes colour in the way that it does.
Use chemical symbols to write an equation for the reaction between the metal and water.
Explain the trend in reactivity of Group 1 elements with water, in terms of electron loss.

F C A

The group 7 elements

You will find out:
> that F_2, Cl_2, Br_2 and I_2 are called halogens
> they react vigorously with alkali metals
> equations for reactions of alkali metals with halogens

Halogens

Group 7 elements are known as the halogens. Halogens have many uses. The first use of chlorine was to bleach textiles in 1785 and it is still used in chemicals for bleaching toilets. Chlorine is used to sterilise water and so prevent diseases like cholera from spreading. You can search for information on why this was possible by looking up the work of John Snow (1813–1858). Clean water has been one of the most important factors in reducing deaths among the world's population. Chlorine was also used as a weapon in the World War I. The effects were devastating.

FIGURE 1: Poolside waterfall – clean, clear and chlorinated.

The halogens

Group 7 elements are called the **halogens**.

Fluorine, chlorine, bromine and iodine are halogens.

Halogens react vigorously with **alkali metals** such as sodium and potassium.

								H hydrogen 1									He helium 2
Li lithium 3	Be beryllium 4									B boron 5	C carbon 6	N nitrogen 7	O oxygen 8	F fluorine 9	Ne neon 10
Na sodium 11	Mg magnesium 12									Al aluminium 13	Si silicon 14	P phosphorus 15	S sulfur 16	Cl chlorine 17	Ar argon 18
K potassium 19	Ca calcium 20													Br bromine 35	
														I iodine 53	

this column has the elements of group 7

FIGURE 2: Group 7 elements in the periodic table. What is another name for them?

The uses of some halogens

FIGURE 3: Chlorine is used to sterilise water.

FIGURE 4: Chlorine is used to make plastics and pesticides

FIGURE 5: Iodine is used to sterilise wounds.

Constructing word equations

If you know the name of the elements and the name of the compound made you can write a word equation. Sodium reacts with bromine to make sodium bromide. The word equation is:

sodium + bromine → sodium bromide

Questions

1 What is chlorine used for?

2 Give one use of iodine.

3 With which type of metal do halogens react vigorously?

Group 7 trends

There is a **trend** in the physical appearance of the halogens at room temperature. Chlorine is a gas, bromine is a liquid and iodine is a solid.

Reactions between alkali metals and halogens

When alkali metals react with halogens there is a vigorous reaction. A **metal halide** is made.

> When lithium reacts with chlorine the metal halide made is lithium chloride.

FIGURE 6: At room temperature chlorine is a green gas and iodine is a grey solid. What colour liquid is bromine?

Q   chlorine   bromine   iodine   halogen   water sterilisation

> When potassium reacts with bromine the metal halide made is potassium bromide.

> When sodium reacts with chlorine the metal halide made is sodium chloride.

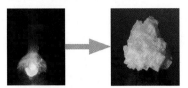

FIGURE 7: What is made when green chlorine gas and sodium metal react?

Constructing word equations

You should be able to write down the word equation of the reaction between a group 1 element and a group 7 element, even if only two of the chemicals are known. For example, the reaction between potassium and chlorine:

potassium + chlorine → potassium chloride

Balancing equations

It is now possible to construct a balanced symbol equation for the reaction of an alkali metal with a halogen, when you know some of the formulae.

Step 1

Write down the symbols for the alkali metal (potassium, K) and the halogen (chlorine, Cl_2). These are the **reactants**.

The '2' in Cl_2 means there are two bonded atoms in the molecule of chlorine. This number cannot be changed!

Step 2

Write down the formula for the product (potassium chloride, KCl). Notice that only one chloride **ion** is joined to one potassium ion. This leaves a 'spare' chloride ion without another ion to join to.

($K + Cl_2 \rightarrow KCl$)

Step 3

Join this spare second chloride ion to a second potassium ion, represented by a large '2' in front of the symbol K, to give 2K. So we have two potassium ions joined to two chloride ions, giving 2KCl:

$2K + Cl_2 \rightarrow 2KCl$

The product is therefore 2KCl.

Questions

4 Which halogen is an orange liquid at room temperature?

5 What is the appearance of iodine at room temperature?

6 What metal halide is made when potassium reacts with bromine?

7 Write down the word equation for the reaction between sodium and chlorine.

Balancing equations

You can construct a balanced symbol equation for the reaction of an alkali metal with a halogen without formula given.

lithium + bromine → lithium bromide

Step 1

Write down the symbols for the reactants.

Step 2

Write down the formula for the product:
($Li + Br_2 \rightarrow LiBr$)

Step 3

The second bromine ion joins to a second lithium ion, (a large '2' in front of the symbol Li, gives 2Li).

$2Li + Br_2 \rightarrow 2LiBr$

Did you know?

Silver bromide is a compound of bromine with silver. It was used in early photography as it turned from cream to a purplish colour when exposed to light.

FIGURE 8: When was silver bromide first used in photography?

Questions

8 Why is the formula of chlorine gas written as Cl_2?

9 Why is KCl not written as KCl_2?

10 Write a balanced symbol equation for the formation of lithium chloride from its elements.

 halogen   alkali metal   balanced equation

Halogens and their order of reactivity

You will find out:

> that F_2 is more reactive than Cl_2

> about the reactivity of halogens

> that Cl_2 displaces halogens from solution

> Fluorine is more reactive than chlorine.

> Chlorine is more reactive than bromine.

> Bromine is more reactive than iodine.

| $_9$F |
| $_{17}$Cl |
| $_{35}$Br |
| $_{53}$I |

Reactivity ↑

FIGURE 9: Halogens dissolve in an organic solvent. Can you say which halogens are dissolved?

FIGURE 10: Halogens reacting with metals.
a Iron reacting in chlorine.
b Aluminium reacting in bromine.
c Aluminium reacting in iodine.

Constructing word equations

When a compound of a metal halide is put in solution it is usually a colourless solution.

Sometimes if a different halogen is reacted with the solution it changes colour. This is because a reaction has taken place.

If you know all the **reactants** and all the **products** of this reaction you can construct a word equation.

> Chlorine reacts with potassium bromide to form potassium chloride and bromine solution.

chlorine + potassium → potassium + bromine
　　　　　bromide　　chloride　(orange solution)

> Chlorine reacts with sodium iodide to make iodine and sodium chloride solution.

chlorine + sodium → sodium + iodine
　　　　　iodide　　chloride　(red–brown solution)

> Bromine reacts with potassium iodide to make potassium bromide and iodine.

bromine + potassium → potassium + iodine
　　　　　iodide　　　bromide

Did you know?

Chlorine gas was used in World War I to attack soldiers in the trenches. This happened before gas masks were available.

Questions

11 Which halogen is the most reactive?

12 Which halogen gives a purple vapour when heated?

13 Which halogen is least reactive?

14 Which gas was used in World War I in the trenches?

Displacement reactions of halogens

The **reactivity** of the halogens increases up the group.

If halogens are bubbled through solutions of metal halides there are two possibilities: no reaction or a **displacement** reaction.

If bromine is bubbled through potassium chloride solution there is no reaction. This is because chlorine is more reactive than bromine. However, if chlorine is bubbled through potassium bromide solution a displacement reaction occurs. If you know the word equation and some of the **formulae** you can construct a balanced symbol equation of a reaction.

fluorine 2. 7

chlorine 2. 8. 7

bromine (outer shell only shown) 7

iodine (outer shell only shown) 7

FIGURE 11: Electronic structures of the halogens.

🔍 reactivity　reduction　negative ion

> Chlorine displaces the bromide to form bromine solution.

chlorine + potassium → potassium + bromine
bromide   chloride   (orange solution)

$Cl_2 + KBr → KCl + Br_2$ unbalanced

$Cl_2 + 2KBr → 2KCl + Br_2$ balanced

> Bromine displaces iodides from solutions.

$Br_2 + 2KI → 2KBr + I_2$ (red–brown solution)

Electronic configurations of group 7

Group 7 elements have similar properties because they all have seven electrons in their outer shell.

 Questions

15 How does the reactivity of halogens change down the group?

16 What will be the product of displacement when chlorine is bubbled through a solution of lithium bromide?

Predicting properties

It is possible to predict the properties of other halogens, such as fluorine or astatine, knowing the properties of the other halogens. This is because the properties follow trends.

Trends of physical properties

Halogen	Melting point in °C	Boiling point in °C	State at room temperature
$_9F$			
$_{17}Cl$	−101	−35	gas
$_{35}Br$	−7	59	liquid
$_{53}I$	114	184	solid
$_{85}At$			

Halogens have similar properties because when they react each atom gains one electron to form a **negative ion** with a stable electronic structure.

> Fluorine has an electronic structure of 2.7. It gains one electron to become 2.8.
>
> Chlorine has an electronic structure of 2.8.7. It gains one electron to become 2.8.8.
>
> An atom of Cl becomes an ion Cl^-.

The nearer the outer shell is to the nucleus, the easier it is for an atom to gain one electron. The easier it is to gain the electron, the more reactive the halogen.

The trend in reactivity is shown by displacement reactions, such as chlorine displacing iodide as iodine from sodium iodide solution.

> $Cl_2 + 2NaI → 2NaCl + I_2$ (red–brown solution)

Knowing this trend allows you to predict the reactions between any halogen and a metal halide of another halogen and construct the balanced symbol equation.

Reduction

If **electrons are gained** the process is called **reduction**.

> $Br_2 + 2e^- → 2Br^-$

You can see why this process is reduction from the ionic equation.

A molecule of bromine gains two electrons. It now becomes two bromide ions. The bromide ion is a negative ion. This is reduction.

Remember!

RIG – Reduction Is Gain.

Questions

17 Predict the state of astatine at room temperature.

18 How many electrons does a bromine atom gain to become a bromide ion?

19 What is the process of electron gain called?

20 Write an ionic equation to show the formation of an iodide ion from a molecule of iodine.

21 Write a balanced equation for the reaction between fluorine and potassium iodide.

Transition elements

Very useful metals

Transition metals are often useful and many have been known since ancient times. Gold and silver are as decorative metals and have had value as currencies for thousands of years.

Liquid mercury has unusual properties and chromium is used in a range of alloys, which are mixtures of metals.

Iron and copper are the most used metals of the transition block because they are strong and conduct electricity.

FIGURE 1: Can you name any of the transition metals in the picture?

Transition elements

																	8
			1 **H** hydrogen 1														4 **He** helium 2
1	2			Key								3	4	5	6	7	
7 **Li** lithium 3	9 **Be** beryllium 4			relative atomic mass **atomic symbol** name atomic (proton) number								11 **B** boron 5	12 **C** carbon 6	14 **N** nitrogen 7	16 **O** oxygen 8	19 **F** fluorine 9	20 **Ne** neon 10
23 **Na** sodium 11	24 **Mg** magnesium 12											27 **Al** aluminium 13	28 **Si** silicon 14	31 **P** phosphorus 15	32 **S** sulfur 16	35.5 **Cl** chlorine 17	40 **Ar** argon 18
39 **K** potassium 19	40 **Ca** calcium 20	45 **Sc** scandium 21	48 **Ti** titanium 22	51 **V** vanadium 23	52 **Cr** chromium 24	55 **Mn** manganese 25	56 **Fe** iron 26	59 **Co** cobalt 27	59 **Ni** nickel 28	63.5 **Cu** copper 29	65 **Zn** zinc 30	70 **Ga** gallium 31	73 **Ge** germanium 32	75 **As** arsenic 33	79 **Se** selenium 34	80 **Br** bromine 35	84 **Kr** krypton 36
85 **Rb** rubidium 37	88 **Sr** strontium 38	89 **Y** yttrium 39	91 **Zr** zirconium 40	93 **Nb** niobium 41	96 **Mo** molybdenum 42	[98] **Tc** technetium 43	101 **Ru** ruthenium 44	103 **Rh** rhodium 45	106 **Pd** palladium 46	108 **Ag** silver 47	112 **Cd** cadmium 48	115 **In** indium 49	119 **Sn** tin 50	122 **Sb** antimony 51	128 **Te** tellurium 52	127 **I** iodine 53	131 **Xe** xenon 54
133 **Cs** caesium 55	137 **Ba** barium 56	139 **La*** lanthanum 57	178 **Hf** hafnium 72	181 **Ta** tantalum 73	184 **W** tungsten 74	186 **Re** rhenium 75	190 **Os** osmium 76	192 **Ir** iridium 77	195 **Pt** platinum 78	197 **Au** gold 79	201 **Hg** mercury 80	204 **Tl** thallium 81	207 **Pb** lead 82	209 **Bi** bismuth 83	[209] **Po** polonium 84	[210] **At** astatine 85	[222] **Rn** radon 86
[223] **Fr** francium 87	[226] **Ra** radium 88																

FIGURE 2: The periodic table with the transition elements shaded grey. What is the symbol for the transition metal called cobalt?

Transition elements are found in the middle part of the **periodic table**.

The periodic table can be used to find the name or **symbol** of a transition element.

Copper and iron are two transition elements.

> The symbol for copper is Cu.

> The symbol for iron is Fe.

🔍 periodic table   lustrous   malleable   ductile   sonorous

Properties of transition elements

All transition elements are **metals** and have typical **metallic properties**. They:

> conduct heat

> are shiny (lustrous)

> conduct electricity

> are sonorous (ring when struck)

> are malleable (can be beaten into a sheet)

> are ductile (can be drawn into a wire).

Questions

1 What are the symbols for manganese and chromium?

2 Give three properties of metals.

Coloured compounds

A compound that contains a transition element is often coloured.

> copper compounds are often blue.

> iron(II) compounds are often pale green.

> iron(III) compounds are often orange/brown.

FIGURE 3: What transition elements do these compounds contain?

Catalysts

A transition element and its compounds are often **catalysts**.

> Iron is used in the **Haber process** to make ammonia, which is used in fertilisers.

> **Nickel** is used in the manufacture of margarine to harden the oils.

A catalyst is an element or compound that changes the rate of a chemical reaction without taking part in the reaction. Catalysts are unchanged during the reaction.

Questions

3 What colour are most compounds of copper?

4 Give one example of the use of a transition metal as a catalyst.

5 Find out from the Internet and books why Haber needed to make more ammonia for fertilisers.

Thermal decomposition

Thermal decomposition is a reaction in which a substance is broken down into at least two other substances by heat.

Many of the transition metals form **carbonates** that are often coloured. When a coloured carbonate is heated it is broken down. This is an example of **thermal decomposition**.

When metal carbonates are heated they give off **carbon dioxide**. Carbon dioxide is tested for by passing it through a colourless solution of **limewater** that turns **milky** if the gas is present.

Precipitation reaction

> Precipitation is a reaction between solutions that makes an **insoluble** solid.

When two solutions react together to make a solid, the reaction is called a **precipitation** reaction.

When a blue solution of copper sulfate is added to a colourless solution of sodium hydroxide, a precipitate (solid) is formed. The precipitate is blue-coloured copper hydroxide.

FIGURE 5: What two solutions react to give a blue precipitate of copper hydroxide?

You will find out:

> about decomposition of coloured carbonates

> coloured solutions

> transition metal hydroxides are different colours

FIGURE 4: What transition element compounds are in solution in these flasks?

Questions

6 What is thermal decomposition?

7 What is the test for carbon dioxide?

8 What is a precipitation reaction?

9 Copper carbonate decomposes to form copper oxide and carbon dioxide. Write down the word equation for this reaction.

Thermal decomposition of metal carbonates

If a transition metal carbonate is heated it **decomposes** to form a metal oxide and carbon dioxide.

On heating:

> $FeCO_3$ decomposes forming iron oxide and carbon dioxide

> $CuCO_3$ decomposes forming copper oxide and carbon dioxide

> $MnCO_3$ decomposes forming manganese oxide and carbon dioxide

> $ZnCO_3$ decomposes forming zinc oxide and carbon dioxide.

The metal carbonates change colour during the reaction.

Reactions with sodium hydroxide solution

Sodium hydroxide solution reacts with a transition metal compound to make a solid of a particular colour. The hydroxide solution is used to identify the presence of transition metal ions in solution.

> Cu^{2+} ions form a **blue** solid.

> Fe^{2+} ions form a **grey/green** solid.

> Fe^{3+} ions form an **orange/brown gelatinous** solid.

These solids are metal hydroxide precipitates.

FIGURE 6: These are metal hydroxide precipitates. The precipitate in the tube on the right is nickel hydroxide. What are the precipitates in the other tubes?

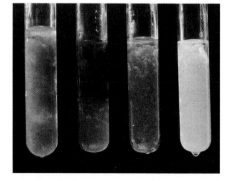

Q precipitate   thermal decomposition   carbon dioxide   limewater

Questions

10 Which gas is given off in the thermal decomposition of metal carbonates?

11 Write a word equation for the thermal decomposition of zinc carbonate.

12 What chemical is used to identify transition metal ions by colour?

13 Which metal gives a blue solid when added to a hydroxide solution?

14 Labels have fallen off two bottles of chemicals believed to be a copper compound and an iron compound. Explain how you could name them using precipitation reactions.

Writing a balanced symbol equation

To write a balanced symbol equation to describe the thermal decomposition of a transition metal carbonate, or a precipitation reaction, the following steps must be followed.

Step 1

Write a word equation to establish the products of the reaction. For example:

copper → copper + carbon
carbonate   oxide   dioxide

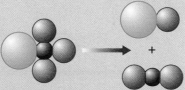

FIGURE 7: Model of the reaction.

Step 2

Assign symbols to the words. For example:

$$CuCO_3 \rightarrow CuO + CO_2$$
$$FeCO_3 \rightarrow FeO + CO_2$$
$$MnCO_3 \rightarrow MnO + CO_2$$
$$ZnCO_3 \rightarrow ZnO + CO_2$$

These are balanced.

Sometimes when writing a symbol equation for precipitation reactions they will need to be balanced.

Step 3

For example:

$$Cu^{2+} + 2OH^- \rightarrow Cu(OH)_2$$
$$Fe^{2+} + 2OH^- \rightarrow Fe(OH)_2$$
$$Fe^{3+} + 3OH^- \rightarrow Fe(OH)_3$$

Questions

15 Write a balanced symbol equation for the thermal decomposition of nickel carbonate, $NiCO_3$.

16 Why are two OH^- ions needed to react with one Cu^{2+} ion?

17 Why are three OH^- ions needed to react with one Fe^{3+} ion?

🔍 precipitation reaction   gelatinous   balanced equation

Metal structure and properties

You will find out:
> about the different properties and uses of metals
> how metals conduct electricity
> about superconductors

'Show your metal...'

Metals are useful materials because they have a wide range of properties. Gold is used for jewellery because it does not corrode and it has an appealing colour and lustre. Copper is used for saucepan bases because of its good thermal conductivity. It is also used for wiring because of its good electrical conductivity.

These metals, however, do not make superconductors. Niobium is a good superconductor. When it is an alloy it is used to levitate trains!

FIGURE 1: In Japan this magnetic levitation bullet train can travel at a speed of nearly 200 miles per hour!

Physical properties of metals

Most metals:

> are **lustrous**
> are **hard**
> have a **high density**
> have a **high tensile strength**
> have a high melting point and a **high boiling point**
> are **good conductors** of heat and electricity.

The uses of a metal depend on its properties.

Property	Iron	Copper
hardness	4 mohs	3 mohs
density	7870 kg/m³	8930 kg/m³
electrical conductivity	103 000 siemen/ cm (S/cm)	588 000 siemen/ cm (S/cm)

FIGURE 2: **a** This is the first bridge made from iron. Iron is used to make steel. This is because it is very strong. These days most bridges are made from steel. **b** Copper is used to make brass. Why do you think copper is used to make the insides of electrical wiring?

What holds a metal together?

A metal is made of particles held together by **metallic bonds**.

FIGURE 3: The structure of a metal. How are the particles of a metal held together?

Questions

1 Why is copper used for electrical wiring but not iron?

2 Suggest why steel is used to make a car chassis.

3 What is the hardness of copper?

conductor   density   lustrous   tensile strength   metallic bonds

Properties of metals

Metals have specific properties that make them suitable for different uses.

A property can be either physical or chemical.

> An example of a physical property is the high **thermal** conductivity of copper. Saucepan bases need to be good conductors of heat. This is why copper is often chosen for the base or the whole of saucepans.

> An example of a chemical property is the resistance to attack by oxygen or acids shown by gold. Copper is also resistant, which is another reason why it is used for saucepans.

a

b

c

FIGURE 4: Other physical properties of metals. **a** Gold is lustrous and shiny and is used in jewellery. **b** Aluminium is used to make aircraft. **c** Brass is sonorous and is used to make bells. What property of aluminium makes it an ideal material for making aircraft?

Other physical properties of metals include being **malleable** or **ductile**.

Metals also have high melting points and high boiling points. This is because of their strong metallic bonds. The bonds between these atoms are very hard to break. A lot of energy is needed to separate them.

Questions

4 Explain why at least four properties of gold make it useful for making jewellery.

5 Explain why copper has a high melting point.

More on properties of metals

Aluminium has a low density and is used where this property is important, such as in the aircraft industry and also in modern cars.

Metallic bonding

A metallic bond is a strong electrostatic force of attraction between **close-packed positive metal ions** and a 'sea' of **delocalised electrons**.

Metals often have high melting points and boiling points. This is because a lot of energy is needed to overcome the strong attraction between the delocalised electrons and the positive metal ions.

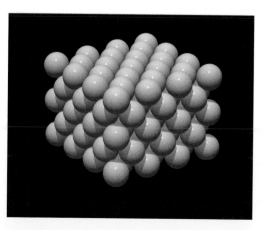

FIGURE 5: Model to show the arrangement of metal ions in a metal. What holds the ions in a metal together?

Questions

6 Explain how metals are bonded.

7 Explain how delocalised electrons in a metal cause it to have a high melting point.

8 Support the theory of the structure of metals by explaining how a metal expands when it is heated.

Q delocalised electrons   malleable   ductile   thermal conductivity

Crystals in metals

A metal has a structure that contains crystals.

The particles in a solid metal are:

> close together

> in a regular arrangement.

Superconductors

At very low temperatures some metals become **superconductors**.

Superconductors can be used to make super-fast circuits and to **levitate** magnets.

The first *Maglev* train (*magnetic levitation*) made for use in England ran between Birmingham International Railway station and Birmingham Airport.

You will find out:

> metals have crystal structures

> metals can become superconductors

> metals conduct electricity

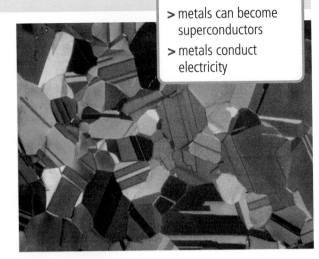

FIGURE 6: A magnified thin section of brass showing its grain structure.

Questions

9 What do metals contain as part of their structure?

10 Describe the arrangement of particles in a metal.

11 When can some metals become superconductors?

12 How can superconductors be used to make trains run faster?

Conductors

When metals conduct electricity, electrons in the metal move. Copper, silver and gold conduct electricity very well, but surprisingly do not become superconductors.

Superconductors

Superconductors are materials that conduct electricity with little or no **resistance**.

The electrical resistance of mercury suddenly drops to zero at −268.8 °C. This phenomenon is called **superconductivity**. The temperature at which it occurs is called the **critical temperature**.

There are two types of superconductor:

> type I, which are metals

> type II, which are alloys.

When a substance goes from its normal state to a superconducting state, it no longer has any **magnetic fields** inside it.

> If a small magnet is brought near the superconductor, it is repelled.

> If a small permanent magnet is placed above the superconductor, it levitates.

FIGURE 7: It looks like magic! ... This small permanent magnet (the globe) is levitating above a superconductor.

Q crystals   metal grains   superconductors

The potential benefits of superconductors are:

> loss-free **power transmission**
> super-fast electronic circuits
> powerful **electromagnets**.

Remember!

Mercury is the only metal that is liquid at room temperature.

13 What happens to mercury at a temperature of −268.8 °C?

14 What is the 'critical temperature'?

15 Describe how a small magnet can be made to 'levitate'.

16 Find out from Internet resources and books how far research has succeeded in making two potential uses of superconductors a reality.

More on delocalised electrons

A metal conducts electricity because delocalised electrons within its structure can move easily.

Difficulties of superconductors

There needs to be a good deal of development work before the true potential of superconductors is realised.

> They work only at very low temperatures; this limits their use.

> Superconductors that function at 20 °C need to be developed.

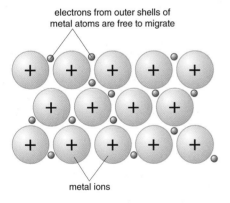

electrons from outer shells of metal atoms are free to migrate

metal ions

FIGURE 8: The structure of a metal. What carries electric charge when a current passes through a metal?

Questions

17 Compare the reasons why a metal is a good electrical conductor and a non-metal is not.

18 Discuss the usefulness of superconductors.

19 Find out one use of superconductors used in healthcare.

Purifying and testing water

You will find out:
> what water is used for
> about some common pollutants in water
> how drinking water is purified

Uses of water

We use water for lots of things – showering, washing the car and cooking at home, but nearly half of the water used in the UK cools power stations.

FIGURE 1: Water to cool power stations doesn't need purifying.

Where do we get water from?

Water as a resource

Water used by industry is:

> a cheap raw material

> used as a coolant

> a valuable solvent.

What is in water before it is purified?

> Dissolved salts and minerals

> **Pollutants**

> Insoluble materials

> **Microbes** (these are killed by **chlorination**).

Pollutants in drinking water

What pollutants may get into tap water?

> **nitrate residues**

> lead compounds

> **pesticide residues**.

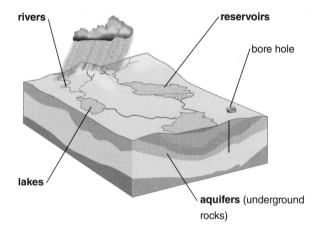

FIGURE 2: Where we get our water from.

Did you know?

London gets drinking water from the River Thames. It has been drunk by six other people (re-purified after each) before it gets to London.

Questions

1 Suggest what happens to cooling water after it has been used.

2 Suggest where aquifers get their water from.

3 Name three pollutants that sometimes get into tap water.

🔍 reservoir   borehole   aquifer   coolant   pesticide residue

Water purification

The water in a river is cloudy and often not fit to drink. To turn it into the clean water in taps it is passed through a **water purification** works.

There are three main stages in water purification:

> **sedimentation** of particles – bits drop to the bottom
> **filtration** of very fine particles – using sand
> **chlorination** – kills microbes.

All these steps cost money so it is important to conserve water. In hot months the water resources tend to have less in them so sometimes bans on using a hosepipe are announced. This is done to **conserve water** so that everybody has access to clean drinking water.

How do pollutants get into drinking water?

Some pollutants get into the water before it has been purified and some get in after water has left the treatment works.

Some older houses still have lead pipes. The lead slowly dissolves into the water.

nitrates from fertiliser run off

pesticides from spraying near to water courses

FIGURE 4: Pathways for pollutants entering the water supply. Where do the nitrates come from?

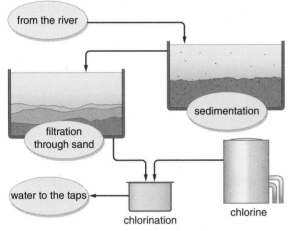

from the river

sedimentation

filtration through sand

water to the taps

chlorination

chlorine

FIGURE 3: The steps in purification of water. Why is the chlorination step necessary?

Questions

4 Which two pollutants come from agriculture?

5 Find out where your local water supply comes from. Will anyone have drunk the water already?

More on water purification

Water from a **bore hole** is usually pure, so it needs little treatment. Water from a river needs much more treatment.

> Sedimentation. Chemicals are added to make solid particles and bacteria settle out.

> Filtration. A layer of sand on gravel filters out the remaining fine particles. Some types of sand filter also remove microbes.

> Chlorination. Chlorine is added to kill microbes.

Some soluble substances remain in the water. Some of these can be poisonous, for example, pesticides and nitrates. These are treated by extra processes.

Seawater has so many substances dissolved in it that it is undrinkable. Techniques such as **distillation** must be used to remove the dissolved substances. Distillation takes huge amounts of energy, so it is very expensive. It is only used when there is not enough fresh water.

FIGURE 5: A sedimentation tank at a water-treatment works.

Did you know?

Many countries with little fresh water rely on desalination processes or reverse osmosis. You can search for information on these processes.

Questions

6 Which stage in distillation makes the process so expensive?

7 Suggest why the chlorination stage of water purification comes last.

Precipitation reactions for testing water

One test to see what is in water is a **precipitation reaction**.

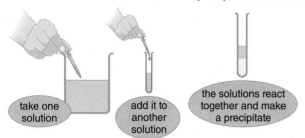

take one solution

add it to another solution

the solutions react together and make a precipitate

FIGURE 6: Making a precipitate. What colour is the precipitate in this reaction?

It is easy to test for some of the **ions** that might be dissolved in water using precipitation reactions.

Test for sulfate ions

Add one or two drops of barium chloride solution to the water in the test tube.

Sulfates give a white precipitate.

Test for chloride, bromide and iodide ions

Add one or two drops of silver nitrate solution to the water in the test tube.

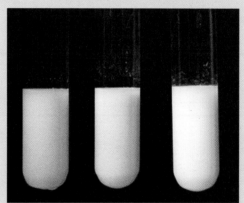

Chlorides give a white precipitate.

Bromides give a cream precipitate

Iodides give a yellow precipitate.

FIGURE 7: Using precipitation reactions to find what ions are present in water.

Word equations

If you add **silver nitrate** to sodium chloride a precipitate of silver chloride is made leaving a solution of sodium nitrate. You can write a word equation knowing these facts.

silver nitrate	+	sodium chloride	→	silver chloride (white precipitate)	+	sodium nitrate

If you add **barium chloride** to magnesium sulfate a precipitate of barium sulfate and a solution of magnesium chloride are made. The word equation is:

barium chloride	+	magnesium sulfate	→	barium sulfate (white precipitate)	+	magnesium chloride

Questions

8 What solution is used to test for chlorides?

9 What colour precipitate do chlorides produce?

🔍 ions   sulfates   chlorides   silver nitrate   barium chloride

Water tests

In a precipitation reaction, two solutions react to form a chemical that does not dissolve. This chemical suddenly appears in the liquid as a solid, a precipitate.

barium + chloride	sodium sulfate	→	barium sulfate + (white precipitate)	sodium chloride

silver + nitrate	sodium bromide	→	silver bromide + (cream precipitate)	sodium nitrate

silver + nitrate	sodium iodide	→	silver iodide + (yellow precipitate)	sodium nitrate

Barium chloride and silver nitrate are used to test ions in water.

Clean water

Clean water saves more lives than medicines do. That is why, after disasters and in developing countries, relief organisations concentrate on providing clean water supplies. In Europe, in 1892 river-borne cholera killed 8500 people in Hamburg in Germany, which took its drinking water from the River Elbe. Downstream in Altona they drank the same water plus Hamburg's sewage. But Altona put its water through simple sand filters, and almost no one died.

Water conservation

Water is a **renewable** resource. However, that does not mean that the supply is endless. If there is not enough rain in the winter, reservoirs do not fill up properly for the rest of the year. In the UK today more and more homes are being built, which increases the demand for water.

Producing tap water does have costs. It takes energy to pump and to purify it – all of which increases global warming.

Remember!

Pb^{2+} ions and SO_4^{2-} ions are doubly charged, so are Mg^{2+} ions.

Questions

10 If there is a major power cut it is possible for water taps to stop working. Suggest why.

11 What would you see if you added silver nitrate solution to potassium iodide solution?

12 What would you see if you added silver nitrate solution to calcium chloride solution?

13 Tap water samples give a positive test for chloride ions. Suggest why.

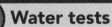

Water tests

Barium chloride is used to test for sulfate ions. Silver nitrate is used to test for halide ions.

The balanced symbol equations for the precipitation reactions are:

$AgNO_{3(aq)}$ + $NaCl_{(aq)}$ → $AgCl_{(s)}$ + $NaNO_{3(aq)}$

$AgNO_{3(aq)}$ + $NaBr_{(aq)}$ → $AgBr_{(s)}$ + $NaNO_{3(aq)}$

$AgNO_{3(aq)}$ + $NaI_{(aq)}$ → $AgI_{(s)}$ + $NaNO_{3(aq)}$

$BaCl_{2(aq)}$ + $MgSO_{4\,(aq)}$ → $BaSO_{4(s)}$ + $MgCl_{2(aq)}$

 ### Questions

14 Write a balanced symbol equation for the reaction between silver nitrate solution and potassium iodide solution. (State symbols are not needed.)

15 Write a balanced symbol equation for the reaction between silver nitrate solution and calcium chloride, $CaCl_2$. (State symbols are not needed.)

balanced equation   precipitate   barium ions   silver ions   renewable

Preparing for assessment: Research and collecting secondary data

To achieve a good grade in science, you not only have to know and understand scientific ideas, but you need to be able to apply them to other situations and investigations. These tasks will support you in developing these skills.

 Tasks

Use the information provided and secondary data to decide where a new reservoir needs to be built.

 Context

Within the UK, March 2011 was the driest since 1961. There was very little rainfall in many southern, central and eastern areas. Reservoir levels fell appreciably in many areas.

Rainfall patterns in March are important in relation to water resources during the following months. Evaporation accelerates during the spring months, and if March is exceptionally dry this may signify the end of the aquifer recharge season and the start of the decline in reservoir levels.

In Somerset, Bristol Water has outlined plans for the possibility of a second reservoir near Cheddar. This is designed to combat the predicted growth in water usage. Bristol Water thinks the reservoir could be a major tourist attraction.

 Planning

Plan how you are going to collect this information. You need to:

1. Write down how you found the information.

2. Write down a list of all the sources of your information.

3. Clearly present the information so it could be used to plan an actual investigation.

✳ General rules

1. You may work with other students on your research but your written work must be done on your own in the classroom.

2. You cannot get detailed help from your teacher.

3. You are not allowed to redraft your work.

4. Your work can be handwritten or word processed.

5. You will need to do this research outside the laboratory.

6. It is expected that you write up this task on your own in two hours.

✳ Research and collecting secondary data

To research:

> where reservoirs are located and what volume they provide per year

> size of population that the reservoirs feed and the amount of water they use

> rainfall and temperature trends that might affect water shortage.

Recommend two ways that a shortfall in water supply for the demand needed can be overcome.

These are the things you will need to consider when using your data.

1. Where will you collect data from about the locations of reservoirs?

2. Where will you collect data from about the volumes of water they hold?

3. Where will you collect data from about the size of population fed by each reservoir?

4. Where will you collect data from about the volume of water each household uses?

5. Where will you collect data from about the temperature each year?

6. Where will you collect data from about the rainfall pattern each year?

7. Where will you collect data from about the ways we can save water?

8. How will you put these pieces of information together to make a judgement about what to do?

9. How will you present your data?

10. Will you present two recommendations or just one?

> What do you need to research?

> Acknowledge the sources of information that you have used, e.g. the website addresses.

> You can answer the questions using the information provided, but it is much better if you do your own research. Remember, there are secondary sources other than the Internet.

> Write the review of the data you collected. Present your data in ways that are clear for your audience to read.

> Recommend at least one course of action.

> Justify why you have recommended this action.

C4 Checklist

To achieve your forecast grade in the exam you'll need to revise

Use this checklist to see what you can do now. It gives you many of the important points you will need to know. Refer back to the relevant pages in this book if you're not sure and to see if there is anything else you need to know. Look across the three columns to see how you can progress.

Remember you'll need to be able to use these ideas in various ways, such as:

> interpreting pictures, diagrams and graphs
> applying ideas to new situations
> explaining ethical implications

> suggesting some benefits and risks to society
> drawing conclusions from evidence you've been given.

Look at pages 278–299 for more information about exams and how you'll be assessed.

To aim for a grade E	To aim for a grade C	To aim for a grade A
know that an atom has a nucleus surrounded by electrons **know** that a nucleus is positive and electrons are negative **know** that the atomic number is the number of protons in an atom **explain**, from a formula, if it is an element or a compound **describe** the main stages of the development of atomic theory	**know** the charges and masses of protons, electrons and neutrons **know** how isotopes are the same but have different mass number **describe** the arrangement of elements in the periodic table **find** the identity of an element from its electronic structure **describe** how the atomic theory changed as evidence was found	**explain** why an atom is neutral in terms of sub-atomic particles **identify** isotopes from data about numbers of protons and neutrons **know** that atoms have a radius of 10^{-10} m and mass about 10^{-23} g **work** out the electronic structure of the first 20 elements, e.g. 2.8.4 **explain** how unexpected results led to theories of a nuclear atom
know that an ion is a charged atom or group of atoms **recognise** an ion, an atom and a molecule from given formulae **compare** how solid and liquid sodium chloride conducts **know** that a solution of sodium chloride conducts electricity **compare** the melting points of NaCl and MgO	**understand** that atoms with eight electrons are electronically stable **explain** that metals form positive ions by losing electrons **explain** that non-metals form negative ions by gaining electrons **work out** the formula of an ionic compound from formulae of ions **describe** the structure of NaCl and of MgO as giant lattices	**explain** ionic bonding using the 'dot and cross' model. **explain** the high melting point of NaCl by its structure and bonding **explain** the conductivity of molten NaCl by its structure and bonding **explain** why the melting point of NaCl is lower than that of MgO **predict** properties of substances with giant ionic structures
know that there are two types of bonding: ionic and covalent **know** that carbon dioxide and water do not conduct electricity **find out**, from the periodic table, elements in the same group **describe** a period of elements as those in a horizontal row **describe** the stages in the idea of the periodic table	**know** that non-metals combine by sharing electron pairs e.g. H_2 **know** that CO_2 and H_2O have weak intermolecular forces **know** that the group is the same as the number of outer electrons **find out** which period an element belongs to from electron shells **describe** evidence Mendeleev used to begin the periodic table	**explain** some covalent bonding using a 'dot and cross' model **explain** double covalent bonds using a 'dot and cross' model **explain** the low melting points of CO_2 and H_2O by their bonding **explain** why structures CO_2 and H_2O do not conduct electricity **explain** how further evidence confirmed the periodic table

To aim for a grade E

explain why group 1 elements are known as the alkali metals and are stored under oil

describe the reaction of group 1 elements with water

construct the word equation for group 1 elements with water

know the flame test colours for lithium, sodium and potassium

To aim for a grade C

predict the reactions of other group 1 elements with water

construct balanced symbol equations of group 1 with water

explain why group 1 elements have similar properties

know how to use a flame test to identify potassium, lithium and sodium compounds

To aim for a grade A

construct balanced symbol equations for Cu^{2+}, Fe^{2+} and Fe^{3+} to make $Cu(OH)_2$ $Fe(OH)_2$ $Fe(OH)_3$

predict the physical properties of other group 1 elements

explain why group 1 elements form similar positive ions

explain the reactivity trend with water to give NaOH and KOH

know why a process is oxidation from its ionic equation

---

know that the group 7 elements are called the halogens

know that fluorine, chlorine, bromine and iodine are halogens

describe the uses of group 7 elements

recognise that group 7 elements react vigorously with group 1

know that the reactivity of group 7 decreases down the group

describe the appearance of Cl_2, Br_2 and I_2

identify the halide when elements from groups 1 and 7 react, e.g. KCl

construct balanced symbol equations of reactions of group 7

know that chlorine displaces bromine and iodine

explain why group 7 elements have similar properties

predict the properties of other group 7 elements

predict if displacement reactions will occur with other halogens

explain why group 7 elements form similar negative ions

explain the trend of reactivity of group 7 by ideas of electron gain

explain why a process is reduction from its ionic equation

---

identify an element as a transition metal from the periodic table

know that transition elements are metals with typical properties

know thermal decomposition is using heat to break substances

recall that carbon dioxide turns limewater milky

know precipitation is a reaction of solutions to make a solid

know that transition elements often have coloured compounds

know transition elements and compounds are used as catalysts

know the thermal decomposition of transition metal carbonates

recall word equations for thermal decomposition → carbon dioxide

describe the use of sodium hydroxide to identify metal ions

construct a balanced symbol equation for thermal decomposition $CuCO_3$, $ZnCO_3$, $MnCO_3$, $FeCO_3$

construct a balanced symbol equation for Cu^{2+}, Fe^{2+} and Fe^{3+} reacting with OH^-

know the formula of ZnO, CuO, FeO, MnO and Na_2O

---

explain why iron is used to make bridges and copper to make wire

know that metals are lustrous, hard, dense and good conductors

interpret data about metals such as density and conductivity

know that particles in metals are held by metallic bonds

know at low temperatures some metals can be superconductors

explain why metals are suited to a given use by looking at data

know that metals have high melting points due to strong metallic bonds

describe how metals conduct electricity by a flow of electrons

describe the potential benefits of superconductors

describe metallic bonding

explain why metals have high melting points and why they conduct electricity

explain some of the drawbacks of superconductors

---

interpret simple data about water resources in the UK

know different water resources are lakes, aquifers and reservoirs

know nitrate residues and lead compounds are found in water

know water contains microbes and pollutants before it is purified

know that barium chloride tests for sulfates in water

interpret data about water resources and its conservation

explain why drinking water may have pollutants including pesticide residues in it

describe the purification of water by filtration and sedimentation

interpret data about water testing with aqueous silver nitrate

know reactions of silver nitrate and halides are precipitation

explain why some dissolved substances stay in purified water

explain the disadvantages of using distillation of sea water for making large quantities of fresh water

construct balanced symbol equations for $BaCl_2$ with sulfates and $AgNO_3$ with halides such as $MgCl_2$

C4 Exam-style questions

Foundation Tier

AO1 **1 (a)** Copy and complete the following table to show the relative charges and masses of the three particles in an atom.

Name of particle	Charge	Mass
proton	1	
		1
electron		

[3]

AO1 **(b)** Copy and complete the sentences using the words below.

electrons   mass   protons
isotopes   atomic   neutrons

In an atom the protons and _____ are found in the nucleus. Atoms of a particular element always have the same number of protons and this is called the _____ number. The total number of _____ and neutrons is known as the _____ number. Atoms that have the same atomic number but different mass numbers are known as _____. Different shells around the nucleus are occupied by

_____.

[6]
[Total: 9]

AO2 **2 (a)** The full symbol for Al is $^{27}_{13}$Al. Copy and complete the table to show information about aluminium atoms.

	Atomic number	Mass number	Number of protons	Number of electrons	Number of neutrons
$^{27}_{13}$Al					

[5]

AO2 **(b)** If the electron arrangement for lithium is 2.1 how many shells do the electrons occupy?

[2]
[Total: 7]

AO1 **3 (a)** Which of these statements is not correct?

A When a metal bonds with a non-metal its atoms lose electrons to make positive ions.

B When a non-metal reacts with a metal the atoms share their electrons.

C A solution of sodium chloride conducts electricity.

D When a non-metal bonds with a metal its atoms gain electrons to make negative ions.

[3]

AO1 **(b) (i)** Sodium chloride is a solid. Describe the structure of sodium chloride. [2]

AO1 **(ii)** Which has the higher melting point, sodium chloride or magnesium oxide? [1]

AO1 **(iii)** Describe the electrical conductivity of solid sodium chloride, molten liquid sodium chloride and sodium chloride in solution. [2]
[Total: 8]

AO1 **4 (a)** When lithium reacts with water two products are made. Which two? [2]

AO1 **(b)** Does potassium react more vigorously or less vigorously with water than lithium? [1]

AO1 **(c)** When potassium reacts with water what colour flame do you see? [1]

AO1 **(c)** Why do the alkali metals all behave in a similar way? [2]
[Total: 6]

AO1 **5** Copy and complete the table to describe the halogens.

	Colour	State
chlorine	green	gas
bromine		
iodine		

[4]

AO3 **6** Jo and Sam tested four samples of water. They tested them with silver nitrate solution and then with barium chloride solution. Here are their results.

	With silver nitrate solution	With barium chloride solution
A	No reaction	White precipitate seen
B	White precipitate seen	No reaction
C	Yellow precipitate seen	White precipitate seen
D	Cream precipitate seen	White precipitate seen

In their next experiment they wanted to use the sample of water with both sulfate and iodide ions present in the water. Say which sample they should use and justify your answer. [3]

AO1 **7**
AO2 Describe the arrangement of elements in the periodic table in terms of their atomic structure. How did the work of Dalton, J. J. Thomson, Rutherford and Bohr contribute to our modern understanding of this arrangement of atoms of elements? [6]

AO1 recall the science   AO2 apply your knowledge   AO3 evaluate and analyse the evidence

✳ Worked Example – Foundation Tier

Avinder and Jo are researching some properties of metals. They know that their metals are **transition** metals.

(a) Shade on this outline part of the periodic table where these transition metals are found. [2]

		H															He
Li																	Ne
Na																	Ar
K																	Kr

> **How to raise your grade!**
> Take note of these comments – they will help you to raise your grade.

> The first two boxes on this row are not transition metal elements. **0/1**

(b) The two metals they have are iron and copper. Write down three physical properties that these metals have. [2]

good conductors of electricity

they have a high density

> Only two correct properties are written. **1/2**

Avinder and Jo look at some of the compounds of copper and iron. They heat some copper carbonate in a test tube. They expect that carbon dioxide will be given off.

(c) How will they test if they are correct? [2]

They pass the gas given off into the solution and it turns milky

> The solution needs to be identified as limewater. **1/2**

Avinder says that the copper carbonate breaks down to make copper oxide and carbon dioxide. [2]

(d) Write a word equation to show this reaction.

copper carbonate → copper oxide + carbon dioxide

> The word equation is correct. **2/2**

(e) Underline the correct answer. When a substance is broken down by heat into at least two other substances this is called: [1]

displacement   precipitation   thermal decomposition

> This is correct. **1/1**

(f) Next Avinder and Jo test three solutions each containing different ions of these metals. They test them with sodium hydroxide solution. Write down the colours of the precipitates they expect to see. [3]

Ions in solution	Cu^{2+}	Fe^{2+}	Fe^{3+}
Colour of precipitate	*blue*	*brown*	*green*

> The colours for the Fe^{2+} and Fe^{3+} ions are the wrong way round. **1/3**

> This student has scored 6 marks out of a possible 11. This is below the standard of Grade C. With more care the student could have achieved a Grade C.

C4 Exam-style questions

Higher Tier

AO1 **1 (a)** Copy and complete the following table to show the relative charges and masses of the three particles in an atom. [3]

Name of particle	Charge	Mass
proton	1	
		1
electron		

AO2 **(b)** Copy and complete the sentences using two of the words below.

electrons   mass   protons
isotopes   atomic   neutrons

In an atom, the protons and _____ are found in the nucleus. Atoms with the same number of protons but different mass numbers are called _____.

If this type of an atom has 6 protons and a mass number of 14, describe how the atom is made up and which element it is. [2]

AO2 **(c)** The full symbol for Al is $^{27}_{13}$Al. Copy and complete the table to show information about aluminium atoms. [2]

	Atomic number	Mass number	Number of protons	Number of electrons	Number of neutrons
$^{27}_{13}$Al					

[Total: 7]

AO2 **2** If the electron arrangement for lithium is 2.1 what is the electron arrangement for potassium? [2]

AO1 **3 (a)** Which of the following statements are correct?

A When alkali metals react their atoms each lose one electron to form ions with a +1 charge.

B When a non-metal reacts with a metal the atoms share their electrons.

C A solution of sodium chloride conducts electricity.

D The halogens react by each of their atoms gaining one electron to form ions with a charge of −1. [3]

AO1 **(b) (i)** Sodium chloride is a solid. Describe the structure of sodium chloride. [2]

AO1 **(ii)** Explain why sodium chloride has a higher melting point than magnesium oxide. [2]

AO2 **(iii)** Use a 'dot and cross' diagram to explain the bonding in lithium fluoride. [2]

[Total: 9]

AO2 **4 (a)** Predict the names of the products formed when caesium reacts with water. [2]

AO2 **(b)** Predict the reactivity of rubidium in water compared to the reaction of sodium in water. [1]

AO2 **(c)** Predict the appearance of rubidium. [1]

AO1 **(d)** Why do the alkali metals all behave in a similar way? [2]

[Total: 6]

AO1 **5** Copy and complete the table to describe the halogens. [4]

	Colour	State
chlorine	green	gas
bromine		
iodine		

AO3 **6** Jo and Sam tested four samples of water. They tested them with silver nitrate solution and then with barium chloride solution. Here are their results.

	With silver nitrate solution	With barium chloride solution
A	No reaction	White precipitate seen
B	White precipitate seen	No reaction
C	Yellow precipitate seen	White precipitate seen
D	Cream precipitate seen	White precipitate seen

In their next experiment they wanted to use the sample of water with both sulfate and iodide ions present in the water. Say which sample they should use and justify your answer. [3]

AO1 **7**
AO2
Describe the arrangement of elements in the periodic table in terms of their atomic structure. Deduce the electronic structure of sulfur using the periodic table. How did the work of Dalton, J. J. Thomson, Rutherford and Bohr contribute to our modern understanding of the structure of atoms? [6]

AO1 recall the science   AO2 apply your knowledge   AO3 evaluate and analyse the evidence

✳ Worked Example – Higher Tier

Avinder and Jo are researching some properties of metals.

(a) Describe the metallic bonding of metals   [2]

Metals are made of metal ions close packed together with a 'sea' of electrons strongly attracted to the ions.

The student has mentioned close-packed ions and sea of electrons.   2/2

(b) Explain how metals conduct electricity.   [1]

The 'sea' of electrons can move freely between the close-packed metal ions.

This is correct.   1/1

Avinder and Jo look at some of the compounds of copper and iron.

They heat some copper carbonate in a test tube. Avinder says that the copper carbonate breaks down using heat, to make two substances.

(c) Write a word equation to show this reaction.   [1]

copper carbonate → copper oxide + carbon oxide

This should be carbon dioxide.   0/1

(d) Next they heat some zinc carbonate $ZnCO_3$.

Write a balanced symbol equation for this reaction.   [1]

$ZnCO_3 + heat → ZnO + CO_2$

The products are correct but heat must not be written as a reactant.   0/1

(e) Next Avinder and Jo test three solutions each containing different ions of these metals.

They test them with sodium hydroxide solution. Write down the colours of the precipitates they expect to see.   [3]

Ions in solution	Cu^{2+}	Fe^{2+}	Fe^{3+}
Colour of precipitate	*blue*	*brown*	*green*

The colours for the Fe^{2+} and Fe^{3+} ions are the wrong way round.   1/3

(f) Construct a balanced symbol equation for the reaction between copper ions, Cu^{2+} and hydroxide ions, OH^-.   [2]

$Cu^{2+} + OH^- → Cu(OH)_2$

The product is correct but there needs to be $2OH^-$ ions to balance.   1/2

This student has scored 7 marks out of a possible 12. This is below the standard of Grade A. With more care the student could have achieved a Grade A.

P3 Forces for transport

Ideas you've met before

Speed and acceleration

Speed = $\dfrac{\text{distance travelled}}{\text{time taken}}$

and is measured in metres per second (m/s).

Distance–time and speed–time graphs show how things move.

If the speed of a car is increasing it is accelerating.

- A car travels 400 m in 20 s. What is its speed?

Forces and motion

Forces can speed things up or slow them down.

The bigger the force the bigger the change in speed.

Force is measured in newtons (N).

Weight is the force acting on an object due to gravity.

- What happens to the speed of a moving object if an unbalanced force acts on it?

Work, energy and power

Work is done whenever a force moves.

Work done = force × distance moved.

Energy is needed to do work.

Power is the rate of doing work or of transferring energy.

- The people push the car but it does not move. How much work are they doing? What must they do to make it move?

Falling safely

All objects near Earth fall due to gravity.

The force of gravity makes falling objects accelerate.

The upward force of air resistance, or drag, reduces their acceleration.

When a parachutist's parachute opens the drag force increases.

- What happens when the forces on a falling object are balanced?

In P3 you will find out about...

> cameras measuring speed and average speed

> interpreting distance–time and speed–time graphs

> calculating acceleration

> the relationship between force, mass and acceleration

> relative velocity of objects moving in parallel

> thinking, braking and stopping distances

> factors that may affect thinking and braking distances

> the importance of thinking, braking and stopping distances in road safety

> alternatives to petrol and diesel as fuels for cars

> evaluating data about fuel consumption and emissions

> the meaning of 'work' in physics and how to calculate work done

> the meaning of power and how to calculate power

> the meaning of momentum and its link with force

> safety features designed to prevent accidents, such as ABS brakes and traction control

> safety features in cars to protect occupants in an accident, including crumple zones, air bags and seatbelts

> how the balance of forces affects motion

> how objects falling through Earth's atmosphere reach a terminal speed

> energy transfer between gravitational potential energy and kinetic energy

> the physics of roller coasters and other theme park rides

Speed

You will find out:
> how to calculate average speed
> how to measure speed
> how speed cameras work

Caught on camera

A speeding motorist is caught on camera driving faster than the speed limit of 50 mph.

The speed camera was invented by a company founded by a rally driver.

FIGURE 1: What does a speed camera do?

Speed and average speed

Speed is a measure of how fast an object is going.

Vehicle speeds are usually measured in km/h. If a car travels at 80 km/h it covers a **distance** of 80 km in a **time** of 1 hour. But its speed changes during a journey so we calculate its **average speed**.

$$\text{average speed} = \frac{\text{distance travelled}}{\text{time taken}} \quad \text{and its units are m/s or km/h.}$$

Examples

> A Formula 1 racing car driver completes a 560 km race in 2 hours.
>
> $$\text{average speed} = \frac{\text{distance travelled}}{\text{time taken}} = \frac{s}{t} = \frac{560}{2} = 280 \text{ km/h}$$

> Alfie takes 6 minutes (360 s) to walk to school, a distance of 1.08 km. What is his average speed in m/s?
>
> 1 km = 1000 m. 1.08 km = 1080 m.
>
> $$\text{average speed} = \frac{\text{distance travelled}}{\text{time taken}} = \frac{1080}{360} = 3 \text{ m/s}$$

FIGURE 2: Why are average speed checks better than a single speed measurement?

Questions

1 Sam ran 100 m in 18 seconds. Priya ran 120 m in the same time. Who ran the fastest?

2 Adi swam 50 m in 2½ minutes (150 s). Calculate his average speed.

3 Why is it impossible to maintain a constant speed during a car journey?

Using a speed camera

On roads, speed is often measured with a **speed camera**.

> As a speeding car passes a camera a photograph is taken.

> A second photograph is taken 0.5 seconds later. There are white lines painted on the road at distances of 1.5 m apart. They show how far the car travels in 0.5 seconds.

> For example, a car passes over six gaps between lines. It travels 1.5 × 6 = 9 m in 0.5 seconds.

> This means the speed of the car was 18 m/s or 65 km/h (about 40 miles per hour).

Measuring average speed

Cameras are used to measure the average speed of a car, through road-works for instance.

> Two cameras are placed a measured distance apart (perhaps 1 km).

> As a car passes each camera its number plate is recorded.

> The two images of the same number plate are paired up.

> Each image carries a date and time stamp so a computer can then work out the car's average speed between the cameras.

Q how speed cameras work   typical speeds

How fast?

average speed $= \dfrac{\text{distance travelled}}{\text{time taken}} = \dfrac{s}{t}$

The speed at a certain point in time is called **instantaneous speed**. If the speed of a car at the start is u and it accelerates uniformly to v the average speed is: $\dfrac{u + v}{2}$

If the average speed of a car is known it is possible to work out

> the distance travelled in a certain time:
distance = average speed × time $= \dfrac{(u + v)t}{2}$

> how long it takes to travel a known distance:
$$\text{time taken} = \dfrac{\text{distance travelled}}{\text{average speed}}$$

So, increasing the average speed decreases the time taken to travel a certain distance.

Remember!
A question may require you to change the subject of the equation.
Practise rearranging the equation until you are really confident.

FIGURE 3: Suggest why it is dangerous for drivers to exceed speed limits.

Example

Zen the robot travels at a speed of 2 m/s. How long will it take Zen to move 30 m?

$$\text{time taken} = \dfrac{\text{distance travelled}}{\text{average speed}} = \dfrac{30}{2} = 15 \text{ s}$$

How far can Zen travel in 40 s?

distance travelled = average speed × time = $2 \times 40 = 80$ m

Speed limits

Different roads have different speed limits. The speed limit can depend on:

> the type of road, such as single or dual carriageway, motorway, straight or winding

> the area the road is in, such as urban or rural.

Questions

4 The speed limit on motorways is 70 mph (112 km/h). Reena drives 150 km in one and a half hours. Does she break the speed limit?

5 Suggest one reason why the speed limit is higher on motorways than on other types of road.

6 Tom averages 48 km/h on his 12 km journey to school. How long does his journey take?

Average speed, distance and time

average speed $= \dfrac{\text{distance travelled}}{\text{time taken}}$

If the average speed of a car doubles the time to travel the same distance halves.

If the average speed doubles the car will travel twice as far in the same time.

If the average speed doubles for twice the time the car will travel four times as far.

Units

Speed is usually measured in m/s or km/h.

The standard unit used in scientific equations is m/s.

1 km = 1000 m   1 hour = 3600 seconds

108 km/h becomes $\dfrac{108000}{(60 \times 60)}$ m/s = 30 m/s

> To change km/h to m/s: multiply by 1000 and divide by 3600.

> To change m/s to km/h: divide by 1000 and multiply by 3600.

Remember!
Always change km/h to m/s before substituting in equations.

Questions

7 A snail travels at a speed of 0.5 mm/s. How far would the snail travel in 1 hour?

8 Top sprinters can run 100 m in about 10 s. What is their average speed in: **a** m/s; **b** km/h?

9 A car reaches a speed of 72 km/h in 40 s starting from rest. How far does it go?

Looking at motion

Looking at moving cars is important for road planners, road safety experts and car manufacturers.

Drawing a graph of distance against time shows how the distance moved by a car from its starting point changes over time.

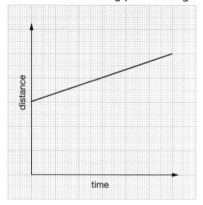

FIGURE 4: Graph of distance against time. Is the speed of the car increasing, decreasing or constant?

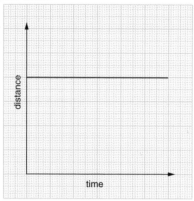

FIGURE 5: Graph of distance against time. The distance does not change over time so the car is stationary. What is its speed?

FIGURE 6: Suggest why motorways are safer than ordinary roads.

Questions

10 Sketch a distance–time graph to show a car that travels a distance of 200 km in 2 hours at a constant speed, stops for half an hour and then travels a further 80 km in 1 hour. Remember to add a scale to the axes.

11 Sketch a possible distance–time graph for one of the cars shown travelling on the motorway in Figure 6.

12 Kate plotted the distance–time graph shown for a bus as it approached her school. Describe the motion of the bus. Give as much detail as you can.

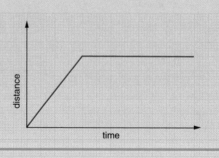

Distance–time graphs

A **distance–time graph** shows how the distance moved by an object changes with time. If the distance *increases* as time increases the line has a positive (+) **gradient**.

If the distance *decreases* as time increases the line has a negative (–) **gradient**. Distance–time graphs allow a collection of data to be shown. It is easier to interpret data when they are plotted on a graph than when they are listed in a results table as shown on the next page.

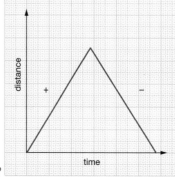

FIGURE 7: Graph of distance against time. How do you know the car returns to its starting point?

FIGURE 8: The distance travelled by the object each second increases as the time increases. The gradient increases. What can you say about the speed of the object?

Q graphs in science   gradient of a graph

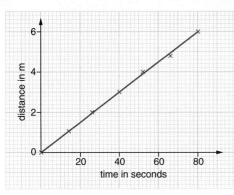

FIGURE 9: Graph showing the duck's movement over time. Gracie draws a **best-fit straight line** through the points on her graph. What can you say about the speed of the duck?

Time in seconds	Distance in m
0	0
12	1
25	2
40	3
51	4
66	5
80	6

Gracie records the movement of a duck over time. She plots her results in a graph.

Questions

13 Suggest why the points on Gracie's graph were not all exactly on a straight line.

14 Sketch a distance–time graph for a car that is slowing down.

15 A car travels a distance of 120 km in 1.5 hours at a constant speed. Sketch a distance–time graph for the car.

More on distance–time graphs

Speed is equal to the **gradient** (steepness) of a distance–time graph.

The gradient of the graph is:

gradient $= \dfrac{AC}{BC} = \dfrac{(20-10)}{(5-0)} = \dfrac{10}{5} = 2$, so speed = 2 m/s

> The larger the gradient (the steeper the line) the higher the speed.

> A **straight line** indicates the speed is constant. A **curved line** shows that the speed is changing.

> If the gradient increases the speed increases. The object **accelerates**.

> If the gradient decreases the speed decreases. The object **decelerates**.

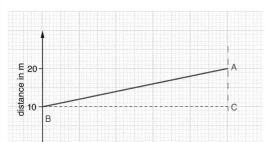

FIGURE 10: Graph of distance travelled by an object over time. How would the graph change if the object went faster?

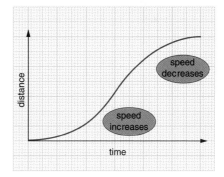

FIGURE 11: Graph of distance against time. What is the relationship between gradient and speed?

Questions

16 Look at the distance–time graph. Describe in words how the speed changes over time.

17 A cheetah chases a small animal at a constant speed and covers 200 m in 10 seconds before stopping suddenly. Sketch a distance–time graph and use it to find the speed of the cheetah.

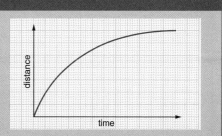

Changing speed

You will find out:
> how to analyse speed–time graphs
> how to calculate acceleration from a speed–time graph
> how to find the distance travelled from a speed–time graph

Quick, quick, slow

Formula 1 racing cars race round bends on Grand Prix race tracks.

To do this they slow down – just a little – as they go into a bend and speed up very rapidly as they turn out of it.

All cars need to speed up and slow down frequently during a journey but not as rapidly as Formula 1 racing cars.

FIGURE 1: Suggest what would happen if a car did not slow down on approaching a bend.

Speed–time graphs

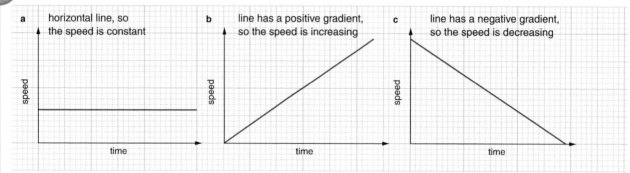

a horizontal line, so the speed is constant

b line has a positive gradient, so the speed is increasing

c line has a negative gradient, so the speed is decreasing

FIGURE 2: Speed–time graphs

A **speed–time graph** shows how the **speed** of an object changes with **time**.

The **gradient** (slope) of a line tells us how the speed is changing.

Remember!
Don't confuse distance–time and speed–time graphs. Always look at the axes carefully.

Questions

1 A car brakes. Does the speed–time graph for the car show a positive or a negative gradient?

2 Sketch a speed–time graph for a tube train travelling between two stations.

More on speed–time graphs

Acceleration

The lines **A**, **B**, **C** and **D** on the graphs show how the speeds of four cars change over time.

> The speeds of cars **A** and **B** are increasing.

> Line **B** is steeper than line **A**. It has a larger positive gradient.

> The speed of car **B** is increasing more rapidly than that of car **A**.

> Car **B** has a larger **acceleration** than car **A**.

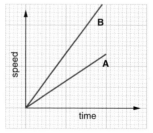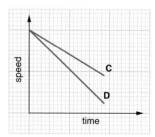

FIGURE 3: Look at the statements for cars A and B. Write four similar statements for cars **C** and **D**.

Distance travelled

Look again at the graphs in Figure 3.

> The speed of car **B** is increasing more rapidly than the speed of car **A**, so car **B** is travelling further than car **A** in the same time.

> The **area** under line **B** is greater than the area under line **A**, for the same time.

Write two similar statements for cars **C** and **D**.

Distance travelled = area under speed–time graph

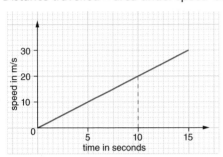

FIGURE 4: Describe the motion.

Distance travelled In the first 10 s = $\frac{1}{2} \times 10 \times 20 = 100$ m

Questions

3 The photograph shows two 100 m runners at the start of a race.

a Which sprinter, a or b, has the greater acceleration out of the starting blocks?

b On the same axes, sketch possible speed–time graphs for sprinter A and sprinter B.

c What is the same for both sprinters?

4 A car accelerates uniformly from rest reaching a speed of 30 m/s in 40 s. How far does it travel?

Interpreting speed–time graphs

Calculating acceleration

Look again at the graph shown in Figure 4.

The gradient (slope) of a speed–time graph is the acceleration. Acceleration is measured in m/s².

gradient $= \frac{(20-0)}{(10-0)} = \frac{20}{10} = 2$ acceleration = 2 m/s²

More about distance travelled

The speed of a car does not usually remain constant for long or vary in the linear way shown in Figure 4.

The graph below shows a car moving between two sets of traffic lights. The speed of the car is not changing uniformly so the acceleration is not constant. The area under the graph can be **estimated** to find the distance travelled.

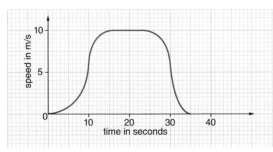

FIGURE 5: How does the acceleration change?

Question

5 Describe in as much detail as possible the motion of a train that has the graph shown here. How far did the train travel?

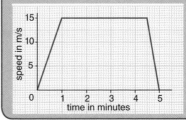

You will find out:

> about acceleration

> about the relationship between acceleration, speed and time

> about velocity and relative velocity

Acceleration

The table shows how the speed of a car changes as it starts to move.

Time in seconds	Speed in m/s
0	0
1	5
2	10
3	15
4	20
5	25

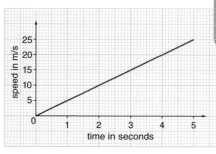

FIGURE 6: Graph of speed against time. What does the graph show?

The speed of the car *increases* by a constant amount (5 m/s) every second.

A change of speed is called acceleration. Acceleration is measured in m/s² (metres per second squared). The car has a constant acceleration of 5 m/s².

This table shows how the speed of a car changes as it comes to a stop.

Time in seconds	Speed in m/s
0	25
1	20
2	15
3	10
4	5
5	0

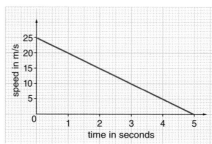

FIGURE 7: Graph of speed against time. What does the graph show?

The speed of the car *decreases* by a constant amount (5 m/s) every second.

The car has a constant **deceleration** of 5 m/s². Its acceleration is –5 m/s².

Calculating acceleration

Acceleration = $\dfrac{\text{change in speed}}{\text{time taken}}$

Example

The speed of a car increases by 20 m/s in 10 s. Calculate its acceleration.

Acceleration = $\dfrac{\text{change in speed}}{\text{time taken}} = \dfrac{20}{10} = 2$ m/s².

Velocity

Direction is important when describing the motion of an object. For example, the rocket moves *upwards* on launch. **Velocity** is the speed of a moving object in a known direction.

Example

The speed of a car is 20 m/s but its velocity is 20 m/s due north.

Remember!

An object at a constant acceleration is different to an object at a constant speed.

FIGURE 8: A space shuttle needs a huge acceleration at lift-off. Why?

Questions

6 Car A accelerates from 0 to 20 m/s in 15 seconds. Car B accelerates from 0 to 30 m/s in 12 seconds. Which car, A or B, has the greater acceleration?

7 The speed of a train increases by 2 m/s in 8 s. Calculate its acceleration.

Relative velocity

If two trains, a and b, are moving at the same speed in *opposite* directions on parallel tracks their velocities will have:

> the same **magnitude** (size)

> opposite signs (one positive, one negative).

If both trains are travelling at 30 m/s:

The velocity of train A relative to train B = (+30) − (−30) = +60 m/s
The velocity of train B relative to train A = (−30) − (+30) = −60 m/s.
The trains pass very quickly.

But if the trains are moving in the *same* direction their **relative velocity** is equal to (+30) − (+30) = 0.

FIGURE 9: Will the trains pass each other quickly or slowly?

Acceleration

During a car journey the speed of the car increases and decreases. It does not stay constant. A change in speed per unit time is called acceleration.

$$\text{acceleration} = \frac{\text{change in speed}}{\text{time taken}}$$

An advertisement for a new car often boasts a rapid acceleration.

A speed of 108km/h is $\frac{108 \times 1000}{60 \times 60}$ = 30 m/s

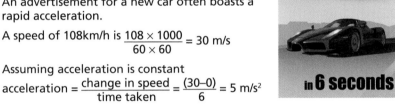

0 to 108 km/h in **6 seconds**

Assuming acceleration is constant

$$\text{acceleration} = \frac{\text{change in speed}}{\text{time taken}} = \frac{(30-0)}{6} = 5 \text{ m/s}^2$$

This means the speed of the car increases by 5 m/s every second.

Suppose the car slows down to a stop in 15 seconds.

$$\text{acceleration} = \frac{\text{change in speed}}{\text{time taken}} = \frac{(0-30)}{15} = -2 \text{ m/s}^2$$

The negative acceleration shows the car is slowing down, or decelerating. What gradient would the speed–time graph show?

Questions

8 Two trains are travelling in opposite directions on parallel tracks. Train A has a speed of 50 m/s. Train B has a speed of 30 m/s. What is their relative speed?

9 The speed of a car increases from 10 m/s to 30 m/s in 8 seconds. Find its acceleration.

10 Luigi is travelling by car at 24 m/s. He brakes and comes to a stop in 30 seconds. Find his deceleration.

11 a Jane is driving at a speed of 72 km/h. Convert this to m/s.

b She accelerates to 108 km/h in 8 seconds. Calculate her acceleration in m/s^2.

Accelerating when speed is constant

A vehicle may go round a roundabout at a constant speed but it is accelerating. This is because the **direction** of its movement is changing. Its velocity is changing. Velocity is the speed of a moving object in a known direction.

Strictly, $\text{acceleration} = \frac{\text{change in velocity}}{\text{time taken}}$

change in velocity
= acceleration × time taken

$\text{time taken} = \frac{\text{change in velocity}}{\text{acceleration}}$

FIGURE 10: Vehicles travelling round a roundabout. Why may the speed of each vehicle be constant when it is accelerating?

Questions

12 A car has a maximum acceleration of 4 m/s^2. How long does it take the car to reach a speed of 36 m/s, starting from rest?

13 Mo and Danny are riding on a carousel at a fairground. They are moving at a constant speed. Why are they accelerating?

Forces and motion

You will find out:
> about when a force acts on an object to speed it up or slow it down
> how mass, force and acceleration are linked
> how to use the equation $F = ma$

Drive safely

Most people are keen to learn to drive as soon as they reach the legal age to drive of 17.

The thrill of passing a driving test and gaining the freedom that it brings takes some beating.

A basic understanding of the forces acting on a moving vehicle can help to make drivers safe.

FIGURE 1: Suggest why it is said that passing the test is really only the first step in learning to drive safely.

What do forces do?

To **accelerate** in a car the driver presses on the accelerator pedal. This increases the **pull** of the engine that provides a forward **force**. If they press the pedal down further, the pull of the engine is greater and the acceleration increases.

Calculating force

$F = ma$ the unit of force, F, is the **newton (N)** when the **mass**, m, is in kg and acceleration, a, is in m/s².

Example

> If car **d** in Figure 3 has a mass of 1000 kg, what force is needed to give it an acceleration of 4 m/s²?
>
> $F = ma = 1000 \times 4 = 4000$ N

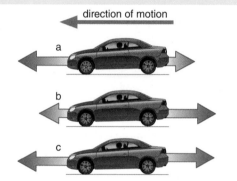

direction of motion

a

b

c

FIGURE 2: In **a** the car accelerates. Describe the motion of the car in **b** and **c**.

d

e

FIGURE 3: These cars have the same engines and so they have the same forward force. Car **d** has a smaller mass than car **e**. Which car has the bigger acceleration?

Questions

1a Meera is playing tennis. What happens to the motion of the tennis ball when she hits it harder?

b She hits a football with the same force. How does the acceleration of the football compare with that of the tennis ball?

2 If car **e** in Figure 3 has a mass of 1300 kg, what force is needed to give it an acceleration of 4 m/s²?

balanced and unbalanced forces for KS4   forces on cars for GCSE

Force, mass and acceleration

If the forces acting on an object are **balanced** it is at rest or has a **constant speed**. If the forces acting on an object are unbalanced it speeds up or slows down. There is a **net**, or **resultant**, force acting.

Using $F = ma$
The **unbalanced** force, $F = ma$
The mass, $m = \dfrac{F}{a}$
The acceleration, $a = \dfrac{F}{m}$

Example

> Ellie pulls a sledge with a force of 15 N in the snow. The sledge accelerates at 3 m/s². Calculate the mass of the sledge.
> $$m = \frac{F}{a} = \frac{15}{3} = 5 \text{ kg}$$

Example

> Professional golfers hit a golf ball with a force of approximately 9000 N. If the mass of the ball is 45 g, the acceleration during the very short time (about 0.5 milliseconds) of impact can be calculated.
> $$a = \frac{F}{m} = \frac{9000}{0.045} = 200\,000 \text{ m/s}^2$$

FIGURE 4: In both these actions the forces are unbalanced How does this affect the speed of the golf ball and the speed of the space shuttle?

Remember!

When using $F = ma$, the units for m must be kg and for a must be m/s²

 Questions

3 Explain why a parachute is used as a space shuttle lands

4 A car with a mass of 1200 kg has a resultant forward force acting on it of 4200 N. Find its acceleration.

More on $F = ma$

The equation $F = ma$ can be used to find mass or acceleration only if the unbalanced, or accelerating, force is known.

The unbalanced force is equal to:

> the forward force – the backward force.

If this is positive the unbalanced force is in the forward direction. The car accelerates.

Example

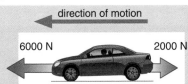

direction of motion

6000 N   2000 N

FIGURE 5: The diagram shows the forces acting on a car of mass 1000 kg. Describe its motion.

> Unbalanced or resultant force,
> $$F = 6000 - 2000 = 4000 \text{ N}$$
> $$a = \frac{F}{m} = \frac{4000}{1000} = 4 \text{ m/s}^2$$

What would happen if the forces in Figure 5 were reversed?

Remember!

Remember: When using $F = ma$, F is the unbalanced or resultant force.

 Questions

5 Explain why there are rules about the mass of golf balls used in competitions.

6 Chloe is driving on the motorway at 30 m/s. She takes her foot off the accelerator and slows down to 27 m/s in 6 s.

a i Calculate Chloe's deceleration.

ii The mass of the car is 1200 kg. Calculate the size of the resistive forces acting on the car.

b The speed of the car returns to 30 m/s.

i What driving force is needed for the car to maintain a steady speed of 30 m/s?

ii Chloe then accelerates and the engine provides a driving force of 840 N. Calculate the car's acceleration.

Road safety

You will find out:

> about thinking, braking and stopping distances

> about the factors that affect thinking, braking and stopping distances

> why a knowledge of stopping distance is important in road safety

A car driver cannot stop a car immediately. It takes the driver time to react to danger. This is called **thinking time**. The higher the speed of a car the larger the distance it travels while a driver thinks.

Thinking distance is the distance travelled between a driver seeing a danger and taking action to avoid it, such as putting their foot on the brake pedal to stop the car.

Braking distance is the distance travelled before a car comes to a stop after the brakes have been applied. Braking distance increases as the speed of the car increases.

stopping distance = thinking distance + braking distance

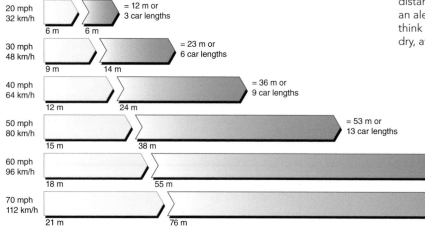

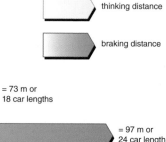

FIGURE 6: How speed affects stopping distance for an average family car, driven by an alert driver on a dry road. How do you think the road conditions, for example wet or dry, affect stopping distance?

A moving vehicle must be able to stop without colliding with another object. Look at Figure 6 to see how the thinking distance, braking distance and overall stopping distance increase with speed.

Example

> Darren is driving at 80 km/h (= 50 mph). What is his stopping distance in metres?
>
> Stopping distance = thinking distance + braking distance = 15 + 38 = 53 m.

Questions

7 Mohammed is driving along a busy road. The brake lights of the car in front come on but Mohammed does not brake immediately. Why not?

8 Lauren is an alert driver whose thinking distance is 18 m. How fast is she going? Using Figure 6, write down her braking distance. What is her stopping distance? What happens to her stopping distance if she halves her speed?

Thinking distance

For an alert driver, thinking time (or **reaction time**) is about 0.7 seconds. Thinking time and therefore thinking distance may increase if a driver is:

> tired

> under the influence of alcohol or other drugs

> distracted or lacks concentration.

If the speed of the car is greater the driver's thinking time stays the same but the thinking distance increases. An increase in thinking distance makes an accident more likely.

Remember!
Be careful not to confuse thinking time and thinking distance.

Braking distance

The braking distance may increase if:

> the road is icy or wet
> the car has poor brakes or bald tyres
> the speed of the car is greater.

An increase in braking distance makes an accident more likely.

Driving safely

The table of stopping distances on the opposite page shows how important it is to:

> keep an appropriate distance from the car in front (at least the thinking distance away)
> have different speed limits for different types of road
> slow down when road conditions are poor.

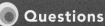

Questions

9 Suggest why road safety campaigns use slogans such as 'Kill your speed not a child!'

10 Two cars are travelling in the same lane along a motorway at a speed of 70 mph. What is the smallest safe distance between them? (Use Figure 6.)

Factors affecting braking distance

> The greater the mass of a vehicle the greater its braking distance.

> The greater the speed of a vehicle the greater its braking distance.

> In a car braking system a disc rotates between two **brake pads**. When the brakes are applied the pads are pushed against the disc. This creates a large force that slows the car down. Worn brakes reduce the friction force.

Worn **tyres** with very little **tread** reduce the grip of the wheels on a slippery road, leading to skidding and an increase in braking distance.

The Ministry of Transport (MOT) test for vehicles includes checks on brakes and tyres.

As the speed of a car increases the graph shows:

> the thinking distance increases linearly (when speed doubles, thinking distance doubles)

> the braking distance increases as a squared relationship (when speed doubles braking distance increases by a factor of 4, when speed trebles braking distance increases by a factor of 9).

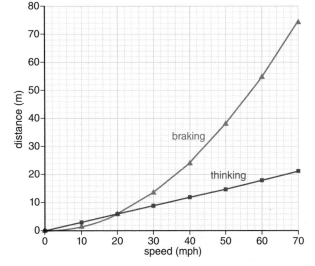

FIGURE 9: How does the graph show that thinking distance and speed are proportional? How does braking distance increase as speed increases?

FIGURE 7: A car disc brake. What happens when the brake pads are worn?

FIGURE 8: The treads on this tyre are worn. Why is this dangerous?

Questions

11 a Estimate the thinking, braking and stopping distances for a car travelling at 45 mph.

b How would these distances change if the car was travelling at 90 mph?

12 Formula 1 drivers use tyres with very little tread when the track surface is dry and tyres with more tread when the surface is wet. Suggest why.

Work and power

You will find out:
> about everyday examples of doing work
> how to calculate work done
> that energy is needed to do work

Push and pull

Work is done whenever a force moves an object. Transport of all kinds involves motion. Different types of transport suit different environments. In big cities, buses or underground trains are convenient, while a cable car is an ideal way to get around in mountainous areas.

FIGURE 1: Why is a cable car an ideal way to get around in mountainous areas?

Work

Work is done when a **force** moves. People and machines do work.

When a person lifts a **mass** or pushes a shopping trolley or pulls a sledge work is done.

The amount of **work done** depends on the:

> size of the force acting on an object in newtons (N)

> distance the object is moved in metres (m).

When a person climbs stairs or jumps in the air the force moved is their **weight**.

Work done = force × distance moved

Work done is measured in **joules** (J)

Example

> Jodie lifts a pile of books weighing 12 N onto a shelf 1.5 m above the ground. How much work does she do?
>
> Work done = force × distance = 12 × 1.5 = 18 J

FIGURE 2: This man is doing work. He is using a force to lift the box. How could he do more work?

Questions

1 What unit is used to measure:

a force

b work

c energy?

2 How much work does Paul do when he lifts a box weighing 250 N off the floor to a shelf 2 m high?

Energy is transferred when work is done. Energy comes from food. The more work is done the more energy is needed.

Energy is also measured in joules (J).

Mass and weight

The force of attraction on a mass due to **gravity** is called weight. force, (F) = mass, (m) × acceleration, (a)

When falling freely $a = g$, the **gravitational field strength** and: weight, $W = mg$.

The weight of an object is found by multiplying its mass by g. On Earth, $g = 10$ N/kg.

Weight is due to gravitational attraction and acts towards the centre of the Earth.

Example

> Sadaf has a mass of 54 kg.
>
> She weighs 54 × 10 = 540 N on Earth.

More on work

Work is only done when a force moves an object in the direction in which the force acts.

The amount of work done increases as:

> the size of the force increases

> the distance moved in the direction of the force increases.

work done = force × distance moved (in the *direction of the force*)
units are joules (J)

Example

If a basketball player weighs 700 N, the work he does against gravity when he jumps 80 cm vertically is:
work done = force × distance moved
= 700 × 0.8 = 560 J

Using the equation for work done

Rearranging the equation

$$force = \frac{work\ done}{distance}$$

$$distance = \frac{work\ done}{force}$$

Example

Jack's car does 300 000 J of work when moving 60 m along a level road. Calculate the pull of the car's engine.

$$force = \frac{work\ done}{distance} = \frac{300\ 000}{60} = 5000\ N$$

FIGURE 3: The basketball player does work in raising his weight off the ground. The woman is walking on a level floor. How much work does she do to move her weight?

Questions

3 Zoe has a mass of 55 kg. What does she weigh?

4 Imran weighs 600 N. How much work does he do when he climbs a vertical rope ladder 5 m high on an assault course?

5 Tom does 3000 J of work in pushing a small van a distance of 10 m. How big is the friction force he has to push against?

Mass and weight

Gravity applies a force of 10 N to each **kilogram** of mass on Earth.

$W = mg$

$m = \frac{W}{g}$ $g = \frac{W}{m}$

Sadaf has a mass of 56 kg. She only weighs 90 N on the Moon. This means the gravitational field strength on the Moon is less than on Earth.

$g = \frac{W}{m}$ so $g = \frac{90}{56} = 1.6$ N/kg.

Car brakes

A car loses all its **kinetic energy** when it stops. The faster it is going the more kinetic energy it possesses. The kinetic energy is transferred mainly into heat by the brakes.

kinetic energy lost = work done by brakes

Example

The brakes in a car produce a force of 5000 N and the car has to lose 200 000 J of kinetic energy in stopping.
work done by brakes = 200 000 J
work done = force × distance moved

$$braking\ distance = \frac{work\ done\ by\ brakes}{braking\ force} = \frac{200\ 000}{5000} = 40\ m$$

Question

6 Bethany has a mass of 60 kg. Calculate her weight
a on Earth
b on planet Zeus where the gravitational field strength is 4 N/kg.

7 A car has 240 000 J of kinetic energy. It stops in 40 m when the brakes are applied. Calculate the braking force.

Power

You will find out:
> about power
> how to calculate power
> how to interpret fuel consumption data

Imagine a tall office block that has two lifts. One lift takes 40 seconds to go up to the tenth floor. The other newer lift takes 25 seconds.

Both lifts do the same amount of work but the new one does it more quickly. The new lift has a greater **power**.

Power is measured in **watts** (W). A large amount of power is measured in **kilowatts** (kW). 1 kW = 1000 W.

Fuel consumption

Some cars are more powerful than others. They travel faster and cover the same distance in a shorter time. More powerful cars need to gain more energy every second and require more fuel.

	Engine capacity in litres	Power in kW	Top speed in km/h	Fuel consumption in litres/100 km
	1.3	75	160	5
	3.6	254	285	15

FIGURE 4: Use the table to help you decide which of these cars is **a** cheaper to run and **b** better for the environment.

The car with the greater engine capacity uses more fuel to travel 100 km.

It also has a greater power and higher top speed.

To sum up, cars have
> different power ratings
> different engine sizes

These are related to **fuel consumption**.

Questions

8 Look at the table above. Which car would go further on a litre of petrol?

9 On a building site, crane A lifts a steel girder on to a building in 60 seconds. Crane B takes 40 seconds to do the same job.

a Why do both cranes do the same amount of work?

b Why is crane B more powerful than crane A?

Calculating power

$$power = \frac{work\ done}{time\ taken}$$

FIGURE 5: Would Anoushka do more work if the stairs were steeper?

Example

Anoushka runs up a flight of 16 stairs, each 0.2 m high, in 8 seconds. She weighs 500 N. She climbs a vertical height of: 16 × 0.2 = 3.2 m.

work done = force × distance moved in the direction of the force
= 500 × 3.2 = 1600 J

$$power = \frac{work\ done}{time} = \frac{1600}{8} = 200\ W$$

If Anoushka walks up the stairs she takes 10 seconds.

Her power = $\frac{1600}{10}$ = 160 W

Car fuels

> Fuel is expensive and a car with a high fuel consumption is expensive to run, particularly over large distances.

> The products of burning fuels **pollute** the **environment**. Car **exhaust gases** are harmful and the production and transport of fuels create many hazards.

The Government wants to reduce car emissions of **carbon dioxide (CO_2)**. Carbon dioxide pollutes local air and is a major source of global **greenhouse gases**, which contribute to climate change. The amount of Vehicle Excise Duty paid is based on carbon dioxide emissions. Vehicle manufacturers and purchasers are made more aware of the environmental impact of vehicles and the use of more fuel-efficient cars is encouraged.

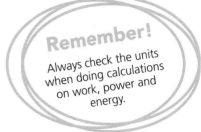

Remember!
Always check the units when doing calculations on work, power and energy.

 Questions

10 Neil can lift 20 weights, each weighing 30 N, through a height of 2 m in 60 seconds. Calculate Neil's power.

11 Sarah has a mass of 55 kg. She can do 24 step-ups in 36 s. The step is 12 cm high. Calculate Sarah's power.

12 Glasgow is 650 km from London. Use the table on the opposite page to compare the amount of fuel used by each car for this journey.

More on power

The power equation can be rearranged to find work done or time taken for a known power value.

work done = power × time

$$\text{time taken} = \frac{\text{work done}}{\text{power}}$$

The Eurostar train provides a high-speed service through the Channel Tunnel from London to Paris reaching a maximum speed of 186 mph (300 km/h) with its engine operating at a power of 2 MW (2 million watts). When Eurostar travels at maximum speed the amount of work done, or energy transferred, in 2¼ hours is calculated by:

To show: $\text{power} = \dfrac{\text{work done}}{\text{time}}$

work done = power × time
= 2 000 000 × (135 × 60)
= 16 200 000 000 J

It does not operate at maximum power for the whole journey.

To show power = force × speed ($P = Fv$)

work done = force × distance

$\text{power} = \dfrac{\text{work done}}{\text{time}}$

$= \dfrac{\text{force} \times \text{distance}}{\text{time}} = \text{force} \times \text{speed}$.

$P = Fv$

Example

Jay's car engine provides a forward force of 6000 N. Calculate its power when travelling at 20 m/s.

$$P = Fv = 6000 \times 20 = 120\ 000\ \text{W}$$

FIGURE 6: The Eurostar takes about 2¼ hours for the 500 km journey to Paris. What is its average speed?

 Questions

13 A 25k W motor raises a 5000 N load through a height of 40 m. How long does it take?

14 Use the data above to calculate the forward force of the Eurostar engine when travelling at maximum speed.

Preparing for assessment: Applying your knowledge

To achieve a good grade in science, you not only have to know and understand scientific ideas, but you need to be able to apply them to other situations and investigations. These tasks will support you in developing these skills.

✳ Surf's up!

Will is very keen on surfing. It is too cold to do much in the winter but as soon as spring comes he and his friends get their boards out and head for the beach. This usually means a car ride and today it is Will's older brother who is driving him and his friends to the beach. This means there will be four people in the car and surfboards strapped to the roof, and this adds to the weight.

Will usually persuades his brother to put some decent music on in the car. He sneaks the volume up a bit as well. On the beach road the wind has been blowing the sand up and some is lying on the road.

Typical of beach roads, the road does not have any pavements and the ground on either side rises quite steeply, so there are pedestrians on the road as well as cars. A couple of lads are play-fighting and one ends up suddenly being pushed further out into the narrow road. Will's brother brakes hard and the car slows down and then slides. It comes to a rest about a metre away from the lad in the road, who gives a sheepish smile and jogs off. A couple of Will's mates laugh nervously. They all feel it was a bit of a close call.

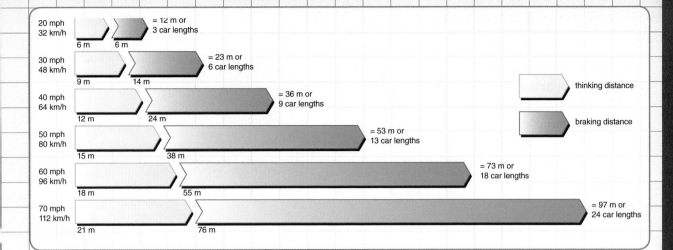

20 mph 32 km/h	6 m / 6 m	= 12 m or 3 car lengths
30 mph 48 km/h	9 m / 14 m	= 23 m or 6 car lengths
40 mph 64 km/h	12 m / 24 m	= 36 m or 9 car lengths
50 mph 80 km/h	15 m / 38 m	= 53 m or 13 car lengths
60 mph 96 km/h	18 m / 55 m	= 73 m or 18 car lengths
70 mph 112 km/h	21 m / 76 m	= 97 m or 24 car lengths

thinking distance

braking distance

✸ Task 1

Will's brother does a lot of driving – most of it is to and from his work, with just him in the car. He is reluctant to admit it but he was a bit surprised by how far the car travelled before it came to a halt. He was not travelling at any great speed because of the road being narrow and the presence of pedestrians. Why did it take longer for Will's brother to stop than he expected?

✸ Task 2

The stopping distance of the car is made up of thinking distance and braking distance. How might the thinking distance have been greater in this case?

✸ Task 3

One of the ways in which the greater braking distance can be explained is by using the concept of kinetic energy. How does this help?

✸ Task 4

The greater stopping distance can also be explained with reference to friction. How does friction affect stopping distance?

✸ Task 5

Will's brother's car is fitted with ABS (an anti-lock braking system).

a Find out how ABS works.

b What difference did ABS make to the braking distance in this case?

✸ Maximise your grade

Answer includes showing that you can...
State **one** reason why it takes Will's brother longer to stop than expected.
State **two** reasons why it takes Will's brother longer to stop than expected.
Explain how distractions in the car affect thinking distance.
Explain the role of friction in stopping a vehicle.
State the factors that affect kinetic energy.
Explain the relationship between kinetic energy and braking distance.
Explain why the car initially starts to slow down, but then starts to slide.
Research anti-lock braking systems and explain how the ABS on Will's brother's car helped to avoid the accident.
As above, but with particular clarity and detail.

Energy on the move

You will find out:
> about kinetic energy
> how to calculate kinetic energy
> about renewable energy sources for vehicles
> how braking distance changes with speed

Dreaming of speed

Many people dream of owning a sleek powerful car like this Porsche with a top speed of 285 km/h (177 mph) and a fuel consumption of 15 litres/100 km (19 mpg) for its 3.6 litre engine.

Beautiful to look at and fun to drive it may be, but it is not environmentally friendly.

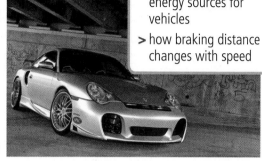

FIGURE 1: What do you think the cars of the future will look like? How do you think they will be powered?

Kinetic energy

Moving objects have **kinetic energy**.

The amount of kinetic energy an object possesses depends on
> its mass
> its speed.

There are different fuels that can be used to gain kinetic energy.

The main fuels for cars and other forms of road transport are **petrol** and **diesel oil**.

Petrol and diesel oil are **fossil fuels**. Some cars use more petrol or diesel oil than others and:
> cause more **pollution**, especially in cities
> cost more to run
> decrease supplies of **non-renewable** fossil fuels.

Scientists are experimenting with types of **renewable** fuels called **biofuels** such as organic wastes.

Solar energy is also being considered as an energy source for vehicles.

Electric cars are becoming more common. These can be
> **battery** driven
> **solar powered** – fitted with solar panels.

What is the major problem associated with the use of solar-powered cars?

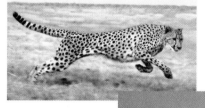

FIGURE 2: The cheetah, the aeroplane and the blades on the wind turbine are all moving. What fuel does each have?

Questions

1 Explain why the cheetah, aeroplane and wind turbine do not always have kinetic energy.

2 Why are scientists developing new types of fuel for cars?

Calculating kinetic energy

> kinetic energy (KE) = $\frac{1}{2}mv^2$

where m = mass of an object in kg

v = speed of an object in m/s

the units of KE are joules (J)

Example

A car has a mass of 1200 kg. Calculate its kinetic energy when travelling at 10 m/s.

Kinetic energy = $\frac{1}{2}mv^2$ = $\frac{1}{2} \times 1200 \times (10)^2$
= $\frac{1}{2} \times 1200 \times 100$ = 60 000J

Fuel consumption

Fuel consumption data are based on ideal road conditions and a car being driven at a steady speed. Values are obtained in urban and non-urban conditions.

Car	Fuel	Engine size in litres	CO_2 in g/km	Consumption in litres/100 km	
				urban	non-urban
Seat Ibiza	petrol	1.6	149	8.4	5.0
	diesel	1.6	112	5.4	3.6

Environmental pollution

Exhaust fumes from petrol- and diesel-fuelled cars cause serious pollution in towns and cities. Battery-driven cars do not pollute the local **environment**, but their batteries need to be recharged. **Recharging** uses electricity from a power station. Fossil-fuelled power stations pollute the local atmosphere and cause **acid rain**. It can be argued that battery-powered cars still cause pollution.

When polluting fossil fuels are no longer available we may have to rely on

> bio-fuelled vehicles

> solar-powered vehicles

> electric vehicles recharged using electricity from **nuclear power stations** or renewable sources.

Questions

3 Carl is running at a speed of 4 m/s. Calculate his kinetic energy if his mass is 70 kg.

4 Look at the table of fuel consumption data.

a Suggest one reason for the difference between urban and non-urban fuel consumption values.

b Write down one other trend that you notice in the data.

5 Give two problems associated with battery-powered cars.

6 Give two advantages of solar-powered cars compared to battery-powered cars.

More about kinetic energy

Kinetic energy (KE) $= \frac{1}{2}mv^2$
so $m = 2 \times KE/v^2$ and $v = \sqrt{(2 \times KE/m)}$

Example

> A charging rhinoceros of mass 800 kg has 90 000 J of kinetic energy. How fast is it running?
> $KE = \frac{1}{2}mv^2$ so $v = \sqrt{\dfrac{(2 \times KE)}{m}} = \sqrt{\dfrac{(2 \times 90\ 000)}{800}} = 15$ m/s

Braking distances

When the car stops its kinetic energy changes into heat in the brakes, tyres and road.

work done by brakes = loss in kinetic energy

braking force × braking distance = change in kinetic energy

If a car has a mass of 1000 kg, its kinetic energy:

> at 20 m/s is $\frac{1}{2}mv^2 = \frac{1}{2} \times 1000 \times (20)^2 = 200\ 000$ J

> at 40 m/s is $\frac{1}{2}mv^2 = \frac{1}{2} \times 1000 \times (40)^2 = 800\ 000$ J.

This is why there are speed limits on roads and stiff penalties for drivers who exceed them.

When the speed of the car doubles

> the kinetic energy quadruples

> the braking distance quadruples.

FIGURE 3: When travelling at the same speed, which vehicle – the car or the lorry – has a greater braking distance? Explain why.

Questions

7 An elephant running at 8 m/s has 64 000 J of kinetic energy. Calculate the elephant's mass.

8 Emma is driving her car at 15 m/s.

a If Emma's car has a mass of 1200 kg, what is its kinetic energy at 15 m/s?

b If the braking distance at this speed is 18 m, calculate the braking force.

Fuel for cars

Most cars use petrol or diesel as fuel. Petrol and diesel are made from oil. Petrol is more **refined** than diesel oil. Petrol cars and diesel cars need different engines.

> Fuel in a petrol engine is ignited by a spark from a spark plug.

> Fuel in a diesel engine is ignited by hot compressed air in a cylinder.

The same amount of diesel oil contains more energy than petrol. A diesel engine is more **efficient** than a petrol engine.

> CO_2 emissions contribute to global warming so must be reduced.

> Diesel cars often have lower CO_2 emissions than petrol cars. But diesel engines produce more of the emissions that affect air quality and can damage our health, such as NO_x and particulates. This is a problem, especially in urban areas.

The table shows data for four Ford Fiesta cars.

engine size in litres	fuel	Fuel consumption in litres/100 km	CO_2 emissions in g/km
1.4	petrol	6.1	145
1.4	diesel	4.4	119
1.6	petrol	6.5	154
1.6	diesel	4.5	116

FIGURE 4: The shapes of the car roof box and the lorry help to streamline the vehicles. What advantage does streamlining give to vehicles?

Streamlining

Friction forces on moving objects such as vehicles can be reduced by **streamlining** their shapes in the following ways:

> shaping car roof boxes

> making high-speed cars wedge-like in shape

> angling lorry deflectors.

A streamlined shape *increases* a car's top speed and *decreases* fuel consumption. Driving with car windows open increases the friction force so *decreases* the top speed and *increases* fuel consumption.

Questions

9 Write down two things about fuel consumption that you notice in the table above.

10 What else, apart from the data in the table, might affect fuel consumption figures?

11 Compare CO_2 emissions for the four cars in the table.

12 Would the top speed of a car be greater or less without a roof box?

How should we power our cars in the future?

Battery driven electric car:

> the batteries takes up a lot of space

> the batteries need recharging frequently causing pollution (see page 205)

> the car has a limited range

> the car is expensive to buy but cost of recharging is low (about £270 for 1200 miles).

With a solar-powered car:

> the Sun does not always shine

> you need to have a back-up such as a battery.

FIGURE 5: This hybrid electric car has solar panels on its roof that convert sunlight into additional power to supplement its battery. It is less environmentally friendly than a car powered only by solar energy, but it is more practical. Can you suggest why?

Q streamlining for KS4   electric cars

We have seen the problems arising from our dependence on oil – resources are finite and it pollutes the environment. The UK is committed to cutting emissions by 80% by 2050.

More and more electric and **hybrid cars** are appearing on our roads, especially in large cities. But this alone will not solve the problem.

Bio-fuelled and solar-powered vehicles clearly reduce pollution at the point of use but still produce pollution in their production in the same way as the manufacture of any other car.

Look at the CO_2 emissions from oil-fuelled cars in the table below. Solar power would lead to an overall reduction in CO_2 emissions but it can only supplement another energy source.

Bio-fuels may reduce CO_2 emissions but we can't be sure. For instance, deforestation leads to an increase in emissions.

So there is no easy answer.

Fuel consumption data

The metric standard method for measuring fuel consumption is litres per 100 kilometres. The UK standard is miles per gallon. A modern fuel efficient car would run at under 8 litres per 100 kilometres, which is equivalent to over 35 miles per gallon.

Fuel consumption units are not easy to interpret. For example, if the measurement in litres/100 km increases, is your engine becoming more or less efficient?

Vehicle model	Fuel	Engine capacity in cm³	Consumption in litres/100 km			CO_2 in g/km	CO in g/km	NO_x in g/km
			urban	non urban	combined			
Ford Focus	petrol	1596	10.6	6.0	7.7	184	0.250	0.060
	diesel	1560	4.5	3.4	3.8	99		0.163
Land Rover Discovery 4	petrol	4999	19.8	10.7	14.1	328	0.145	0.032
	diesel	2993	11.3	8.3	9.3	244	0.303	0.177
Jaguar S-type	petrol	2967	15.8	7.3	10.3	249	0.414	0.026
	diesel	2720	9.3	5.5	6.8	179	0.334	0.186

Questions

13 Why does the data in the table above suggest that diesel is better than petrol as a fuel?

14 Which car has the best fuel consumption figures?

15 What do you notice about CO_2 emissions and fuel consumption?

16 Comment on CO and NO_x emissions for the cars in the table.

More about fuel consumption data

Factors that affect the fuel consumption of a car are:

> the amount of energy required to increase its kinetic energy

> the amount of energy required for it to do work against friction, e.g. the shape of the car

> its speed; kinetic energy = ½ mv^2

> the way in which it is driven such as excessive acceleration and deceleration, constant braking, speed changes

> road conditions, such as a rough surface.

Questions

17 Fuel consumption is usually much better on a long journey using motorways than when driving on minor roads. Suggest why.

18 Suggest why fuel consumption data are unlikely to be achieved in normal driving conditions.

19 Discuss what we can do to reduce the pollution caused by cars.

20 Evaluate the data in the table with respect to fuel consumption and emissions.

fuel consumption data   cars of the future

Crumple zones

You will find out:
> about momentum
> how seatbelts, crumple zones and air bags act in a crash
> how forces can be reduced in a crash

Car safety

To stop a car safely energy must be absorbed. Sometimes people say, 'They don't make cars like they used to', when they see the damage caused to a modern car by a fairly minor accident. But modern cars are built with crumple zones at the front and rear so that the car absorbs the maximum amount of energy with the minimum injury to the driver and passengers.

FIGURE 1: Why may damage to a car in an accident help to protect the driver and passengers from serious injury?

Momentum

A moving car has **momentum**.

momentum = mass × velocity   Unit: kgm/s

Example

> Calculate the momentum of a car of mass 1000 kg travelling at 20 m/s.
>
> momentum = mass × velocity = 1000 × 20 = 20 000 kgm/s

In an accident a car stops suddenly. Its momentum becomes zero. The rapid change in momentum means there is a large force on the occupants which can cause serious injury.

Car safety

Modern cars have safety features that absorb energy when a vehicle stops suddenly to reduce or avoid injury. Safety features include:

> **brakes** that get hot
> **crumple zones** at front and rear that change shape
> **seatbelts** that stretch a little
> **air bags** that inflate and squash.

A seatbelt and an air bag work together. On **impact**:

> the air bag inflates
> the seatbelt stretches and slows the forward motion of the driver's body.

In this way the head and thorax (chest) of the driver are protected.

FIGURE 2: An air bag and seatbelt in action during a car accident. What parts of the man are protected?

Questions

1 Jo has a mass of 50 kg. She is running at 5 m/s. Calculate her momentum.

2 In addition to absorbing energy, in what other way does a seatbelt act to avoid serious injury?

3 Why are accidents at high speed more likely to cause serious injury than accidents at low speed?

Using the momentum equation

momentum = mass × velocity

$$mass = \frac{momentum}{velocity} \qquad velocity = \frac{momentum}{mass}$$

Example

> Ellie has a mass of 50 kg. How fast is she moving if her momentum is 150 kgm/s?
>
> $velocity = \frac{momentum}{mass} = \frac{150}{50} = 3\,m/s$

Reducing injury in a crash

To minimise injury, **forces** acting on the people in a car during a crash must be made as small as possible.

$$force = \frac{change\ in\ momentum}{time}$$

So by spreading the change in momentum over a *longer* time the force on the people in the car is *smaller*. This means they are less likely to be seriously injured.

> Crumple zones absorb some of the car's energy by changing shape or 'crumpling'.

> A seat-belt is designed to stretch a little so that some of the person's kinetic energy is converted to elastic energy.

An air bag absorbs some of the person's KE by squashing up around them.

All these safety features:

> change shape

> absorb energy

> reduce injuries

> reduce momentum to zero more slowly so reduce the force on the car's occupants.

Questions

4 a Dilip's momentum is 700 kg m/s when he is travelling in a car at 10 m/s. What is his mass?

b Use momentum to explain why he is more likely to be injured in an accident if the car is travelling at 30 m/s.

5 Why is a *driver's* air bag especially important?

More about momentum

momentum = mass × velocity

force = mass × acceleration
F = ma (Newton's second law)

acceleration, $a = \dfrac{\text{change in velocity}}{\text{time}} = \dfrac{(v - u)}{t}$

force, $F = \dfrac{m(v - u)}{t} = \dfrac{(mv - mu)}{t} = \dfrac{\text{change in momentum}}{\text{time}}$

$\text{Force} = \dfrac{\text{change in momentum}}{\text{time}}$

Change in momentum = force × time;
$\text{time} = \dfrac{\text{change in momentum}}{\text{force}}$

Example

A car of mass 900 kg is moving at 12 m/s when it collides with a wall. The force on the car is 3600 N. Calculate the stopping time.

Change in momentum = mass × change in velocity
= 900 × 12 = 10 800 kg m/s.

$\text{Stopping time} = \dfrac{\text{change in momentum}}{\text{force}} = \dfrac{10\ 800}{3600} = 3\text{s}$

Reducing injury

Force can be reduced by reducing the **acceleration**:

> by increasing stopping or collision time

> by increasing stopping or collision distance.

How do safety features work?

> Crumple zones increase the time between first impact and the car stopping.

> A seatbelt stretches a little and slows a person down more slowly.

> An air bag inflates on impact, slowing a person down more slowly and protecting them from protruding objects.

> A metal **crash barrier** is made from a material that changes shape readily on impact so the car travels further before stopping and takes longer to come to a halt. It also helps to stop the car bouncing back across the road into other traffic.

> An **escape lane** on a steep hill allows a vehicle with failed brakes to stop more slowly by running into an area with a rough surface that may slope upwards.

FIGURE 3: This escape lane on a steep hill has deep sand or a rough surface and an incline. Can you suggest why?

Questions

6 Explain why there is a greater risk to the driver and front passenger if a passenger in the back breaks the law by failing to wear a seatbelt.

7 Consider each of the safety features mentioned above. List them in order of importance in terms of saving lives and reducing injuries. Give reasons for your choice.

car safety features   force and momentum for GCSE

Safety features

You will find out:

> about some typical safety features of cars

> how safety features can make driving safer

> how ABS brakes reduce braking distances

Modern cars are built with many features designed to improve safety.

Features designed to prevent crashes include:

> **ABS brakes (anti-lock** braking system)
> **traction control** (stops wheel spin)
> **electric windows**
> **paddle shift controls** on steering column

Features designed to protect occupants in the event of a crash include:

> crumple zones
> **safety cage** (protects in a roll over accident)
> air bags
> seat belts
> collapsible steering column
> **side impact beams** in doors

All safety features must be kept in good repair if they are to continue to be of benefit.

> Seatbelts must be replaced after a crash in case the belt fabric has been overstretched.

> The safety cage must be examined for possible damage after a crash.

> Seat fixings should be checked frequently to make sure they are secure.

Can seat belts be bad for users?

The wearing of seat belts is compulsory but some people believe they can make accidents worse:

> wearing a seat belt makes a driver feel safe so he/she drives more daringly

> it may not be possible to undo a seat belt to escape from a burning car

> there is a risk of chest injuries

What do you think?

FIGURE 4: A car safety cage. How does it make the car safer?

Remember!
ABS brakes don't stop a car more quickly. They give improved control and prevent skidding.

Questions

8 How do electric windows contribute to road safety?

9 How does a collapsible steering column contribute to road safety?

10 State two risks and two benefits of wearing a seat belt. Suggest why the wearing of seat belts is compulsory.

Crash testing

You might think that **computer modelling** would have replaced **crash testing** using **dummies**. But actually crashing cars gives more information about where safety features need to be introduced or improved. The tests are usually impacts upon a solid concrete wall, but can also be vehicle–vehicle tests.

For example, 4 x 4 cars such as Land Rovers have become very popular but tend to roll over in a serious accident. Crash tests have led to improvements to the safety cage.

Crash test dummies now come in all shapes and sizes, including pregnant women and babies.

FIGURE 5: Why are 4 x 4 cars more likely to roll over in an accident?

Questions

11 Suggest why crash testing using dummies is important although expensive cars are destroyed.

12 Why is it important to use dummies of different shapes and sizes?

FIGURE 6: A simulated crash with a dummy with no seatbelt. How does wearing a seatbelt help to protect a person in a crash?

Making driving safer

Primary safety features which help to prevent a crash include:

> ABS brakes which give a vehicle stability and maintain steering when braking hard or going into a skid:

– Hard continuous pressure activates anti-lock brakes which then work by automatically pumping on and off to avoid skidding

– The driver gets the maximum braking force without skidding and can still steer the car.

– The driver does not necessarily stop more quickly.

> Traction control which stops the wheels on a vehicle from spinning when it accelerates rapidly. It gives maximum grip and stability on the road during acceleration.

Secondary safety features, which protect occupants in the event of a crash, include a safety cage, crumple zones, seatbelts, side impact beams and air bags.

Other safety features include:

> A **cruise control** system accelerates the car to a fixed speed irrespective of the load in the vehicle or the gradient of the road. It is less tiring on long motorway trips and avoids 'lead-foot syndrome', where the driver rests their foot too hard on the accelerator pedal and inadvertently speeds up.

> Electric windows open and close quickly at the push of a button leaving the driver to concentrate on driving.

> Paddle shift controls allow the driver to operate gears, lights, stereo and wipers without taking their hands off the steering wheel or their eyes off the road.

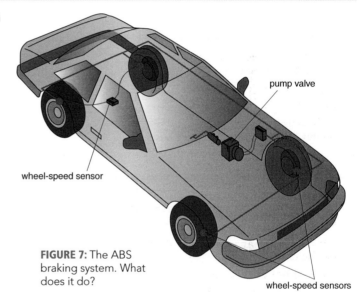

pump valve

wheel-speed sensor

FIGURE 7: The ABS braking system. What does it do?

wheel-speed sensors

Questions

13 What is the difference between primary and secondary safety features?

14 a Describe how ABS brakes work.

b ABS brakes stop a car of mass 1200 kg moving at 30 m/s in a distance of 75 m. Find the braking force of the car.

15 Consider the risks and benefits of wearing seatbelts. Decide whether you think this should be a personal choice rather than a legal requirement.

Falling safely

You will find out:
> about the motion of falling objects
> how air resistance slows down moving objects
> about terminal speed

Free-fall

Free-fall parachutists fall for several kilometres before opening their parachutes.

They accelerate rapidly at first but soon reach a constant speed when the forces acting on them balance. When they open their parachutes they decelerate rapidly and gently float to the ground.

FIGURE 1: Why does a parachutist fall more slowly after their parachute opens?

Falling objects

If an object is dropped it gets faster as it falls. It is pulled towards the centre of Earth due to **gravity**. The Earth's gravity gives all objects a downward force called weight.

There is a story that a famous scientist called Galileo dropped a small cannon ball and a large one from the Leaning Tower of Pisa in Italy to demonstrate the effect of gravity. The balls hit the ground at the same time. This showed that all objects **accelerate** at the same rate (about 10 m/s^2), regardless of their mass.

But if a ball and a feather are dropped, the ball hits the ground first.

Objects that have a large area of cross-section such as parachutes, shuttlecocks or feathers, fall more slowly. The frictional **force** that causes the objects to accelerate more slowly is called **air resistance** or **drag**.

Air resistance has a bigger effect on a ball of low mass, such as a table tennis ball, than on a ball of larger mass, such as a golf ball.

The force due to air resistance gets bigger the faster an object falls. When it is equal to the weight of the object the forces on it are balanced. It falls at a constant speed – **terminal speed**.

On the road, any object moving though air, such as the cyclist, is slowed down by air resistance.

On the Moon, where there is no atmosphere, there is no drag force on falling objects. They all fall with the same acceleration.

FIGURE 2: Why does a shuttlecock fall more slowly than a ball?

Did you know?

Apollo 15 landed safely on Earth after its mission to the Moon even though one of its parachutes failed.

FIGURE 3: What force slows the cyclist down?

Questions

1 If you drop a ball it accelerates. What happens if you throw it upwards?

2 Why would the cyclist be able to travel faster if there was no air present?

Balanced and unbalanced forces

Moving objects:

> *increase* speed if the forward force is *bigger* than the backward force

> *decrease* speed if the forward force is *smaller* than the backward force

> maintain a steady speed if the forward force is **equal** to the backward force.

Parachutists

The speed of a **free-fall** parachutist changes as they fall to Earth. This is because the upward air resistance force on them changes.

Terminal speed is the constant speed reached by a falling object. It occurs when forces acting on an object are **balanced**.

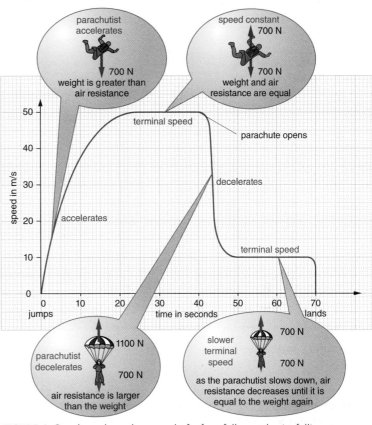

parachutist accelerates

700 N
weight is greater than air resistance

speed constant 700 N

700 N
weight and air resistance are equal

terminal speed

parachute opens

decelerates

accelerates

terminal speed

1100 N
parachutist decelerates
700 N
air resistance is larger than the weight

slower terminal speed
700 N
700 N
as the parachutist slows down, air resistance decreases until it is equal to the weight again

speed in m/s (y-axis): 0, 10, 20, 30, 40, 50

time in seconds (x-axis): 0 jumps, 10, 20, 30, 40, 50, 60, 70 lands

Questions

3 What would happen to the speed of a parachutist if their parachute failed to open?

4 Why do free-fall parachutists fall with their body in a horizontal position?

5 What can you say about the forces on a car when it is accelerating?

FIGURE 4: Graph to show the speed of a free-fall parachutist falling to Earth. What happens to the forces acting on the parachutist when the terminal speeds are reached?

Terminal speed

> As the speed of a free-fall parachutist increases, they displace more air **molecules** every second so the air resistance, or drag force increases. This reduces their acceleration.

> When their weight is equal to the drag force, the forces on them are balanced so they travel at a constant speed – the terminal speed.

> When the parachute opens the upward force on them increases suddenly as there is a much larger surface area, displacing more air molecules every second.

> They decelerate, displacing fewer air molecules each second, so the drag force decreases.

> Eventually they reach a new slower terminal speed when their weight is equal to the drag force once more.

> This means they can land safely.

The same principle applies to a car being driven along a road. Terminal speed is reached when the driving force from the engine equals the total backward force.

FIGURE 5: What happens to the speed of a parachutist when their parachute opens?

Question

6 Ali and Charlie are free-fall parachutists. Ali weighs 500 N and Charlie weighs 800 N. Use your ideas about forces to explain who reaches the greatest terminal speed.

7 Sketch a graph with force diagrams, as in Figure 4, for a cyclist who starts from rest and reaches terminal speed.

8 What can the cyclist do to increase his terminal speed?

Frictional forces

FIGURE 6: Friction forces on a car. What do friction forces do?

All **friction** forces act in the opposite direction to the direction of motion and slow an object down. This leads to energy loss and inefficiency.

There are ways of reducing friction forces.

> Friction forces between moving parts of a machine can be reduced by **lubricating** (oiling) them.

> Friction forces on moving objects such as vehicles can be reduced by **streamlining** their shapes in the following ways, as seen on page 202:

• shaping car roof boxes

• making high-speed cars wedge-like in shape

• angling lorry deflectors.

• closing car windows when driving

Streamlining is designed to reduce the drag force acting on a vehicle. Less energy is wasted. This:

> allows its top speed to increase

> improves fuel consumption.

Air passes over a streamlined vehicle more easily than if it had sharp corners and a square shape.

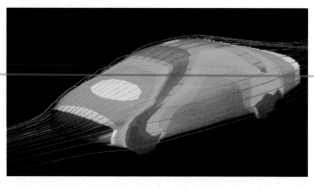

FIGURE 7: Computer simulation used in the design of a new car. How do you know it is streamlined?

Questions

9 Suggest why dolphins and sharks have a streamlined shape.

10 How do racing cyclists make themselves more streamlined?

Falling objects

In general all objects fall with the same acceleration. But this is only true if the effect of air resistance is very small.

When falling towards Earth through the atmosphere there is always a drag force (air resistance) as the falling object displaces air molecules.

The size of the air resistance force on a falling object depends on:

> its **cross-sectional area** – the larger the area the greater the air resistance

> its speed – the faster it falls the greater the air resistance.

Air resistance only has a significant effect on motion when it is large compared to the weight of the falling object.

Acceleration due to gravity

The acceleration due to gravity (g) is the same for
any object at a given point on the Earth's surface.
The variation from place to place is small so we take it as 10 m/s² when using it in calculations.

Air resistance has an effect on the actual acceleration of a falling object, as seen in Figure 8.

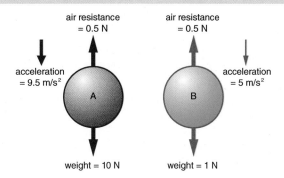

FIGURE 8: Use the equation F=ma to check that the values given for the acceleration of balls A and B in Figure 8 are correct.

Questions

11 Explain why free-fall parachutists are not actually in free-fall unless they jump from an aeroplane at a very high altitude.

12 a What happens to the air resistance force as the balls in Figure 8 continue to fall?

b How does this affect their acceleration?

c Which one, A or B, would reach terminal velocity first?

More about g

Gravitational field strength is the force on each kilogram of mass due to gravity. It is measured in N/kg.

On Earth we say the gravitational field strength, g = 10 N/kg.

The acceleration due to gravity is the acceleration of a freely falling object due to the force of gravity pulling it down. Acceleration is measured in m/s².

On Earth we say the acceleration of free-fall, g = 10 m/s². (A more precise value is 9.81 m/s².)

> g is unaffected by atmospheric changes.

> g varies slightly with position on the Earth's surface, from about 9.78 m/s² at the equator to approximately 9.83 m/s² at the poles. This is because the Earth is not quite spherical. Its radius at the poles is a tiny bit less than its radius at the equator so you are a little bit nearer the centre of the Earth at the poles.

> Very small local variations are caused by height above sea level; g is slightly different on the top of a mountain or down a mineshaft.

Questions

13 Will the acceleration due to gravity be larger or smaller down a mine-shaft than at the Earth's surface? Explain.

14 Discuss how the bounce of a ball dropped at the top of Mount Everest would be different from dropping the same ball at sea level.

 acceleration due to gravity for KS4   variation of g for KS4

The energy of games and theme rides

Understanding the science of theme park rides

The rides at theme parks are designed to thrill and frighten people. They cause rapid energy changes that distort a person's 'gravity'. People on the rides experience G-forces similar to those experienced by astronauts on lift-off. Roller coasters can be quite gentle or terrifying, depending on the steepness of the track – and a person's 'scare factor'.

FIGURE 1: What happens to your kinetic energy as you accelerate rapidly?

Gravitational potential energy

An object held above the ground has **gravitational potential energy**. The amount of gravitational potential energy an object has increases with:

> its **mass**

> its height above the ground.

The vertical-drop roller coaster shown starts with a 60 m climb (to gain gravitational potential energy) and then is stationary for 3 seconds hanging over the edge of a 60 m vertical drop. After being released each carriage reaches a speed of 110 km/h and the people in the carriage experience a force of 4.5 G (this means a person feels 4.5 times their normal weight).

FIGURE 2: A vertical-drop roller coaster. Why does the roller coaster start with a climb to the top?

Did you know?

The world's first vertical-drop roller coaster called 'Oblivion' opened in 1998 at Alton Towers, Staffordshire.

Questions

1 What sort of energy do the riders possess waiting at the top of a roller coaster?

2 Sam and his Dad are riding on 'Oblivion'. Who has the greater gravitational potential energy when they are held at the top of the drop?

Energy transfers

The riders at the top of a roller coaster ride have a lot of gravitational potential energy. The gravitational potential energy is quickly changed to kinetic energy as the carriages descend.

A bouncing ball converts gravitational potential energy to **kinetic energy** and back to gravitational potential energy. If a ball is dropped from a height of 2 m it does not return to its original height because energy is transferred to other forms such as **thermal energy** and **sound energy**.

Q gravitational potential energy for KS4   the G force for KS4

FIGURE 3: A person swinging experiences constant changes in energy. What are the main energy changes?

B the ball has *just* reached the ground so it has kinetic energy but no gravitational potential energy

D the ball has gravitational potential energy but no kinetic energy at the top of the bounce

2.00 m

1.20 m

A   B   C   D

A the ball has gravitational potential energy but no kinetic energy

C the ball squashes and has **elastic potential energy** which is converted to kinetic energy as it leaves the ground

FIGURE 4: Stages of energy transfer when a ball is dropped. Where is the kinetic energy of the ball greatest?

Calculating gravitational potential energy

To calculate the gravitational potential energy (GPE) of an object the following equation is used:

$GPE = mgh$

where m = mass in kg
       h = vertical height moved in m
       g = gravitational field strength in N/kg (on Earth, g = 10 N/kg)

The unit for GPE is the joule (J).

If the ball in Figure 4 of mass 50 g is dropped from a height **A** of 2.00 m and bounces up to a height **D** of 1.2 m, the energy wasted in the bounce is:

GPE lost between **A** and **D**
    $= mgh = 0.05 \times 10 \times (2.00 - 1.20) = 0.40$ J

The initial gravitational potential energy is 1.0 J, so 40% of the ball's energy is wasted. The energy **efficiency** of the bounce is 60%.

Questions

3 Look at Figure 4:

a What types of energy does the bouncing ball possess when it is 1 m above the ground?

b Suggest why the GPE of the ball at D is less than its GPE at A.

4 Jake is at the top of a vertical-drop roller coaster, 60 m above the ground. He has a mass of 70 kg. Calculate his GPE.

Using gravitational potential energy equation

$GPE = mgh$

so $m = \dfrac{GPE}{gh}$ $h = \dfrac{GPE}{mg}$

Example

A ski jumper gains 84 000 J of gravitational potential energy as he climbs to the top of a 120 m jump. What is his mass, assuming g = 10 N/kg?

mass, $m = \dfrac{GPE}{gh} = \dfrac{84\,000}{10 \times 120} = 70$ kg

FIGURE 5: When a skydiver reaches terminal velocity, what happens to their kinetic energy?

Energy and terminal speed

When a skydiver reaches terminal speed their kinetic energy ($\frac{1}{2} mv^2$) has a maximum value and remains constant. The gravitational potential energy lost as they fall is used to do **work** against friction (**air resistance**). This increases the thermal energy of the surrounding air particles. When terminal speed is reached:

    change in gravitational potential energy
    = work done against friction

Question

5 The skydiver in Figure 5 is in a streamlined position. How could he slow down?

6 Using the data for a vertical-drop roller coaster on the opposite page calculate the speed of each carriage. Why does each carriage only reach a speed of 110 km/h?

7 How much work is done against friction when a skydiver of mass 60 kg falls 100 m at terminal velocity?

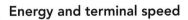

You will find out:

> more about gravitational potential energy and kinetic energy

> how a roller coaster works

Water-powered funicular railway

A water-powered **funicular railway** has a carriage that takes on water at the top of the hill giving it extra gravitational potential energy.

As the carriage travels down the hillside it transfers gravitational potential energy to kinetic energy and pulls up another similar carriage with an empty water tank on a parallel rail.

Kinetic energy

A moving object has kinetic energy. The amount of kinetic energy an object has increases if:

> its mass increases

> its speed (how fast it is moving) increases.

FIGURE 6: The water-powered Lynton and Lynmouth funicular railway. Which carriage has more kinetic energy, the 'up' or the 'down' carriage?

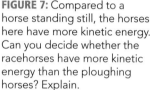

FIGURE 7: Compared to a horse standing still, the horses here have more kinetic energy. Can you decide whether the racehorses have more kinetic energy than the ploughing horses? Explain.

Questions

8 Give **two** ways of increasing the gravitational potential energy of an object.

9 Give **two** ways of increasing the kinetic energy of an object.

How a roller coaster works

A traditional roller coaster works by using a motor to haul a train up in the air, giving it a lot of gravitational potential energy. The train is then released, converting gravitational potential energy to kinetic energy as it falls. Each peak is lower than the one before because some energy is transferred to heat and sound due to friction and air resistance.

The principle of **conservation of energy** tells us that:

gravitational potential energy at top = kinetic energy at bottom + energy transferred (to heat and sound) due to friction

> At the peaks the train has a lot of gravitational potential energy (high up) and little kinetic energy (moves slowly).

> At the bottom the train has little gravitational potential energy (low down) and a lot of kinetic energy (moves fast).

FIGURE 8: What type of energy is the leading train gaining on this roller coaster?

Relationship of speed and kinetic energy

kinetic energy = $\frac{1}{2} mv^2$

Ignoring friction, as the train falls:

> loss of gravitational potential energy = gain in kinetic energy = $\frac{1}{2} mv^2$

> if the gravitational potential energy is doubled, the kinetic energy doubles

> if the kinetic energy doubles the speed increases but does not double. (It increases by $\sqrt{2}$.)

> If mass doubles, kinetic energy doubles (kinetic energy $\propto m$).

> If speed doubles, kinetic energy quadruples (kinetic energy $\propto v^2$).

● Questions

10 Why do the heights of the peaks on roller coaster rides decrease progressively?

11 On long roller coaster rides the trains are given a short lift by a motor part-way through the ride. Why is this done?

12 Sam and his Dad are on a roller coaster moving at 20 m/s. Sam's mass is 45 kg, half that of his Dad.

a How does Sam's kinetic energy compare with his Dad's?

The roller coaster slows down to 10 m/s.

b What is Sam's kinetic energy now?

c What is his Dad's kinetic energy now?

Falling roller coaster

Ignoring friction, as the train falls,

loss of GPE = gain in KE

$mgh = \frac{1}{2} mv^2$.

$h = \dfrac{v^2}{2g}$

This is independent of the mass of the falling object.

Example

> A roller coaster train reaches a speed of 18 m/s when falling from rest. How far did it fall?
>
> $h = \dfrac{v^2}{2g} = \dfrac{(18 \times 18)}{20} = \dfrac{324}{20} = 16.2$ m
>
> It falls through a *vertical* height of 16.2 m.

● Questions

13 Meena rides on the roller coaster shown.

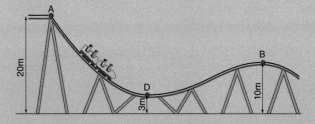

a She is not moving at A. How fast is she going at B?

b How fast is she going at D?

c Why are your answers likely to be too large?

Preparing for assessment: Analysis and evaluation

To achieve a good grade in science, you not only have to know and understand scientific ideas, but you need to be able to apply them to other situations and investigations. These tasks will support you in developing these skills.

✳ Tasks

> Analyse and evaluate an investigation to find out if there is a simple relationship between drop height and bounce height.

> Find out if different balls bounce in the same way.

✳ Context

A ball is held above the ground and released. Its gravitational potential energy (GPE) gets less and its kinetic energy increases as it falls.

As it hits the ground its kinetic energy is transferred to elastic potential energy (EPE) as the ball squashes. As the ball regains its shape, the EPE is transferred to kinetic energy and it bounces upwards.

When the ball reaches the top of its travel, all the kinetic energy has been transferred to GPE again.

It never bounces up to the same height it was released from as some energy is released as heat and sound to the surroundings.

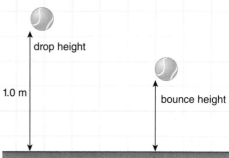

drop height

1.0 m

bounce height

✳ How to do this experiment

1. Fix a metre rule to a vertical wall or door frame. (You may want to use two metre rules later.)

2. Hold a ping-pong ball so that its base is level with the 1 m mark on the rule. (Remember to read the scale at eye level.)

3. Drop the ball so that it falls straight down without touching the wall or metre rule.

4. Note the reading of the base of the ball on the metre rule at the highest point of its bounce.

5. Repeat steps 3 and 4 twice more and find the mean value of the bounce height.

6. Repeat the experiment six more times using a range of heights.

7. To find out if other balls bounce in the same way, replace the ping-pong ball with one made from a different material and do the whole experiment again.

Take care – avoid using hard balls such as golf or cricket balls.

Results, analysis and evaluation

A group of students found these results when they tested a ping-pong ball and a tennis ball.

ping-pong ball			
drop height in cm	bounce height in cm 1	bounce height in cm 2	mean bounce height in cm
110	74.4	74.2	
100	70.0	69.0	
90	65.8	66.2	
80	61.0	60.7	
70	54.6	54.4	
60	47.2	47.0	
40	32.1	31.7	

tennis ball			
drop height in cm	bounce height in cm 1	bounce height in cm 2	mean bounce height in cm
110	70.8	71.3	
100	63.8	64.2	
90	58.4	58.8	
80	50.2	50.6	
70	45.5	45.9	
60	39.1	38.8	
50	33.2	32.7	

1. Complete the final column in each table.

2. Use the same set of axes to plot graphs of mean bounce height (*y*-axis) against drop height (*x*-axis).

> Think about the scale for the axes. Use sensible divisions but make the graph as large as possible.

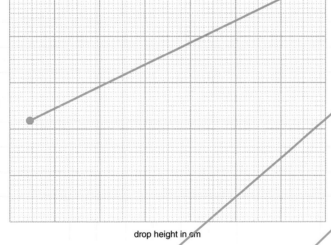

mean bounce height in cm

drop height in cm

> Should these be straight lines or smooth curves? Never draw dot to dot.

> What shape graph would give a simple relationship?

> What measurement(s) were hard to do? How could the students have done it better?

3. Draw lines of best fit through the points.

4. Is there a simple relationship between the drop height and the bounce height? Explain your answer by referring to the shape of the line graph.

5. Do all balls bounce in the same way? Explain your answer by referring to the two graphs.

6. How could the students improve their experiment?

7. Were the results precise? How do you know?

> Do not confuse precision with accuracy; precision is about the repeatability of results.

8. Use ideas about forces to explain the shape of the graphs.

P3 Checklist

To achieve your forecast grade in the exam you'll need to revise

Use this checklist to see what you can do now. It gives you many of the important points you will need to know. Refer back to the relevant pages in this book if you're not sure and to see if there is anything else you need to know. Look across the three columns to see how you can progress.

Remember you'll need to be able to use these ideas in various ways, such as:

> interpreting pictures, diagrams and graphs
> applying ideas to new situations
> explaining ethical implications
> suggesting some benefits and risks to society
> drawing conclusions from evidence you've been given.

Look at pages 278–299 for more information about exams and how you'll be assessed.

To aim for a grade E	To aim for a grade C	To aim for a grade A
use the equation: average speed = $\frac{\text{distance}}{\text{time}}$ **recall** that 1 km = 1000 m and change units from km to m **understand** how speed and average speed cameras work **draw and interpret** graphs of distance against time	**manipulate** the equation distance = average × speed time **interpret** the gradient (steepness) of a distance–time graph as speed	**draw and interpret** distance–time graphs for non-uniform speed **calculate** speed from a distance–time graph for uniform speed
describe trends from simple speed–time graphs **recognise** that acceleration involves a change in speed **calculate** acceleration when change in speed is given **understand** that velocity is speed in a certain direction	**draw and interpret** speed–time graphs for uniform acceleration **calculate** distance travelled from a simple speed–time graph **describe** acceleration as change in speed per unit time **use** the equation acceleration = $\frac{\text{change in speed}}{\text{time taken}}$ **calculate** the relative velocity of objects moving in parallel	**describe, draw and interpret** speed–time graphs **calculate** acceleration from graph for uniform acceleration **manipulate** the equation for acceleration
recognise situations where forces change speed of an object **use** the equation force = mass × acceleration to calculate force **describe** thinking, braking and stopping distances **calculate** stopping distance **explain** why these distances are important for road safety	**manipulate** the equation force = mass × acceleration ($F = ma$) **explain** factors that may increase thinking and braking distances **interpret** charts of thinking and braking distances **explain** why stopping distances are important in road safety	**calculate** the accelerating force and manipulate $F = ma$ **explain** everyday situations that change braking distance **interpret** graphs of thinking and braking distance against speed **explain** the effect of speed on thinking and braking distances
recall examples in which work is done **recall** that the joule (J) is the unit for work and energy **use** the equation work done = force × distance **describe** power and **recall** that it is measured in watts (W) **recognise** that power and engine size affect fuel consumption	**use** the equation weight (W) = mass × gravitational field strength (mg) **manipulate** the equation work done = force × distance **use** the equation power = $\frac{\text{work done}}{\text{time}}$ **interpret** fuel consumption data including cost and environmental issues	**manipulate** the equation $W = mg$ **manipulate** the equation power = $\frac{\text{work done}}{\text{time}}$ **use and derive** the power equation in the form power = speed × time

To aim for a grade E

understand that kinetic energy depends on mass and speed

recall bio-fuels and solar energy as alternatives to fossil fuels

describe how electricity is used in battery and solar-powered cars

recognise that a car's shape can affect fuel consumption

To aim for a grade C

use and apply the equation $KE = \frac{1}{2} mv^2$

describe arguments for and against battery powered cars

interpret fuel consumption data including emissions

To aim for a grade A

manipulate the equation $KE = \frac{1}{2} mv^2$

explain how solar and bio-fuelled vehicles reduce *and* produce pollution so *may* lead to a reduction in CO_2 emissions

explain the factors that affect car fuel consumption figures

evaluate and compare fuel consumption and emissions data

---

calculate momentum using momentum = mass × velocity

recall that a sudden change in momentum causes a large force

recognise the risks and benefits of wearing seatbelts

recall safety features that are intended to prevent accidents

recall safety features that are intended to protect occupants

manipulate the equation momentum = mass × velocity

calculate force using force = change in momentum /time

explain how spreading the change in momentum over a longer time reduces the likelihood of injury

explain how airbags, crumple zones and seatbelts work

describe how test data is used to develop safety features for cars

manipulate the equation force = change in momentum/time and use Newton's second law of motion ($F = ma$) to explain it

explain how forces are reduced in a collision

evaluate the effectiveness of given safety features

describe how ABS brakes work

---

recognise the effect of frictional forces on movement and how they can be reduced

explain how falling objects reach a terminal speed

understand why falling objects do not experience drag when there is no atmosphere

explain in terms of the balance of forces how moving objects change speed

recognise that the acceleration due to gravity (g) is the same for any object at a given point on Earth's surface

explain in terms of the balance of forces why objects reach a terminal speed

understand that the acceleration due to gravity (g) is unaffected by atmospheric changes and how it varies over Earth's surface

---

recognise why objects have gravitational potential energy (GPE)

recognise everyday examples in which objects use GPE

use the equation for gravitational potential energy, $GPE = mgh$

interpret examples of energy transfer between GPE and KE such as a roller coaster

describe the effect of changing mass and speed on KE

understand what happens to the GPE of a body falling at terminal speed

manipulate the equation $GPE = mgh$

use and apply the relationship $mgh = \frac{1}{2} mv^2$

show that for an object falling to Earth this relationship becomes $h = v^2/2g$

P3 Exam-style questions

Foundation Tier

1 Darren completed the London Marathon, a distance of 42.195 km (26 miles 385 yards), in 6 hours.

AO2 **(a)** Calculate his average speed for the race. [3]

AO2 **(b)** Why is this an *average* speed? [1]

[Total: 4]

2 The table shows the shortest stopping distances for a car on a dry road, with good brakes and a good reaction time for the driver.

speed in m/s	thinking distance in m	braking distance in m	stopping distance in m
10	7	8	
20		32	
30	21		93

AO2 **(a)** Fill in the blanks in the table. [4]

AO1 **(b)** How would the data change if the road was wet? [1]

AO1 **(c)** How would the data change if the car's brakes were poor? [1]

AO1 **(d)** How would the data change if the driver was tired? [1]

AO3 **(e)** Use the information in the table to explain why we have speed limits on our roads. [2]

[Total: 9]

3 Olivia got in her car and started the engine. She accelerated to 10 m/s in 20 s.

AO2 **(a)** Calculate her acceleration. [3]

AO2 **(b)** Olivia's car has a mass of 800 kg. What forward force was needed to give this acceleration? [3]

AO2 **(c)** Charlie's car has a mass of 1200 kg. Would he need the same, a bigger or a smaller forward force to produce the same acceleration as Olivia? [1]

[Total: 7]

4 Beth is a gymnast. She jumps onto a beam. The beam is 1.5 m above the ground. Beth weighs 400 N.

AO2 **(a)** How much work does Beth do in jumping onto the beam? [3]

AO1 **(b)** What sort of energy has she gained when she stands on the beam? [1]

AO2 **(c)** What happens to this energy when she jumps off the beam? [1]

[Total: 5]

5 Amy runs up a flight of steps 6 m high in 7.5 s. Her mass is 50 kg.

AO2 **(a)** How much work does she do? [3]

AO2 **(b)** What power does she develop? [3]

[Total: 6]

AO3 **6** Carl has bought a battery powered electric car to help to protect the environment. Meera says Carl should have bought a solar-powered car.

Consider the two options and decide which is better giving reasons for your choice.

The quality of written communication ✎ will be assessed in your answer to this question. [6]

AO1 recall the science AO2 apply your knowledge AO3 evaluate and analyse the evidence

✳ Worked Example – Foundation Tier

(a) Modern cars have many safety features. Some, like electric windows, are intended to prevent accidents.

(i) Suggest how electric windows may prevent an accident.   [1]

You just have to push a button so you can keep your eyes on the road. It's quicker than wind down ones.

(ii) Give another example of a safety feature intended to prevent accidents.   [1]

airbag

(b) Other safety features are intended to protect a car's occupants in the event of an accident. Seat-belts must be worn by everyone travelling by car. They are made from a material that is slightly stretchy.

(i) Suggest how seat-belts protect a car's occupants if there is an accident.   [2]

They stop you going through the windscreen.

(ii) Explain why seat-belts have to be replaced after a crash.   [1]

In case they have been broken in the accident.

How to raise your grade!
Take note of these comments – they will help you to raise your grade.

Two correct points here but only one is required. It is sometimes sensible to give two options provided they are not contradictory as you have two chances of getting the mark.
1/1

It is a safety feature but does not **prevent** accidents. ABS brakes or traction control, for example, would have scored the mark.   **0/1**

The answer is true but the question is worth two marks so two correct points are expected. Seat-belts also stop you being thrown out of the car or bounced around inside it. Rear seatbelts protect front seat occupants from being hit from behind.   **1/2**

This answer is too vague to score the mark. Reading the whole question carefully gives a hint of what is expected. The seat-belt fabric is likely to have been overstretched in a crash so will be permanently damaged.   **0/1**

This student has scored 2 marks out of a possible 5. This is below the standard of Grade C. With more care the student could have achieved a Grade C.

P3 Exam-style questions

Higher Tier

1 Daniel rode his bike for one minute before stopping.

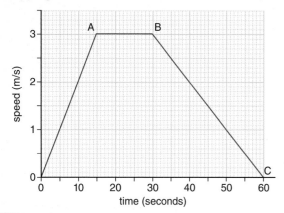

AO2 **(a)** Calculate Daniel's initial acceleration. [3]

AO2 **(b)** Calculate how far he travelled. [3]

AO2 **(c)** If Daniel had decelerated at the same rate as he accelerated, for how long would he have moved? [1]

AO2 **(d)** Would he have travelled the same distance? Explain you're answer. [2]

[Total: 9]

2 Mollie is a free-fall parachutist. She falls out of an aircraft.

AO1 **(a) (i)** What is her initial acceleration? [1]

AO2 **(ii)** Explain why her acceleration decreases as she falls. [2]

AO2 **(iii)** What can Mollie do to decrease her acceleration even more? [1]

AO1 **(b)** Eventually Mollie reaches terminal velocity. Explain, in terms of forces, why this happens. [2]

AO1 **(c)** Mollie is losing gravitational potential energy as she falls. What happens to this energy? [1]

[Total: 7]

3 Beckie is doing step-ups to keep fit. She does 80 step-ups in two minutes. The step is 25 cm high. Beckie has a mass of 60 kg.

AO2 **(a) (i)** What does Beckie weigh? [1]

AO2 **(ii)** How much work does Beckie do in two minutes? [3]

AO2 **(iii)** What is Beckie's personal power when doing step-ups? [2]

AO2 **(b)** Andy can do 80 step-ups in 2½ minutes but is more powerful than Beckie. Explain. [1]

[Total: 7]

AO2 **4 (a)** A car of mass 1200 kg is travelling at 20 m/s. Calculate its change of momentum when it stops. [3]

AO2 **(b)** Crumple zones are a safety feature on modern cars. Use ideas about momentum to explain why they reduce the forces acting in the event of a collision. [3]

[Total: 6]

AO3 **5** Ben is buying a new car. He has chosen the make and model but has not decided how to fuel it. Use the information in the table to help Ben to decide. Explain how you came to this conclusion and what other factors should be considered.

Toyota Auris		
fuel	combined fuel consumption in litres/100km	CO2 emissions in g/km
petrol	5.8	135
diesel	4.7	125
hybrid (petrol + electric)	3.8	89
hybrid (petrol + solar panels on roof)	3.9	89

The quality of written communication ✎ will be assessed in your answer to this question. [6]

AO1 recall the science   AO2 apply your knowledge   AO3 evaluate and analyse the evidence

✪ Worked Example – Higher Tier

Kate's car has a mass of 1200 kg. At a certain moment the car is travelling at 20 m/s and the forces acting on the car are as shown.

6500 N   1700 N

(a) (i) Calculate the acceleration of the car.   [4]

$$F = ma$$
$$6500 - 1700 = 1200 \times a$$
$$4800 = 1200 \times a$$
$$a = 4800 / 1200 = 4\ m/s^2$$

(ii) Calculate the kinetic energy of the car.   [3]

$$KE = \tfrac{1}{2}mv^2$$
$$KE = \tfrac{1}{2} \times 1200 \times 20^2 = 144\,000\,000\ J$$

(b) Kate comes to a steep hill. She maintains the same forward force from the car's engine but her acceleration decreases. Use ideas about forces to explain why.   [3]

There is a bigger force backwards so the resultant force is smaller. This means she can't accelerate as fast as F = ma.

(c) The speed of Kate's car drops to 10 m/s. How does its kinetic energy compare with the value calculated in (a) (ii)? Explain.   [2]

The speed is halved so the KE is halved.

A good answer. The correct equation has been selected and the *resultant* force has been calculated and substituted in $F = ma$. This has been processed correctly, with the correct unit.   4/4

Again the correct numbers have been substituted in the equation but the whole expression has been squared instead of just *v*. All the working has been shown clearly.   2/3

The resultant force must be smaller so the acceleration is less (from $F = ma$). The source of the extra force backwards has not been explained though. As the car goes up the hill there is a backwards force due to its weight. The steeper the hill the bigger the extra force backwards.   0/3

The speed, *v*, is halved but the kinetic energy is quartered as $KE = \tfrac{1}{2}mv^2$.   1/2

This student has scored 7 marks out of a possible 12. This is below the standard of Grade A. With more care the student could have achieved a Grade A.

P4 Radiation for life

Ideas you've met before

Electrostatics

Electrical insulators can be charged by rubbing (friction).

There are two kinds of electric charge – positive and negative.

Like charges repel, unlike charges attract.

Static electricity can make your hair stand on end.

- Why is the girl's hair standing on end?

Current electricity

Electric charges on conductors move, producing an electric current.

Resistors can be used to change the current flowing in a circuit.

Care must be taken when using mains electricity as it has a high voltage (230 V).

Fuses and circuit-breakers switch off the current if a fault occurs.

- What happens to the wire inside a fuse if the current gets too high?

Ultrasound

Ultrasound is sound with a pitch too high for humans to hear (above 20 0000 Hz).

Sound cannot travel in a vacuum.

Reflection of sound is called an echo.

Ultrasound is used in medicine for body scans.

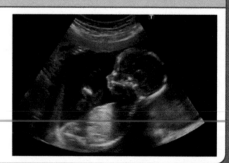

- Why can sound not travel in a vacuum?

Radioactivity

Alpha, beta and gamma radiation come from the nucleus of a radioactive atom.

Nuclear radiation can be measured using a Geiger counter.

Radioactive materials must be handled with care.

Background radiation is always around us.

Nuclear power stations produce electricity.

- Apart from nuclear power stations, give two more uses of radioactive materials.

In P4 you will find out about...

> how static electricity is due to the transfer of electrons

> how static electricity can be a nuisance and sometimes dangerous

> some uses of static electricity

> electric circuits and resistance

> the safe use of electricity in the home

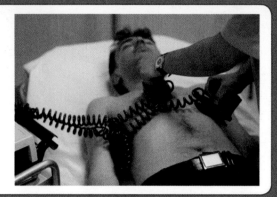

> the properties of longitudinal and transverse waves

> how ultrasound is used for diagnosis in medicine, for example, in body scans

> how ultrasound is used for treatment in medicine, for example, to break down kidney stones

> why ultrasound is preferred to X-rays for certain scans

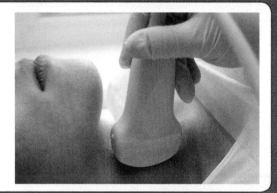

> the properties of alpha, beta and gamma radiation

> the ionising effect of nuclear radiation

> the meaning of half-life and how to calculate it

> X-rays and gamma rays, which are part of the electromagnetic spectrum and have a very short wavelength

> what happens to a nucleus when an alpha particle or a beta particle is emitted

> how alpha sources are used in some smoke alarms

> how radioactive tracers are used to locate blockages in industry and in medicine

> the use of gamma rays to treat cancer

> how measuring the activity of radioactive carbon can lead to an approximate age for certain objects

> how radioactivity is used to date rocks

> fission and fusion

Sparks

Lightning strikes

Lightning occurs when a cloud becomes charged and an electric current passes between the cloud and Earth.

Static electric charges can build up on insulating materials. These charges can cause electric shocks or explosions.

You will find out:

> how insulating materials can become charged

> how charged objects can attract other objects

> that there are two types of electric charge

> about electron movement

FIGURE 1: Cloud to ground lightning striking a mountain peak. What causes lightning?

Insulating materials

Metals are good electrical **conductors**. They allow electric **charges** to move through them.

Materials such as wood, glass and polythene are **insulators**. They do not allow electric charges to pass through them.

Charge can build up on an insulator. An insulator can be charged by **friction**.

If a polythene rod is charged by rubbing it with a duster it attracts small pieces of paper.

Other materials can be charged by friction:

> when a balloon is rubbed on a sweater it becomes charged and sticks to the wall

> when a plastic comb is used to comb hair both the comb and hair can become charged

> some types of dusting brushes are designed to become charged and attract dust.

There are two kinds of electric charge, positive and negative. When rubbed with a duster:

> acetate and Perspex become **positively** charged

> polythene becomes **negatively** charged.

FIGURE 2: A polythene rod becomes charged by friction and attracts discs of paper. Would a Perspex rod attract discs of paper if charged?

Did you know?

The floor tiles in an operating theatre are made of a conducting material because some anaesthetics are explosive.

Questions

1 Which one of the following is not an insulator?
candle wax   glass   iron   rubber   wood

2 Explain why static charges do not build up on a conductor.

Positive and negative charges

An **atom** is a small positively charged **nucleus** surrounded by negatively charged **electrons**. In a stable, neutral atom, there are the same numbers of positive and negative charges.

Electrostatic effects are caused by the transfer of negatively charged **electrons**.

The law of electric charge states that like charges **repel** and unlike charges **attract**.

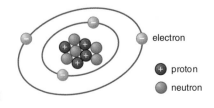

electron
proton
neutron

FIGURE 3: The charges in a neutral atom balance. This atom has four electrons. How many protons does it have?

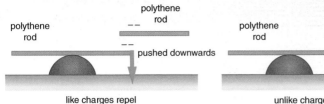

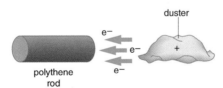

FIGURE 4: Like charges repel and unlike charges attract. What would happen if two Perspex rods were used?

Van de Graaff generator

A **Van de Graaff generator** collects electric charge on a metal dome.

> A person places their hands on the dome when it is uncharged.

> The dome is switched on and it and the person become charged.

> All the person's hairs gain the same charge.

> Like charges repel, so the hairs move away from each other.

Remember!
It is only electrons that move when an atom becomes charged.

FIGURE 5: When the girl puts her hands on the Van de Graaff generator she becomes charged. Why does her hair stand on end?

Questions

3 Copy and complete the table.

+	−	attract
+	+	
−	−	
−	+	

4 A neutral carbon atom has six protons in its nucleus. How many electrons does it have?

Moving electrons

When a polythene rod is rubbed with a duster electrons are transferred from the duster to the polythene, making the polythene rod negatively charged.

When an acetate rod is rubbed with a duster electrons are transferred from the acetate to the duster, leaving the acetate rod positively charged.

In general an object has:

> a negative charge due to an excess of electrons

> a positive charge due to a lack of electrons.

Atoms or molecules that have become charged are **ions**.

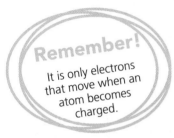

FIGURE 6: Why does the polythene rod become negatively charged when it is rubbed with a duster?

Question

5 A polythene rod is charged by rubbing it with a duster. What charge, if any, does the duster gain?

6 What happens when a conductor becomes positively charged?

7 What has happened to an atom if it becomes a positive ion?

Electrostatic shocks

If you touch a charged object you can get an electrostatic **shock**.

If you are wearing a sweater or top made from a synthetic material and pull it off very quickly the friction can charge it.

If you do this in the dark you may even see **sparks**.

A person gets an electrostatic shock if they become charged and then become **earthed**.

For example, a person can become charged if they walk on a nylon carpet or vinyl floor because:

> the floor is an insulator

> they become charged as they walk due to friction.

The person can become earthed by touching water pipes or even another person.

FIGURE 7: Explain what happens to Laura's hair when she takes her hat off.

Questions

8 Jake got an electric shock when he touched the car door after a journey. How did the car become charged?

9 Why can a person become earthed by touching water pipes?

FIGURE 8: Birds sitting on high-voltage power lines. Can you suggest why they do not get an electric shock?

Did you know?

A microscopic coating of oil prevents 'frictional charging'.

When static electricity is dangerous

Static electricity is dangerous in conditions where there are explosive materials.

When inflammable gases or vapours are present or there is a high concentration of oxygen, a spark from static electricity could ignite the gases or vapours and cause an explosion.

> When cleaning oil tankers their tanks are first filled with an inert gas such as nitrogen to avoid a spark that could cause an explosion.

> Mobile telephones must not be used on petrol station forecourts to prevent sparks that could cause an explosion.

If a person touches something at a high **voltage**, large amounts of electric charge may flow through their body to earth.

Current is the rate of flow of charge. The table on the next page shows that even small currents can be fatal.

FIGURE 9: How are explosions prevented in the tanks of oil tankers during cleaning?

dangers of static electricity   causes of electric shocks

Electric current in mA (contact time is 1 second)	Effect on the body
1	tingling sensation
10–20	'can't let go', muscles keep on contracting
100–300	ventricular fibrillation (heart attack), fatal in some cases

The voltage that produces a given current depends on the **resistance**. The resistance of the body varies.

If a person is barefoot and sweaty, resistance is low and the current is greater for a given voltage.

When static electricity is a nuisance

There are times when static electricity is a nuisance but not dangerous.

> Dust and dirt are attracted to insulators, such as a television screen.

> Clothes made from synthetic materials often 'cling' to each other and to the body.

FIGURE 10: Factory workers in China assembling television and stereo parts. Can you suggest why they wear a wristband to discharge static electricity?

Questions

10 Why are high oxygen levels dangerous in a situation where electric charges are allowed to build up?

11 Explain why clothes made from synthetic materials may 'cling'.

Safety measures

Electric shocks can be avoided in the following ways.

> If an object that is likely to become charged is connected to earth, any build-up of charge immediately flows down the earth wire.

> When refuelling an aircraft bonding cables are used to connect the fuel tanker to the aircraft. By interconnecting all exposed non-current carrying metal objects together, they should remain near the same voltage. This prevents dangerous static discharges in aircraft fuel tanks and hoses.

> In a factory where machinery is at risk of becoming charged, the operators stand on insulating rubber mats so that charge cannot flow through them to earth.

> Shoes with insulating soles are worn by workers if there is a risk of charge building up so that charge cannot flow through them to earth.

FIGURE 11: Why are the fuel tanker and aircraft connected by a bonding cable (orange cable) during refuelling?

Anti-static devices

> Sprays, liquids and dusters made from conducting materials carry away electric charge. This prevents a build-up of charge that could be dangerous or a nuisance.

> 'Dryer sheets' containing oil are used in tumble dryers to prevent static charging.

Questions

12 Explain why aircraft tyres are made from a type of rubber that conducts electricity.

13 Why is there a risk of static charge building up in a factory using machinery?

14 When a jar of coffee is filled in a factory it passes below a large funnel, which delivers the correct amount. But it was noticed that some of the coffee 'missed' the jar.

a Explain why.

b Suggest what could be done to stop it happening.

Uses of electrostatics

You will find out:
> about some uses of electrostatics
> how electrostatic dust precipitators work

Missing a beat...

Static electricity can be a nuisance but it can also be of potential benefit to us all.

When a person suffers a heart attack their heart stops beating. A defibrillator uses static electricity to restart their heart. It is a procedure that does not always work but it has saved many lives.

Portable and implanted defibrillators allow a more rapid response and help to save even more lives.

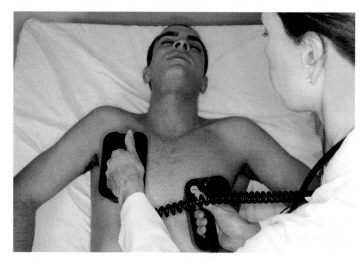

FIGURE 1: A defibrillator. What is it used for?

Uses of static electricity

There are many uses of static electricity.

> A **paint sprayer** charges paint droplets. They spread out. This gives an even coverage.

> **Crop sprayers** on farms work in a similar way. The fertiliser or insecticide particles are charged to give even coverage.

> A **photocopier** and laser printer use charged particles to produce an image.

> **Electrostatic dust precipitators** are used to remove the dust or soot in smoke. Charged plates are placed inside factory chimneys to attract soot particles.

> A **defibrillator**, which works by discharging a charge, delivers a controlled electric shock through a patient's chest to restart their heart.

FIGURE 2: How do photocopiers use static electricity?

Did you know?

Defibrillators use a high voltage so operators must take care to avoid an electric shock.

Questions

1 Why do paint sprayers charge paint droplets?

2 Give two other uses of static electricity.

3 What is a defibrillator used for?

Electrostatic dust precipitators

Gases emitted from the chimneys of many factories and power stations contain harmful particles that pollute the atmosphere. A dust precipitator is used to remove harmful smoke particles from a chimney.

> Metal plates and a grid (or rods) are put into the chimney.

> They are connected to a high-voltage supply so that the grid becomes negatively charged.

> The plates are earthed and gain the opposite charge to the grid.

> As the dust particles pass the grid (or rods) they become negatively charged.

> The dust particles are attracted to the plates because unlike charges attract.

> At intervals the plates are vibrated and the dust falls down to a collector.

FIGURE 3: What does a dust precipitator do?

Questions

4 Write down two possible consequences of not using electrostatic dust precipitators in an industrial area with lots of chimneys.

5 Explain what would happen if the wires in a dust precipitator were positively charged and the metal plates negatively charged.

Did you know?

Dust precipitators in large power stations can remove 40 tonnes of soot per hour.

More on dust precipitators

Some power stations burn coal to produce electricity. The smoke from their chimneys must be cleaned before it is released into the air.

> A dust precipitator inside a chimney contains wires in a grid that are given a large negative charge.

> As the dust particles pass close to the wires they gain electrons, becoming negatively charged.

> The negative charge on the grid induces the opposite charge on the earthed metal plate.

> Opposite charges attract so the dust particles are attracted to the plate.

> They stick to the plate and are removed at intervals.

> The dust is used to make building blocks.

Question

6 A similar method to a dust precipitator is used for fingerprinting. Paper is put near a charged wire. A black powder is used in place of smoke. Suggest how it works.

7 Explain, in terms of electron movement, how a dust precipitator works.

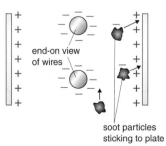

FIGURE 4: How does a dust precipitator work?

You will find out:

> about the application of electrostatics to paint spraying
> how a defibrillator can be used to restart a person's heart

Paint spraying

> The spray gun is charged.

> All the paint particles become charged with the same sign of charge as the gun.

> Like charges repel so the paint particles move away from each other giving a fine spray. The article being painted gets a fine, even coat of paint.

> The object to be painted is given the opposite charge to the paint.

> This means the paint is attracted to the object.

> The paint can get into awkward nooks and crannies, or 'shadows' of the object, giving an even coat with less waste of paint.

nozzle is charged up positively

object to be painted is negatively charged

FIGURE 5: Why is the bicycle frame given the opposite charge?

What does a defibrillator do?

Defibrillation is a procedure that restores a regular heart rhythm by delivering an electric shock through the chest wall to the heart. Once the heart resumes its pumping action, blood can once more flow throughout the body.

> Two **paddles** are charged from a high-voltage supply, typically 300 –1000 V.

> They are then placed firmly on the patient's chest to ensure a good electrical contact. A **conducting gel** is usually applied to the patient's chest first.

> Electric charge is passed through the patient to make their heart contract. Once the heart has been artificially restarted it is hoped that it will continue to contract normally.

FIGURE 6: Paddles on a defibrillator. What are they used for?

> Great care is taken to ensure that the operator does not receive an electric shock; for example, the paddles have insulating plastic handles.

A typical shock from a defibrillator supplies about 400 J of energy in a few milliseconds (1 millisecond = 0.001 seconds).

If a defibrillator is switched on for 5 milliseconds (0.005 seconds), the power can be calculated from:

$$\text{power} = \frac{\text{energy}}{\text{time}} = \frac{400}{0.005} = 80\ 000 \text{ W}$$

Portable defibrillators for home use

Rapid treatment is essential if defibrillation is to be successful. Up to 80 000 people die in the UK each year due to sudden cardiac arrest, with 80 per cent of such incidents happening at home. The chances of surviving decrease by 10 per cent for each minute that passes before treatment is received.

In recent years, small portable machines have been developed for use in the home.

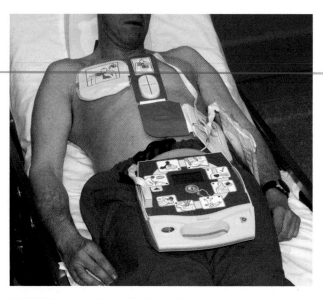

FIGURE 7: A portable defibrillator. How does this machine help to save lives?

🔍 paint sprayers   portable defibrillators

An implantable cardiac defibrillator

An implantable defibrillator is a small (6 cm × 6 cm × 1.5 cm) battery-powered device.

It is implanted under the skin in the upper part of the chest wall and is connected to the heart. It monitors heart rhythm and senses if there is about to be a severe disturbance in heart rhythm. If necessary it delivers an electrical impulse to the heart to stop abnormal rhythm and allow normal rhythm to resume. It can deliver up to four consecutive discharges of 25 J to 30 J.

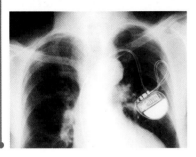

FIGURE 8: An implantable defibrillator. What is its main advantage?

Questions

8 Explain why paint spreads out into a fine spray from the nozzle of a spray gun.

9 How could the current through a patient be increased in defibrillation?

10 A defibrillator passes 96 mC (0.096 C) of charge through a patient in 2 ms (0.002 seconds). What is the average current (charge flow per second) through the patient?

11 Why is everyone warned to 'stand clear!' before a defibrillator is used?

12 Suggest why a larger current than those referred to in the table on page 233 can be used in a defibrillator.

13 What is one disadvantage of an implanted defibrillator?

More on paint spraying

If the object to be painted is not charged the paint still moves on to it, but:

> the object becomes charged from the paint, gaining more and more of the same charge

> further paint droplets are repelled away from the object.

This can be avoided by **earthing** the object to prevent a build up of charge.

The above method can be used when painting small objects but on the production line of a factory or in a vehicle-repair shop electrostatic paint sprayers are used.

> The object to be painted, such as a car, is given the opposite charge to the paint. So if the paint is positively charged, having lost electrons, the car is given a negative charge so gains electrons.

> Opposite charges attract so the paint is attracted to the car and gets into the 'shadows' giving good coverage.

The advantages of electrostatic paint sprayers are:

> reduction in wasted paint

> an object receives an even coat of paint

> paint covers awkward places ('shadows').

FIGURE 9: Paint spraying. What are the advantages of using electrostatic paint sprayers?

FIGURE 10: How do crop sprayers give a large even coat of pesticide?

Questions

14 When disinfectant is sprayed on an area, the droplets of disinfectant are sometimes given a positive charge by a spray gun. Suggest why this is done.

15 What are the advantages of painting a car using electrostatics?

16 Explain, in terms of electron movement, how an electrostatic paint sprayer works.

Safe electricals

Electricity is important in our lives

People use electrical appliances, from computers to washing machines, every day but they often forget that electricity can be dangerous if not used correctly.

A modern domestic wiring system that includes fuses and circuit breakers is designed to protect people and property if an electrical fault occurs.

In the UK, more than 28 000 fires a year are caused by electrical faults, leading to over 2500 deaths or serious injuries.

FIGURE 1: Incorrect wiring in electrical appliances can lead to fire. What other potentially fatal event might occur if you touched this appliance?

Electric circuits

A closed loop, with no gaps, is required for a circuit to work. Charge cannot flow across a gap in a circuit.

Electric current is the rate of flow of electric charge.

A resistor is added to a circuit to limit the amount of current in it. Adding a large resistor to a circuit reduces the rate of flow of charge so the current is smaller.

A **variable resistor**, or rheostat, is made of a length of wire. The length of wire in the circuit can be changed. It allows the **current** to have a range of values.

> Longer wires allow less current to pass.

> Thinner wires allow less current to pass.

The resistance of a wire is measured in **ohms (Ω)**.

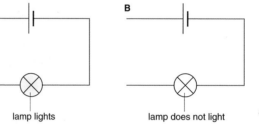

A — lamp lights
B — lamp does not light
C — resistor — adding a resistor to the circuit decreases the current and the lamp is dimmer

FIGURE 2: Electric circuits. Why doesn't the lamp light in **B**?

FIGURE 3: Carbon resistors. What does a resistor do when it is connected in an electric circuit?

FIGURE 4: A variable resistor. What would happen to the brightness of the lamp if this was added to the circuit?

Questions

1 In Figure 2, which of the circuits, A, B or C, has:

a the largest current **b** the smallest current?

2 How could the lamp in circuit **C** in Figure 2 be made brighter?

3 Will a variable resistor added to circuit **A** have its longest or shortest length of wire connected in the circuit when the lamp is brightest?

Electric current

In an electric circuit, charge is carried by negatively charged electrons. The electrons are free to move. They flow in the opposite direction to the conventional current.

If a circuit has a large resistance it is hard for the charge to move. In this case the rate of flow of charge (or current) is small.

A variable resistor (or rheostat) can be used to change the resistance and current in a circuit. The slider alters the length of wire in the circuit.

The longer the length of wire:

> the bigger its resistance
> the smaller the current
> the dimmer the lamp.

If thinner wire is used it has a bigger resistance so the current is smaller for the same length of wire.

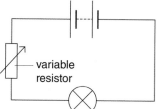

FIGURE 5: This circuit has a variable resistor. How does it act as a dimmer switch?

The current in a circuit is measured in **amperes (A)** using an **ammeter**. An ammeter is connected in **series**.

Voltage

Voltage (or **potential difference**) is measured in **volts (V)** using a **voltmeter**. A voltmeter is always connected in **parallel**.

> For a fixed resistor, as the voltage across it increases, the current increases.

> For a fixed voltage, as the resistance increases, the current decreases.

$$\text{resistance} = \frac{\text{voltage}}{\text{current}} = \frac{V}{I}$$

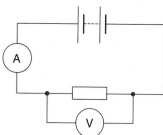

FIGURE 6: Measuring the voltage (potential difference) across a resistor using a voltmeter. Is a voltmeter connected in series or parallel?

Example

In the circuit in Figure 6 the voltmeter reads 5.0 V and the ammeter reads 0.2 A:

resistance of the lamp, $R = \frac{V}{I} = \frac{5.0}{0.2} = 25\ \Omega$

Questions

4 How would the brightness of the lamp in Figure 5 change if the variable resistor was replaced by one of the same length wire but thinner?

5 The voltage across a resistor is 12 V and the current is 0.6 A. What is the value of its resistance?

Linking resistance, voltage and current

Resistance, $R = \frac{\text{voltage}}{\text{current}} = \frac{V}{I}$

So, $V = IR$ and $I = \frac{V}{R}$

Current in a circuit

The current in a circuit must not be allowed to get too high.

> Electrons are 'pushed' around a circuit by the **battery**. They bump into the atoms in the resistor. This makes the atoms vibrate more so the resistor gets hotter.

> The increased atomic vibrations impede the electrons' motion more so the resistance increases.

> The **filament** in a lamp connected in a circuit becomes so hot it emits light.

Example

A piece of wire has a resistance of 3 Ω and melts if the current through it exceeds 5 A.

The maximum voltage across the wire, V, possible without melting the wire is:
$V = IR = 5 \times 3 = 15$ V.

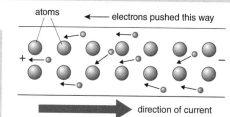

FIGURE 7: The movement of electrons in a wire carrying a current. Why does the wire get hot?

Question

6 What is the voltage across a 6 Ω resistor when the current in it is 1.5 A?

7 What current passes through a 4 Ω resistor when the voltage across it is 12 V?

Q charge flow in a circuit   resistance

Mains electricity in the home

The mains electrical supply to a house is usually from a power station. There are two wires connecting a house to the power station called **live** and **neutral**.

These two wires and an earth wire can be seen in a plug. If the appliance is working properly, there should be no current in the earth wire.

Colour coding for mains cables

The cable used to connect an appliance to the mains has three wires inside it covered in plastic. These are coloured brown, blue and green/yellow striped and they must be connected correctly:

> brown is connected to the live terminal (L)

> blue is connected to the neutral terminal (N)

> green/yellow stripe is connected to the earth terminal (E).

Earthing a conductor

If a conductor is earthed it is connected to the earth or ground. This means any charge in it flows safely down to the ground and a person touching it does not get an electric shock. It cannot become live.

Fuses and circuit-breakers

Fuses are always included in mains electricity circuits. They are also used in plugs, as seen in Figure 8. Fuses stop the electric current if a fault occurs. They prevent wires and equipment overheating and possibly catching fire. A wire fuse has to be replaced when the fault has been corrected.

Circuit-breakers are resettable fuses. They break the circuit if a fault occurs. Once the fault has been corrected they can be reset at the flick of a switch.

Double-insulated appliances do not need an earth wire.

FIGURE 8: Wiring inside a plug. Make sure you know which colour of wire connects to which terminal. What is F?

FIGURE 9: A circuit-breaker. Why is a circuit-breaker better than a wire fuse?

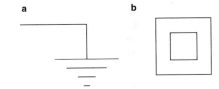

FIGURE 10: **a** Earth symbol and **b** double-insulation symbol. Why don't double-insulated appliances need an earth wire?

Questions

8 Which wire, live, neutral or earth, normally does not carry a current?

9 Why are most electric wires in the home covered in plastic?

Mains electricity

A cell or battery has two terminals, positive and negative. The electrons always move in the same direction, from negative to positive.

Mains electricity is different in two ways:

> it is **alternating** (the electrons vibrate to and fro)

> it is at a much higher voltage (230 V).

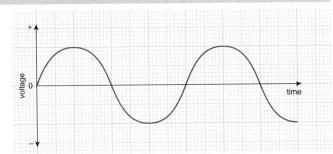

FIGURE 11: Graph to show variation in the voltage from the mains supply. How does the mains supply differ from that produced by a cell or battery?

mains electricity   double insulation

Live, neutral and earth wires

> The live wire carries a high voltage into and around the house.

> The neutral wire completes the circuit, providing a return path for the current to the local electricity sub-station, where it is earthed.

> The **earth wire** is a safety wire. It is connected to the metal case of an appliance to prevent it becoming charged if touched by a live wire. It provides a low-resistance path to the ground. There is normally no current in it.

How a fuse works

If an appliance develops a fault, such as a live wire touching the case:

> There is suddenly a large current in the live and earth wires.

> A fuse melts when the current becomes too large, stopping the flow of charge.

> As the fuse is in the live wire, the appliance is now disconnected from the mains supply. It is safe to touch.

> The flex cannot now overheat and cause a fire.

> Further damage to the appliance is prevented.

A fuse in a plug or in a main fuse box contains a cartridge that must be replaced after a fault has occurred and been corrected. Fuses of values 5 A and 13 A are commonly available for use in three-pin plugs.

Double insulation

A double-insulated appliance does not need an earth connection. It has a plastic case with no electrical connections to it so the case cannot become live.

Electrical power

The rate at which an appliance transfers energy is its **power** rating: power = voltage × current

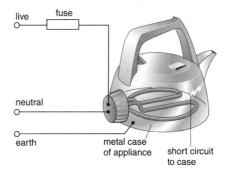

FIGURE 12: The arrangement of wires in a metal-cased appliance. Why is the earth wire connected to the metal case?

 Questions

10 Give two ways in which electricity from a battery and a mains supply differ.

11 A vacuum cleaner working from a 230 V mains supply normally has a current of 7 A.
a Should its plug be fitted with a 5 A or a 13 A fuse?
b Calculate its power.

Calculating the correct fuse rating

Example

power = voltage × current

current = $\dfrac{\text{power}}{\text{voltage}}$

mains voltage = 230 V
power of kettle = 2500 W

current = $\dfrac{2500}{230}$ = 10.9 A

Therefore a 13 A fuse is required.

> A current can become too large if live and neutral wires touch or if a live wire touches an earthed case.

> The type and thickness of fuse wire (the length is usually fixed) are chosen so that it melts when the current in it reaches the required value.

> A fuse is connected in the live wire so that if a fault occurs the dangerous live wire is disconnected

Protecting people

Earth wires and fuses protect us from electric shock.

As soon as the metal case of an appliance becomes 'live', a large current flows in the earth and live wires and the fuse 'blows'.

Circuit-breakers

Resettable fuses (circuit-breakers) have largely replaced the wire fuses in the main fuse box as a wire does not need to be replaced to restore power. A circuit-breaker can be reset at the flick of a switch.

 Questions

12 How do earth wires and fuses protect the user?

13 Why is it important to fit a fuse that has the correct value in a plug connected to an appliance?

14 Sara's 230 V iron has a power rating of 1900 W. Calculate the current when working normally. Should the plug contain a 5 A or 13 A fuse?

Ultrasound

You will find out:
> about longitudinal waves
> how ultrasound is a longitudinal wave
> about the frequency of ultrasound

Echo location

The limit of human hearing is about 20 kHz (20 000 Hz). Some animals can hear sounds of much higher frequencies.

Bats emit pulses of sound between 30 and 100 kHz and find their way by listening to the echoes. This high-frequency sound is called 'ultrasound'.

Ultrasound has many applications in industry and medicine to investigate objects internally without causing damage.

FIGURE 1: Bats use echo location to orient themselves. Why can't we hear the sound bats make?

Sound waves

All **sound** is produced by vibrating **particles** that form a **wave**.

> **Wavelength** (λ) is the distance occupied by one complete wave.

Sound and **ultrasound** are **longitudinal** waves. The **vibrations** of the particles are along the direction of the wave. This sets up a **pressure** wave with **compressions** and **rarefactions**.

These compressions and rarefactions make a person's eardrums vibrate and signals are then sent to the brain.

Longitudinal waves cannot travel through a **vacuum**. This is why sounds are not heard from space.

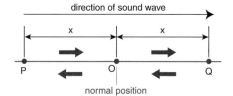

FIGURE 2: Particle movement in a sound wave. An individual particle vibrates between P and Q in the same direction as the wave moves.

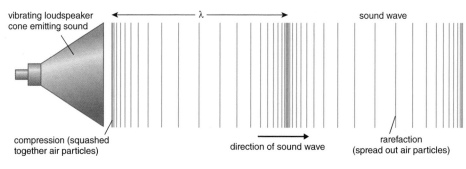

FIGURE 3: A sound wave showing rarefactions, compressions and wavelength. Does the wavelength get bigger or smaller if the compressions and rarefactions are further apart?

Questions

1 Compressions in a sound wave are 12 cm (0.12 m) apart. What is the wavelength of the sound?

2 Why wouldn't anyone hear you if you screamed in space?

What is ultrasound?

Frequency (f) is the number of complete waves in a second. Frequency is measured in **hertz (Hz)**. A frequency of 50 Hz means there are 50 complete waves in 1 second.

In the range of sounds that humans can hear, the higher the frequency:

> the higher the **pitch**

> the smaller the wavelength.

Wavelength is the distance between two successive points on the wave having the same displacement and moving in the same direction.

A sound wave can be represented as a graph so that it is easier to see the motion.

Ultrasound is sound of a higher frequency than humans can hear.

The upper **threshold** of human hearing is about 20 000 Hz. It becomes lower as a person gets older.

Ultrasound is a longitudinal wave. It travels as a pressure wave that is a series of compressions and rarefactions.

> At the centre of a compression the pressure is greater than when no wave is present (particles are squashed together).

> At the centre of a rarefaction the pressure is smaller than when no wave is present (particles are spread out). See Figure 3.

A pressure wave can only exist in a medium.

> The denser the medium the faster a wave travels. For example, sound travels faster in water than in air.

Material	Speed of sound in m/s
air	330
water	1500
steel	5000

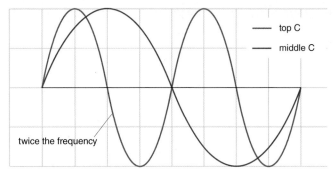

twice the frequency

FIGURE 4: A cathode ray oscilloscope (CRO) trace of two sound waves. Middle C has a frequency of 256 Hz. What is the frequency of top C?

Questions

3 Could a person hear the following sound frequencies?

a 50 000 Hz

b 5000 Hz.

4 If the frequency of a wave doubles, what happens to its wavelength?

Transverse and longitudinal waves

In a **transverse** wave the vibrations of the particles are at right angles to the direction of the wave.

In a longitudinal wave the vibrations of the particles are parallel to the wave direction.

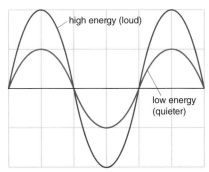

vibrations from side to side at

right angles to direction of wave

wave direction

FIGURE 5: A transverse wave. Give an example of a transverse wave.

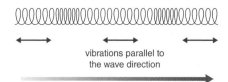

vibrations parallel to the wave direction

wave direction

FIGURE 6: A longitudinal wave. Give an example of a longitudinal wave.

Amplitude

Amplitude (A) is the maximum distance a particle moves from its normal position. (x in Figure 2.)

The louder a sound wave is or the more powerful the ultrasound wave:

> the more **energy** it carries

> the larger its amplitude.

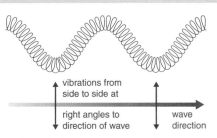

high energy (loud)

low energy (quieter)

FIGURE 7: Comparing a loud sound with a quieter sound. What happens to the height of the trace on the CRO during the quieter sound?

Questions

5 What would you notice if the amplitude of a light wave was increased?

6 What evidence do you have that light travels faster than sound?

You will find out:

> how ultrasound is used in medicine
> about the reasons for using ultrasound rather than X-rays

Using ultrasound in medicine for diagnosis

Ultrasound allows a doctor to 'see' inside a patient without surgery.

Uses of ultrasound for **diagnostic** purposes include:

> to check the condition of a foetus

> to investigate heart and liver problems

> to look for tumours in the body

> to measure the speed of blood flow in vessels when a blockage of a vein or artery is suspected. See Figure 12.

Did you know?

Ultrasound is used by dentists. The vibrations shake the dirt and plaque off teeth!

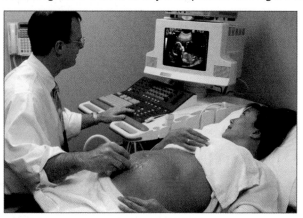

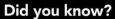

FIGURE 8: An ultrasound scan being carried out on a pregnant woman. The image of the uterus and foetus is shown on the computer screen. The technician checks for abnormalities in growth and development of the foetus. What else might the scan show?

Questions

7 Give two uses of ultrasound scanning.

8 Explain how ultrasound is used to clean jewellery.

Using ultrasound in medicine for therapy

An ultrasound body scan is used to establish the exact position of a problem in a patient, making surgery easier.

Using ultrasound for therapy

Kidney stones

A high-powered ultrasound beam is used to break down kidney and other stones inside the body. Even large stones can be broken down into fragments that are then excreted from the body in the normal way. Before this method was introduced stones required major surgery.

Cataract surgery

A surgeon uses ultrasound to break up the opaque lens of a patient suffering a cataract (loss of transparency in the lens of the eye). The defective lens is surgically removed and replaced by an artificial one.

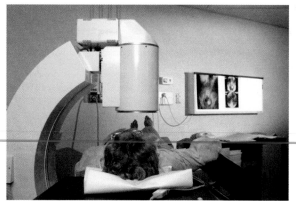

FIGURE 9: A patient undergoing ultrasound treatment to break down kidney stones. What is the main advantage of this approach?

Questions

9 Why is a *high-powered* ultrasound beam needed to break down kidney stones?

10 Why is ultrasound used in cataract surgery?

How a body scan works

A pulse of ultrasound is sent into a patient's body. At each boundary between different tissues or organs some ultrasound is **reflected** and the rest is **transmitted**. The returning **echoes** are recorded and used to build up an image of the internal structure.

Foetal ultrasound scanning is routine during pregnancy. A three-dimensional (3-D) scan uses computer technology and gives a more detailed image than a conventional two-dimensional (2-D) scan.

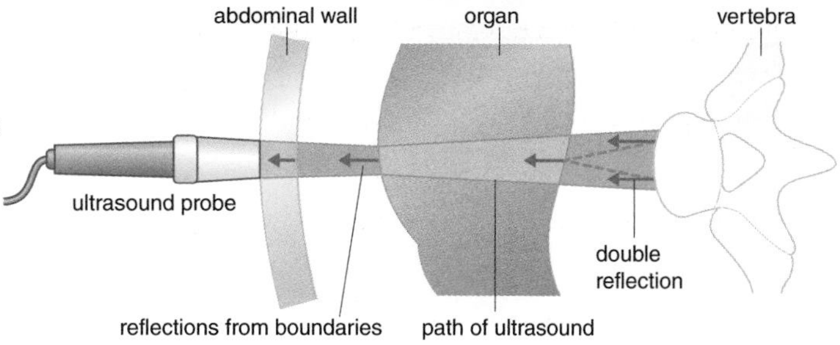

A **gel** is placed on a patient's body between the ultrasound probe and their skin. Without gel virtually all the ultrasound would be reflected at the skin and a good image of internal structure would not be obtained.

When ultrasound is reflected from different interfaces in the body, reflections from different layers return at different times from different depths. The depth of each structure is calculated using:

speed $= \dfrac{\text{distance}}{\text{time}}$

given the standard speed of ultrasound for different tissue types and remembering that the distance is there and back – twice the depth.

The proportion of ultrasound reflected at each interface depends on:

> the **densities** of the adjoining tissues

> the speed of sound in the adjoining tissues.

If the tissues are very different (for example, blood and bone) most of the ultrasound is reflected, leaving very little to penetrate further into the body.

Doppler ultrasound scanner

This probe detects movement. See Figures 8 and 12. It is held against an artery or a pregnant woman's abdomen. If ultrasound waves hit a moving object the frequency of the echoes alters. This allows speed of blood flow or foetal heart rate to be measured.

Advantages of ultrasound compared to X-rays

> Ultrasound differentiates between soft tissues better than **X-rays**. X-rays show bones very well.

> Ultrasound does not damage living cells. X-rays can kill cells or bring about changes in them that can lead to the development of cancers.

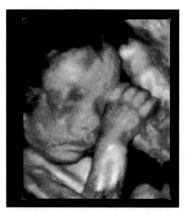

abdominal wall   organ   vertebra

ultrasound probe

double reflection

reflections from boundaries   path of ultrasound

FIGURE 10: How ultrasound is reflected during a body scan. What does the proportion of ultrasound reflected at a tissue boundary depend on?

FIGURE 11: Coloured image of a 3-D ultrasound scan of the face of a foetus at approximately 25 weeks gestation. What is the advantage of 3-D scanning compared to 2-D scanning?

FIGURE 12: Portable equipment to measure speed of blood flow in a vein or artery. What problem might this diagnose?

Questions

11 Suggest why a pregnant woman is scanned using ultrasound but is never X-rayed.

12 The speed of ultrasound in soft tissue is 1500 m/s. The time delay for an echo from ultrasound in soft tissue was 0.0002 seconds. At what depth was it reflected?

13 The difference in time between the return of ultrasound pulses from either side of the head of a foetus is 140 µs (140×10^{-6} seconds). The standard speed of ultrasound in the head is 1500 m/s. Calculate the size of the foetal head.

Preparing for assessment: Applying your knowledge

To achieve a good grade in science, you not only have to know and understand scientific ideas, but you need to be able to apply them to other situations and investigations. These tasks will support you in developing these skills.

✳ A first look...

Ben and Louise are sitting in their local hospital waiting room.

It is Louise who is to be examined, though she is not injured or unwell. She is pregnant and her doctor has calculated that this is her 16th week of pregnancy. It is routine for pregnant women to have an ultrasound scan at this stage, so she and Ben have taken time off work for the appointment.

They are both excited and a little anxious. This is their first child and they are not quite sure what the procedure is. Louise had three cups of herbal tea shortly before they came to the hospital because she had been told that the scan gives a clearer image if her bladder is full. She has changed to drinking herbal tea because since being pregnant she has gone off the taste of ordinary tea!

A few minutes later they are called into a room that has a bed and some equipment nearby. Mary, the operator, asks Louise to lie on the bed and then she dims the lights. She moves a computer screen into view, lifts Louise's blouse and squirts some clear gel on to Louise's belly. Mary then rests a probe on the gel on Louise's skin to make a good contact. It feels cold but Louise soon forgets this as an image comes into view on the screen.

Louise had imagined that the image of their baby would be rather fuzzy and difficult to understand, especially if you were not used to making sense of medical images. She thought it would have to be explained to her. In fact, she gasped in amazement as a tiny figure appeared. It was easy to see the shape of its body, backbone and even its heart beating. It was active as well, its little arms and legs moving around.

Mary turns the probe to get a cross-sectional view. She then uses a tracker ball to trace round the perimeter of the skull on the image and then zooms in on the spine to take a closer look.

It was hard to believe that this was all being done with sound waves. The probe resting on Louise's tummy was producing a stream of high-frequency sound waves, which passed through her body, including her womb and the baby's body. Where there was a difference in density, some of the waves were reflected and the probe picked up the reflections.

All too soon it was over. Mary had the information she needed, the baby was doing well and there were other people waiting. Ben and Louise could have spent much longer gazing at the screen but Mary gave them a printout of one of the images to take away with them. Thankfully the ultrasound scan had not shown anything wrong with the baby and now she or he seemed even more real.

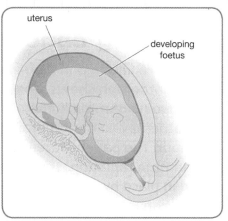

The foetus in the womb.

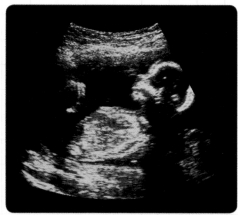

Image of a 16-week-old foetus during an ultrasound scan.

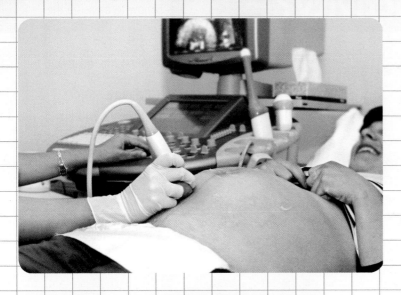

✺ Task 1

If ultrasound uses sound waves why could Ben, Louise and Mary not hear the waves?

✺ Task 2

How does an ultrasound machine detect if a wave has been reflected from near the surface or deeper down in the body? Why does it need this information?

✺ Task 3

Hospitals make extensive use of X-rays. Find out why these are not used to get images of unborn babies.

✺ Task 4

What does 'difference in density' mean? Look at the diagram of a baby in the womb. Where are the changes in density going to be?

✺ Task 5

Mary said that the measurement of the circumference of the skull was to check that Louise was 16 weeks pregnant. How did that help?

✺ Task 6

Ultrasound scans are not done to give parents pictures of babies – that is an added bonus. Mary is a highly trained operator – what features do you think she is looking for in particular?

✺ Maximise your grade

Answer includes showing that you can...
State the reason why ultrasound cannot be heard by humans.
State an effect on the body of exposure to X-rays.
Explain why ultrasound is used instead of X-rays to monitor the progress of a baby in the womb.
State the relationship between skull circumference and number of weeks since fertilisation.
Explain what is meant by the term density.
Describe differences in density of body parts and explain where such changes may occur.
Explain how the reflection of ultrasound allows distance to be calculated and hence produce a three-dimensional image.
Explain how information from the ultrasound scan can be used by a trained operator to identify abnormalities in the development of a baby in the womb.
As above, but with particular clarity and detail.

F

C

A

What is radioactivity?

You will find out:
> how to measure the activity of radioactive materials
> about the radiation given out by radioisotopes
> about the half-life of radioisotopes

Marie Curie's achievements

Marie Curie (1867–1934) was a famous Polish scientist who researched radioactivity.

In 1903 she won the Nobel Prize for physics with her husband Pierre for their work on the three types of radiation emitted – alpha, beta and gamma.

In 1911 she won the Nobel Prize again, for chemistry, for the discovery of two new elements, radium and polonium.

She died of leukaemia caused by radiation in 1934. Pierre was killed when he was run over by a cart in Paris in 1906.

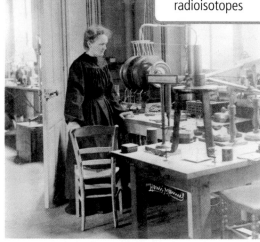

FIGURE 1: Marie Curie measuring the activity of a radioactive material between 1897 and 1899. Why did she win the Nobel Prize twice?

Measuring radiation

A **Geiger–Müller tube** and **ratemeter** (together called a **Geiger counter**) are used to detect the rate of **decay** of a **radioactive** substance. Each 'click' sound or number on the display screen represents the decay of one **nucleus**. **Radiation** is emitted when a nucleus decays.

Activity is measured by counting the average number of nuclei that decay every second. This is also called the **count rate**.

Activity is measured in counts per second or **becquerels (Bq)**.

$$\text{activity} = \frac{\text{number of nuclei that decay}}{\text{time taken in seconds}}$$

The activity of a radioactive substance decreases with time. This is shown by the count rate falling.

Nuclear radiation ionises materials. This damages living cells. Exposing living organisms to nuclear radiation should be avoided.

FIGURE 2: A Geiger counter is a Geiger–Müller tube and ratemeter. What does a Geiger counter measure?

 Questions

1 The activity of a radioactive source is 200 Bq. How many counts would be recorded in 10 seconds?

2 Sam records a count of 4000 in 25 seconds from a radioactive source. What is its activity?

Radioactive decay

Radioactive substances decay naturally and give out **alpha** (α), **beta** (β) and **gamma** (γ) radiation. The Curies and Becquerel identified and named the three different types of radiation.

Radioactive decay is a **random** process. It is not possible to predict when a nucleus will decay. Decay is independent of physical conditions such as temperature.

However, there are so many **atoms** in even the smallest amount of radioisotope that an average count rate can be calculated. Compare this with throwing dice. In a large number of throws, on one-sixth of all the throws each number will turn up.

Look at Figure 3. It shows how the activity of a radioactive material changes with time. The activity falls rapidly at first but changes less and less as time goes on. The graph line gets closer and closer to the time axis but never reaches it.

The **half-life** of a radioisotope is the average time it takes for half the nuclei present to decay. This is the time for the activity to halve.

Radioisotopes

A **radioisotope** has nuclei that are **unstable** because its nuclear particles are not held together strongly enough.

The largest stable nucleus is an isotope of lead. This is why lead is often found near radioactive rocks.

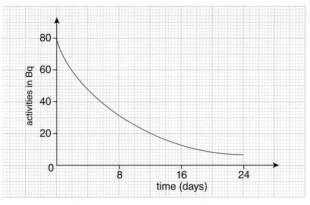

FIGURE 3: Radioactive decay. Estimate the half-life of the radioactive material.

FIGURE 4: Pitchblende, a radioisotope that is a form of uranium oxide and which emits alpha particles. What can you say about its nuclei?

Questions

3 What can you say about all the isotopes of the elements above lead in the periodic table?

4 Sita records the count rate from a radioactive source. She takes four readings. They are 138 Bq, 149 Bq, 133 Bq and 142 Bq. Why are the readings different?

More about half-life

The rate of radioactive decay:

> is different for different radioisotopes

> depends on the number of nuclei of the radioisotope present; the more nuclei present the greater the rate of decay.

Example

In an experiment to find the half-life of a radioisotope, the following results are obtained and a graph of activity against time is drawn.	
Time in minutes	**Activity in Bq**
0	100
1	48
2	26
3	12

The initial activity is 100 Bq. The average time for the activity to halve from 100 to 50 Bq, 50 to 25 Bq and so on is about 1 minute. What is its half-life?

The half-lives of different radioactive isotopes vary from a fraction of a second to millions of years.

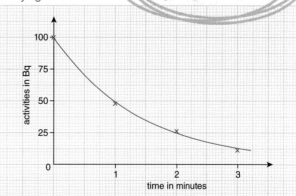

FIGURE 5: Graph of activity against time

Remember! The half-life of a radioisotope cannot be changed.

Question

5 The activity of a radioactive sample took 4 hours to decrease from 100 Bq to 25 Bq. What is its half-life?

6 A radioactive substance has a half-life of 3 hours. What fraction of it remains after 9 hours?

You will find out:
> about alpha particles and beta particles
> about alpha particle and beta particle decay
> how to write and balance nuclear equations

Radioactive decay

Nuclear radiation **ionises** materials.

When nuclear radiation strikes atoms of the material it is passing through, it knocks negatively charged **electrons** out of the atoms. The atoms have lost electrons so are positively charged; they are now positive ions.

Other particles may gain these electrons so are negatively charged; they are now negative ions.

The presence of ions can cause problems.

What is an alpha particle?

Experiments carried out by Marie Curie and other scientists around 1900 showed that an alpha particle:

> is positively charged

> has a relatively large mass

> has **helium** gas around it.

An alpha particle is a helium nucleus.

It consists of two **protons** and two **neutrons**.

FIGURE 6: Coloured image of alpha particles emitted from radium. What is an alpha particle?

What is a beta particle?

Experiments showed that a beta particle:

> is negatively charged

> has a very small mass

> travels very fast (at approximately one-tenth the speed of light).

A beta particle is a fast-moving electron.

When an alpha particle or a beta particle is emitted from the nucleus of an atom the remaining nucleus is a different **element**.

Gamma radiation is an **electromagnetic wave**. When emitted it removes excess energy from a nucleus. The emission of gamma radiation does not change the composition of the nucleus; it remains the same element.

Did you know?

Radium is the most powerful radioactive substance known.

It emits one million times more radiation than uranium.

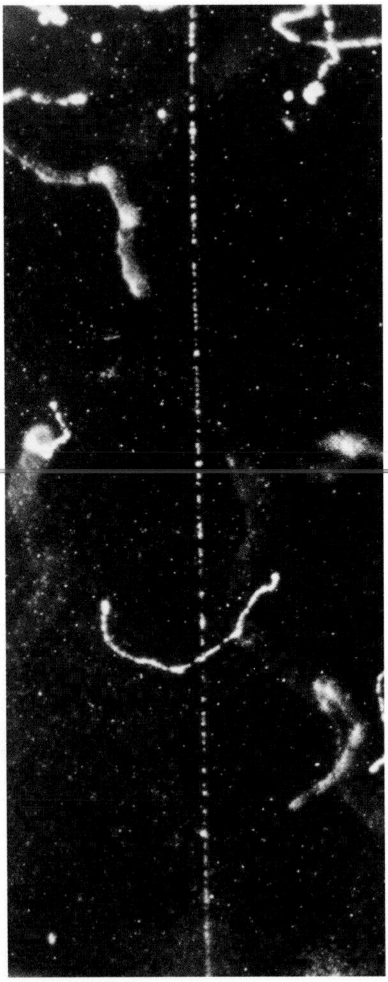

Remember!

Alpha, beta and gamma radiation aren't radioactive. It is the emitting **sources** that are radioactive.

FIGURE 7: The track of a beta particle moving from top to bottom in a cloud chamber used to detect beta particles. What is a beta particle?

Questions

7 If an alpha particle is emitted from a nucleus what happens to the mass of the nucleus?

8 What is the difference between an alpha particle and a helium atom?

9 Why doesn't the emission of gamma radiation change the composition of a nucleus?

Nuclear equations

Alpha particles are very good ionisers. They are the largest particles emitted in radioactive decay. This means they are more likely to strike atoms of the material they are passing through, ionising them.

Describing a nucleus

A **nucleon** is a particle in a nucleus. Protons and neutrons are nucleons.

The nucleus of an atom can be represented as:

$$^{A}_{Z}X$$ where
A = mass number (or nucleon number)
Z = atomic number (or proton number)
X = chemical symbol for the element

FIGURE 8: Alpha particle tracks emitted from polonium-212 in a cloud chamber. Radioactive materials emit alpha particles at one or more specific energies. Most of the particles here have the same energy. Suggest how you can tell this.

Z is the number of protons in the nucleus, so the number of neutrons is (A − Z).

For example, the carbon **isotope** $^{14}_{6}C$ has six protons and eight neutrons in its nucleus.

Alpha decay

An alpha particle or helium nucleus contains four nucleons: two protons and two neutrons.

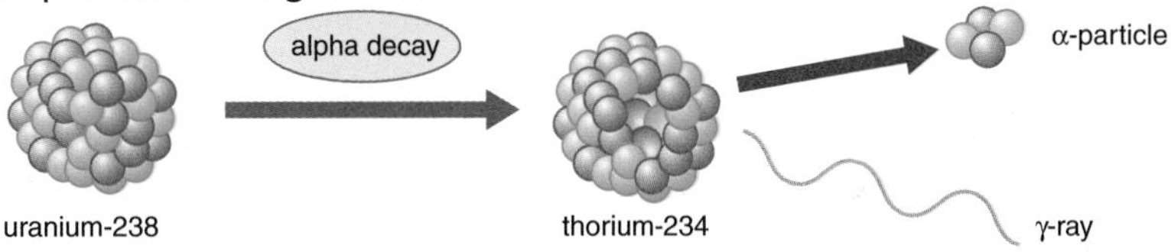

When an alpha particle is emitted from a nucleus:

> the nucleus has two fewer protons
> the nucleus has two fewer neutrons
> the mass number, or nucleon number, decreases by four
> the atomic number, or proton number, decreases by two
> a new element is formed that is two places lower in the **periodic table** than the original radioisotope.

This can be shown as a **nuclear equation**:

$$^{238}_{92}U \longrightarrow \ ^{234}_{90}Th + \ ^{4}_{2}He$$

alpha particle

> the **mass number** is conserved (238 = 234 + 4)
> the **atomic number** is conserved (92 = 90 + 2).

uranium-238 → thorium-234 + α-particle + γ-ray

FIGURE 9: Alpha decay.

Beta decay

When a beta particle is emitted from a nucleus:

> the nucleus has one more proton (p)
> the nucleus has one less neutron (n)
> the mass number, or nucleon number, is unchanged
> the atomic number, or proton number, increases by one
> a new element is formed that is one place higher in the periodic table than the original radioisotope.

A beta particle is an electron but there are no electrons in the nucleus.

In beta decay a neutron changes into a proton and an electron.

Charge is **conserved**: $n^0 \rightarrow p^+ + e^-$

Beta decay can be shown as a nuclear equation:

$$^{14}_{6}C \longrightarrow \ ^{14}_{7}N + \ ^{0}_{-1}e$$

beta particle

carbon-14 → nitrogen-14 + β-particle

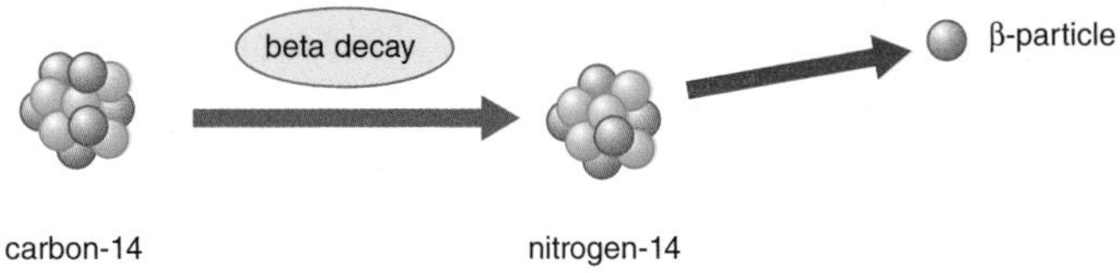

FIGURE 10: Beta decay. How is the nitrogen nucleus different from the carbon nucleus?

Questions

10 Which are better ionisers, beta particles or gamma radiation? Explain your choice.

11 Complete the nuclear equation:
$^{226}Ra \rightarrow \ _{86}Rn + \ ^{4}_{2}He$

12 How many protons and neutrons are in the nucleus of an atom of $^{235}_{92}U$?

13 What happens to the nucleus of an atom during beta decay?

Uses of radioisotopes

You will find out:
> about background radiation
> how background radiation is caused
> how tracers are used

Radon risks

Granite rocks contain small amounts of radioactive uranium. When uranium decays it emits radon gas that is also radioactive.

In granite areas, such as Devon and Cornwall, where houses were traditionally built from granite, there is concern about the health risks to the inhabitants who may be harmed by breathing in radon gas.

FIGURE 1: How might radon gas get inside some of these houses?

 ## Background radiation and tracers

Background radiation is **ionising** radiation that is always present in the environment. The level of background radiation varies from place to place and from day to day. But it is low level and does not cause harm. Background radiation mainly comes from rocks, especially **granite**, and **cosmic rays**.

Tracers

Radioisotopes are used as **tracers** in industry, research and medicine.

Tracers are used to:

> detect leaks or blockages in underground pipes
> find the route of underground pipes
> track dispersal of waste
> monitor the uptake of fertilisers in plants
> check for a blockage in a patient's blood vessel.

Smoke alarm

One type of **smoke detector** uses a source of **alpha** particles to detect smoke. It is sensitive to low levels of smoke.

FIGURE 2: Burying an oil pipeline underground. What is used to detect leaks in an underground pipe?

FIGURE 3: A smoke alarm is used to detect smoke. What type of radiation is emitted by the radioactive source?

Questions

1 Suggest how the uptake of a fertiliser in a plant is monitored.

2 Why should all houses have smoke alarms fitted?

3 What is background radiation?

Using radioisotopes

How are tracers used in industry?

To locate a leak or blockage in an underground gas pipe using a tracer:

> a very small amount of a suitable radioisotope is put into the pipe

> a **detector** is passed along the ground above the path of the pipe to track the progress of the radioisotope

> an increase in **activity** is detected in the region of the leak or blockage and little or no activity is detected after this point.

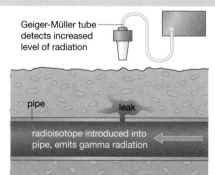

Geiger–Müller tube detects increased level of radiation

pipe

leak

radioisotope introduced into pipe, emits gamma radiation

FIGURE 4: Using a radioisotope to detect a leak in a pipe. Why does the reading on the Geiger–Müller tube fall rapidly after the leak has been passed over?

Q background radiation in the UK   uses of radioisotopes for GCSE

How a smoke alarm works

Many smoke detectors contain a radioisotope such as americium-241 that emits alpha particles.

Without smoke:

> The alpha particles hit air particles and ionise them.

> The positive **ions** and negative **electrons** move towards the negative and positive plates, respectively.

> This creates a tiny **current** that is detected by electronic circuitry in the smoke alarm.

With smoke:

> Some alpha particles hit the smoke particles so there is less ionisation of air particles.

> The smoke detector senses the drop in current and sets off an alarm.

Americium-241 has a long **half-life** of about 430 years. This is important as:

> the source does not need to be replaced frequently

> a decrease in ionisation current is due to the presence of smoke and not the decay of the source reducing the number of ions present.

Although most background radiation is naturally occurring, some comes from man-made sources and waste products from, for example, hospitals and industry.

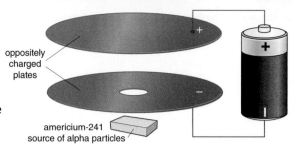

FIGURE 5: The radioisotope americium-241 emits alpha particles inside a smoke alarm. How does the detector work?

oppositely charged plates

americium-241 source of alpha particles

Questions

4 Give two causes of background radiation.

5 There is a blockage in an underground gas pipe. Describe how the location of the blockage can be found.

6 Why is a beta source not suitable for use in a smoke detector?

More on background radiation and tracers

Most background radiation is from natural sources such as rocks and cosmic rays.

Human activity contributes less than 1 per cent to the level of background radiation. Examples are:

> waste products from hospitals

> man-made radioisotopes obtained by firing particles such as neutrons at stable nuclei, making them unstable and causing them to decay.

> waste products from nuclear power stations and other industries

Industrial tracers

> A tracer is a radioactive material that can be introduced into an underground pipe to track its path or to locate a leak or blockage without digging up large areas.

> Alpha and **beta** radiation would not pass through the pipe and soil to be detected above ground.

> The tracer *must* be a radioisotope that emits **gamma** radiation.

> A detector such as a **Geiger counter** can then be moved over the surface to trace the path and intensity of the gamma radiation.

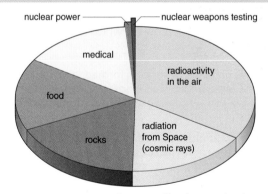

nuclear power — nuclear weapons testing

medical

radioactivity in the air

food

rocks

radiation from Space (cosmic rays)

FIGURE 6: Sources of background radiation. Which is the largest source?

Question

7 Suggest why background radiation varies **a** in different regions and **b** at different times.

8 Why must a gamma source be used to detect a leak in an underground pipe?

9 It is thought that the mouth of a river is silting up from one of two sites upstream. How can a tracer be used to find out which site is the cause?

You will find out:
> how radioactivity is used to date rocks
> how radioactive carbon is used to date old materials

Dating rocks

Some rock types such as granite contain traces of **uranium**. All uranium **isotopes** are radioactive. These uranium isotopes go through a series of decays, eventually forming a **stable** isotope of **lead**.

By comparing the amounts of uranium and lead present in a rock sample, its approximate age can be found.

FIGURE 7: Granite rocks. How can the approximate age of these rocks be found?

 Questions

10 Why is granite radioactive?

11 How can the approximate age of a rock sample be found?

Dating rocks and radiocarbon dating

Use of uranium-238

Uranium-238 decays, with a very long half-life of 4500 million years, to form thorium, which is also unstable.

$$^{238}_{92}U \rightarrow \, ^{234}_{90}Th(+\, ^{4}_{2}He) \rightarrow \, ^{234}_{91}Pa(+\, ^{0}_{-1}e) \rightarrow \, ... \rightarrow \, ^{206}_{82}Pb$$

A series of unstable isotopes is formed, all with relatively short half-lives, until a stable isotope, lead-206, is formed. (Lead is the element with the highest **atomic number** that has stable isotopes.)

The **ratio** of uranium to lead in a sample of rock indicates the age of the rock. The proportion of lead increases as time increases. If there are equal quantities of $^{238}_{92}U$ and $^{206}_{82}Pb$, the rock is 4500 million years (one half-life) old.

Use of uranium-235

Uranium-235 has a half-life of approximately 100 million years – shorter than that of uranium-238 – and it is used to date younger rocks.

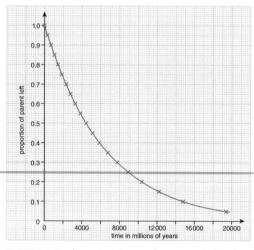

FIGURE 8: Radioactive decay of uranium-238. What stable isotope does it form?

FIGURE 9: Uranium-235 was used to date this rock sample brought back from the Moon by an Apollo mission. It was formed from solidified lava. The Moon rocks, and hence the Moon, were found to be about 4400 million years old. Why was uranium-235 used rather than uranium-238?

Q dating rocks   radiocarbon dating

Radiocarbon dating

Carbon-14 is a radioactive isotope of carbon. Carbon is present in all living things. By measuring the proportion of carbon-14 present in an archaeological find, its approximate age can be found.

FIGURE 10: The Turin shroud was originally thought to have been worn by Jesus Christ. However, recent radiocarbon dating has suggested the shroud is only about 500 years old. Why do some scientists think it is a fake?

Questions

12 Which of the following could *not* be dated using carbon-14?
wool jumper   wooden axe   iron nail
nylon shirt   cotton sheet

13 Why is lead always found with radioactive rocks?

14 The ratio of $^{238}_{92}$U to $^{206}_{82}$Pb in a rock sample is 1 : 4.

a How old is the rock?

b Why does this suggest the rock did not originate on Earth?

More on radiocarbon dating

In 1947 a shepherd discovered the Dead Sea Scrolls in caves at Qumran in Jordan. **Radiocarbon dating** was used to estimate their age. They were found to be about 2000 years old and are likely to be genuine.

> Cosmic rays enter Earth's atmosphere and collide with atoms that release **neutrons** from their **nuclei**. The energetic neutrons collide with nitrogen atoms forming **carbon-14**. The amount of carbon-14 in the atmosphere has not changed for thousands of years.

> Only a very small fraction of the carbon present in living things is carbon-14 (about 1 in 1012 atoms). (It is mainly stable carbon-12.)

> Plants absorb carbon dioxide in **photosynthesis**. Animals and humans eat plants and take in carbon-14.

> When a living thing such as a tree or an animal dies gaseous exchange with the air stops; no more carbon-14 is produced.

> The carbon-14 present decays to nitrogen with a half-life of about 5700 years. $^{14}_{6}C \rightarrow {}^{14}_{7}N (+{}^{0}_{-1}e)$ The activity of the object gradually decreases.

> By looking at the ratio of carbon-12 to carbon-14 in a sample from the dead organism and comparing it to the ratio in a living organism, the age of a dead organism can be estimated.

> Carbon-14 decays very slowly so the method is not suitable for dating organisms that are a few hundred years old.

FIGURE 11: A section of The Scroll of Isaiah, one of the Dead Sea Scrolls. Why are they likely to be genuine?

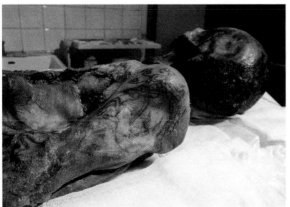

FIGURE 12: The preserved remains of 'Ice Maiden', a young woman found in 1993 in ice in central Asia. The body has been dated to approximately 2500 years old. She is so well preserved that tattoos can still be seen on her skin. Suggest why she is well preserved.

Questions

15 Why is carbon-14 dating not suitable for dating old rocks?

16 Explain how carbon-14 dating could be used to estimate the age of the remains of an animal believed to be a mammoth.

🔍 preserved remains in ice   Turin shroud   Dead Sea scrolls

Treatment

You will find out:
> how nuclear radiation is used in medicine
> about beta and gamma radiation
> about X-rays

Treating cancers

Nuclear radiation and high energy X-rays are used to treat cancers. The photograph in Figure 1 shows a patient undergoing gamma knife radiotherapy to treat a brain tumour. The patient receives a dose of gamma radiation. Their head is held in place by a metal frame. Radiation is targeted at a specific area of the brain and has a minimal effect on surrounding areas. It is non-invasive and provides access to areas of the brain that cannot be reached by other techniques.

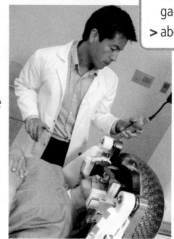

FIGURE 1: Why is it vital that patients do not move their heads during treatment?

Using radiation

Comparing X-rays and gamma rays

> **X-rays** and **gamma rays** are both **electromagnetic waves**.

> They have similar **wavelengths**.

> Both are **ionising radiation**.

> Although exactly the same they are produced in different ways.

Both X-rays and gamma rays are used in medicine for:

> **diagnosis** (finding out what is wrong)

> **therapy** (treatment).

They are both very penetrating and can pass into the body to treat internal organs.

Nuclear radiation and X-rays can damage living cells so great care must be taken to limit exposure to such radiation when it is used for diagnosis or therapy.

Producing radioisotopes for medical use

The radioisotopes used in medicine need to have special properties. For example, they must:

> give out gamma or beta radiation because these can pass through skin

> have a suitable half-life

> not be **toxic** to humans.

They are made artificially to suit their use by placing materials into a nuclear reactor.

Such man-made radioisotopes can be produced with different properties that make them ideal for different uses such as in hospitals to diagnose and treat patients.

Questions

1 What is the difference between diagnosis and therapy?

2 What do X-rays and gamma rays have in common?

3 How are radioisotopes for medical use obtained?

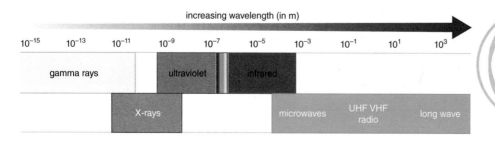

increasing wavelength (in m)

| 10^{-15} | 10^{-13} | 10^{-11} | 10^{-9} | 10^{-7} | 10^{-5} | 10^{-3} | 10^{-1} | 10^{1} | 10^{3} |

gamma rays · ultraviolet · infrared · X-rays · microwaves · UHF VHF radio · long wave

Remember! X-rays are NOT emitted from the nucleus of an atom.

FIGURE 2: The electromagnetic spectrum. What do you notice about the wavelengths of gamma rays and X-rays?

What sort of radiation is used in medicine?

Radiation emitted from the nucleus of an unstable atom can be alpha (α), beta (β) and/or gamma (γ).

Radiation	Alpha	Beta	Gamma
Ionising power	very strong	medium	weak
Range in air	about 5 cm	about 1 m	large, its intensity reduces with distance
What stops it?	paper	aluminium	greatly reduced by thick lead

> Alpha radiation is absorbed by the skin so is of no use for diagnosis or in therapy.

> Beta radiation passes through skin but not bone so its medical applications are limited.

> Gamma radiation is a very penetrating nuclear radiation. When nuclear radiation passes through a material the material absorbs some ionising radiation. Ionising radiation damages living cells, increasing the risk of **cancer**.

Nuclear radiation can be used to kill cells and living organisms. Cancer cells can be destroyed by exposing the affected area of the body to large amounts of radiation. This is called **radiotherapy**.

Cobalt-60 is a gamma-emitting radioactive material that has been widely used to treat cancers.

Producing an X-ray image

When X-rays pass through the body the tissues **absorb** some of this ionising radiation. The amount absorbed depends on the thickness and the **density** of the absorbing material.

The soft tissues in the body all have similar densities. They absorb some radiation but bone, having a much greater density, absorbs virtually all of it so hardly any radiation reaches the film. This is why X-rays are good at showing bones but do not distinguish clearly between tissues such as liver and kidney.

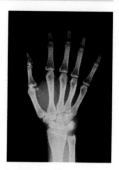

FIGURE 3: X-ray image. Why are the bones clearer than the soft tissue?

Artificial radioactivity

Materials can become radioactive when their nuclei absorb extra neutrons in a nuclear reactor. Neutrons are uncharged so they are easily captured by many nuclei. The nuclei become unstable and emit nuclear radiation.

Questions

4 Alpha radiation is of no use in medicine. Why?

5 Why must exposure to nuclear radiation be avoided?

6 Which type of nuclear radiation is used to treat a cancer deep inside the body?

Gamma radiation and X-rays

How X-rays are produced

> A hot **cathode** (negatively charged) emits electrons that are attracted by a highly positive target.

> When the fast-moving electrons hit the target some of their kinetic energy is used to emit X-rays but most of it is converted to heat.

Advantages of using X-rays instead of gamma rays

> X-rays are produced only when needed.

> An X-ray machine allows the rate of production of the X-rays to be controlled.

> The energy of X-rays can be changed. For example, the higher the tube voltage the higher the energy of the electrons, and so the higher the energy of the X-rays.

> Some X-rays can have a much higher energy than some gamma rays.

You cannot change the rate of production or energy of the gamma radiation emitted from a particular radioactive source.

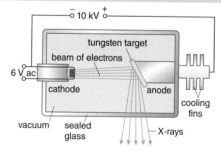

FIGURE 4: An X-ray tube used to produce X-rays. Why are cooling fins needed?

Question

7 Suggest why there is a vacuum in an X-ray tube.

8 Give two advantages of using X-rays rather than gamma rays as a source of radiation.

Using gamma radiation

Gamma radiation is used to treat cancer. It damages and destroys cancerous cells. Large doses of radiation are sometimes used in place of surgery. However, it is more common to use radiation after surgery. This helps to make sure all cancerous cells are removed or destroyed.

If any cancerous cells are left, they can multiply and cause secondary cancers at different sites in the body.

Cobalt-60 is a radioactive source that has been commonly used in medicine. The side effects of a treatment can be unpleasant but it can slow down the growth of a cancer or completely cure it.

Sterilising hospital equipment

Gamma radiation kills **bacteria**. It is used to **sterilise** hospital equipment to prevent the spread of disease. See Figure 5.

This is a much simpler process than old-fashioned methods that used high-pressure steam treatments.

Radiographer

A **radiographer** is a person who carries out procedures using X-rays and nuclear radiation. They must ensure that they are not exposed to X-rays or nuclear radiation. They can do this by leaving the room while the radiation source is switched on.

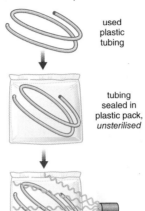

FIGURE 5: Sterilising medical equipment using gamma radiation. Why must equipment be sterilised before use on patients?

used plastic tubing

tubing sealed in plastic pack, *unsterilised*

source of gamma radiation

tubing ready for use again, *sterilised*

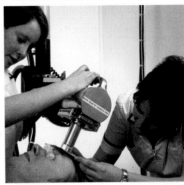

FIGURE 6: Radiographers position an X-ray generator over a patient's forehead for treatment of a skin cancer. There is a lead block on the patient's eye. Can you suggest why? The lead is covered in cling film because it is toxic to skin.

Questions

9 Why is it important to try to remove all cancerous cells during a treatment?

10 Suggest how radiographers protect themselves from exposure to radiation.

Using tracers

A radioactive tracer is used to investigate inside a patient's body without surgery. Gamma (and sometimes beta) emitters are introduced into the body to be used as tracers.

Radiation from a radioisotope used as a tracer is emitted inside the body. The radiation can penetrate tissues and leave the body to be detected.

A radioactive tracer used in medicine should have a short half-life so that it does not remain active in the body, emitting radiation for long periods. Technetium-99m is a commonly-used tracer.

> It emits only gamma radiation, the most useful radiation in medical diagnosis, without exposing the patient to useless alpha or beta radiation.

> It has a half-life of 6 hours, long enough for its passage through the body to be traced but short enough not to remain in the body for too long.

X-rays are not suitable as tracers as they are produced in an X-ray tube and cannot act as a radiation source inside the body.

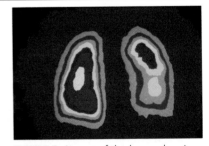

FIGURE 7: A scan of the lungs showing a pulmonary embolism on the left lung. This occurs when a blood clot blocks the flow of blood in a blood vessel. The different colours indicate the amount of radiation detected. Why is there no tracer in the dark area?

Q sterilising equipment using gamma radiation

> Iodine-123 is an artificially produced radioisotope of iodine and emits gamma radiation. It is used to investigate the **thyroid** gland. The thyroid absorbs **iodine**, so a small amount of radioactive iodine is given to the patient orally. It is absorbed into the bloodstream and collects in the thyroid. The radiation given out is monitored over a period of about 24 hours and is compared with the result from a 'normal' thyroid gland.

Questions

11 Suggest two essential properties that make a radioisotope suitable for use as a tracer.

12 Technetium-99m is the radioisotope most widely used as a tracer in medicine. Suggest why.

Using radiation for diagnosis and therapy

Treating cancer

High-powered gamma radiation can be used to destroy a **tumour** inside the body, such as a brain tumour. A dose large enough to destroy the tumour would also destroy the healthy tissue it passed through.

Two techniques are commonly used to limit damage to healthy tissue:

1. A radiation source is slowly rotated around the patient, with the tumour at the centre of the circle, so that a wide beam of gamma rays is focused on the tumour.

2. Three sources of radiation, each providing one-third of the required dose, are arranged around the patient's head. Each source is focused on the tumour.

Diagnosis using tracers

> A radioactive tracer with a short half-life, emitting beta or gamma radiation, is mixed with food or drink and swallowed, or it is injected into the patient's body.

> The tracer is allowed to spread through the body.

> Its progress through the body is monitored on the outside using a radiation detector, such as a gamma camera connected to a computer.

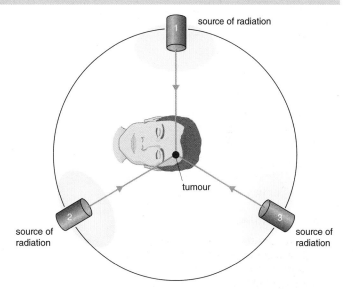

FIGURE 8: Treating a brain tumour with three fixed sources of gamma radiation. Why is a single fixed source of radiation not used?

Questions

13 Why is a syringe used to inject technetium-99m into a patient surrounded by lead?

14 Suggest why technetium-99m is used in diagnosis but cobalt-60 is used in therapy.

15 Why must the radioisotope used as a tracer in medicine
a have a short – but not very short – half-life?
b emit beta or gamma radiation?

Q using tracers   using radiation to treat cancer

Fission and fusion

You will find out:
> how electricity is generated in nuclear power stations
> how uranium releases energy
> about radioactive waste

Fission – friend or foe?

Fission is the splitting of a large nucleus such as uranium to release energy. Fission was used with devastating effects in the nuclear bombs that were dropped on the cities of Hiroshima and Nagasaki in 1945.

These days, controlled fission is used in nuclear power stations to produce electricity.

FIGURE 1: An atom bomb explosion. In what way was this worse than a conventional bomb?

Power stations

A **power station** makes electricity. It uses an energy source such as coal, oil, gas or nuclear to:

> heat water

> produce steam

> turn a **turbine**

> generate **electricity**.

A nuclear power station uses **uranium** as a fuel to heat water.

The process that gives out energy in a nuclear reactor is called nuclear **fission**. The rate of energy production by nuclear fission is kept under control.

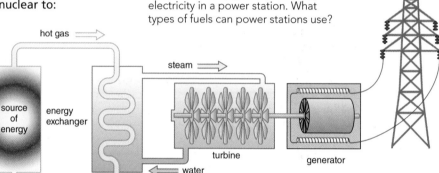

FIGURE 2: The stages of producing electricity in a power station. What types of fuels can power stations use?

Radioactive waste

Nuclear fission produces **radioactive waste**. This is a major problem since the waste products have to be handled carefully and disposed of safely.

Did you know?

Only 1 kg of uranium-235 produces about the same amount of energy as 2 million kg of coal!

Questions

1 What use is made of the uranium fuel in a nuclear power station?

2 What is the difference between a coal-fired and a nuclear power station?

3 Why is it difficult to dispose of radioactive waste?

How a nuclear power station works

Natural uranium consists of two isotopes, uranium-235 and uranium-238. The fuel used in a nuclear power station contains a greater proportion of the uranium-235 isotope than occurs naturally. The fuel is called 'enriched uranium' and forms the **fuel rods**.

Fission occurs when a large unstable nucleus is split up and there is a release of energy in the form of heat.

> The decay of uranium in a nuclear reactor starts a **chain reaction** that produces heat.

> The heat is used to boil water to produce steam.

> The pressure of the steam acts on the turbine blades which turn.

FIGURE 3: Fuel rods in a nuclear power station. What material do they contain?

Q how power stations work for kids   atomic bombs in World War 2

> The rotating turbine turns the **generator** that produces electricity.

Chain reaction

A chain reaction can carry on for as long as any of the uranium fuel remains. This allows large amounts of energy to be produced.

> In a nuclear power station the chain reaction is controlled to produce a steady supply of heat.

> A nuclear bomb involves an uncontrolled chain reaction.

How does uranium release energy?

In a nuclear power station, uranium-235 nuclei are bombarded with neutrons. A typical fission of uranium-235 can be shown as:

$$^{235}_{92}U + ^{1}_{0}n \rightarrow ^{90}_{36}Kr + ^{143}_{56}Ba + 3(^{1}_{0}n) + \gamma \, rays$$

The extra neutrons emitted cause a chain reaction and produce a large amount of energy.

Graphite rods placed between the fuel rods slow down fast-moving neutrons emitted during fission. The graphite acts as a **moderator**. Slow-moving neutrons are more likely to be captured by other uranium nuclei, which maintains the chain reaction.

Controlling nuclear fission

> **Boron control rods** are placed between the fuel rods.

> Boron absorbs neutrons so fewer neutrons are available to split more uranium nuclei.

> The boron rods can be raised or lowered in the reactor, which controls the fission rate.

> Enough neutrons are always left to keep the process operating.

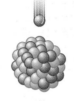

a neutron is absorbed by the nucleus of a uranium-235 atom

the nucleus is now less stable than before

it splits into two parts and releases energy

several neutrons are also produced – these may go on to strike the nuclei of other atoms causing further fission reactions

this is called a chain reaction

FIGURE 4: Uranium-235 undergoes a chain reaction that produces a large amount of energy. What keeps the chain reaction going?

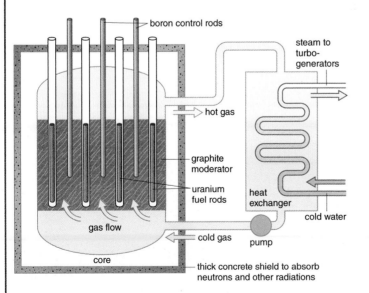

FIGURE 5: A gas-cooled nuclear reactor. Why is it important to be able to control the number of fission reactions in a reactor?

Questions

7 The boron control rods in a reactor are raised. Explain how this affects the energy produced.

8 What is the purpose of a moderator in a nuclear reactor?

nuclear fission   radioactive waste   artificial radioactivity

You will find out:

> about nuclear fusion
> how scientists work

Nuclear fusion

Fusion is the energy source for stars as it:

> produces large amounts of heat energy

> happens at extremely high temperatures.

'Cold fusion' – is it true?

Fusion needs very high temperatures of several million degrees Celsius. One group of scientists, led by Pons and Fleischmann, did once claim to have achieved fusion at room temperature – so-called '**cold fusion**'. Their claim has always been disputed because other scientists have not been able to repeat their findings.

FIGURE 6: Pons and Fleischmann in their lab. What did they claim to have done?

Remember!

• Fission is the splitting of large, heavy nuclei such as uranium.
• Fusion is the joining of small, light nuclei such as hydrogen and helium.

Questions

9 What is the difference between nuclear fission and nuclear fusion?

10 Why do we doubt the scientists who claim to have achieved 'cold fusion'?

Energy for the future?

> Nuclear fusion happens when two light nuclei **fuse** (join) together releasing large amounts of heat energy. For instance, hydrogen nuclei can fuse together to form helium.

> Nuclear fusion happens at very high temperatures in stars. Attempts to produce fusion on Earth started in the 1950s with **fusion bombs** (hydrogen or H-bombs).

Fusion requires extremely high temperatures, of the order of 10 million degrees Celsius. Such high temperatures have proved difficult to achieve and to manage safely on Earth and so its use for large-scale power generation remains a dream.

Projects such as the JET project have tried for many years to achieve fusion to produce energy safely and economically. For a long time the input energy was greater than the output energy. Now scientists have achieved equality, but only for a short time and a much greater output is needed for commercial use. Such research is very expensive so it is carried out as an international joint venture to share costs, expertise and ultimately the benefits.

FIGURE 7: The first hydrogen bomb was detonated in the Pacific in 1952. It is believed it was 1000 times more powerful than the atomic bomb that destroyed Hiroshima. How was it different to an atomic bomb?

Q hydrogen bomb   nuclear fusion

Fusion would be preferred to fission as an energy source because:

> there is a plentiful supply of hydrogen for fusion in sea water

> waste products of fusion are less harmful.

Is cold fusion possible?

The sensational announcement in 1989 by Pons and Fleischmann that they had achieved 'cold fusion' on a laboratory bench appeared to herald clean inexpensive energy for the future. The experimental details and data were shared between scientists. This can lead to greater knowledge and understanding by joint consideration of new ideas. So far no-one has been able to reproduce Pons and Fleischmann's results.

FIGURE 8: JET nuclear fusion experiment, Culham, Oxford, designed to produce temperatures of several million degrees. Why is it so complicated?

Questions

11 Why do scientists want to achieve fusion on Earth?

12 Why do scientists share details of their work with other scientists?

More about fusion

> In stars fusion happens at extremely high temperatures and pressures.

> So far attempts to replicate these conditions safely on Earth have been unsuccessful.

> Scientists are still trying to solve the safety and practical challenges presented.

> These reactions produce large quantities of energy.

> Fusion bombs are started with a fission reaction, which creates the exceptionally high temperatures needed for fusion reactions to start.

Examples of fusion reactions include:

$$^{1}_{1}H + ^{2}_{1}H \rightarrow ^{3}_{2}He$$

$$^{3}_{2}He + ^{3}_{2}He \rightarrow ^{4}_{2}He + 2^{1}_{1}H$$

More about 'cold fusion'

'Cold fusion' is still not accepted as a realistic method of energy production. Occasional reports have continued to trickle in from various small research groups, but in each case the results have proved erratic or impossible for other groups to replicate. If it ever proves possible it would revolutionise energy production.

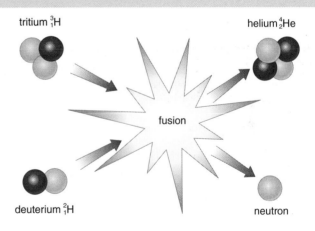

tritium $^{3}_{1}H$

helium $^{4}_{2}He$

fusion

deuterium $^{2}_{1}H$

neutron

FIGURE 9: Write the nuclear equation for this fusion reaction.

Questions

13 Why has 'cold fusion' not become an acceptable method of energy production?

14 Discuss the advantages that would be gained by producing energy using fusion in the future compared with building more fission power stations.

Preparing for assessment: Planning and collecting primary data

To achieve a good grade in science, you not only have to know and understand scientific ideas, but you need to be able to apply them to other situations and investigations. These tasks will support you in developing these skills.

✸ Tasks

> Plan an investigation to see how the length of wire connected in a circuit affects the current in a lamp.

> Once your plan has been approved, perform the investigation, record your results and write a simple conclusion.

✸ Context

A dimmer switch is a useful electrical component that lets you adjust light levels from nearly dark to fully lit by simply turning a knob or sliding a lever. The simplest dimmer switch works by altering the resistance in the lighting circuit. This alters the current in the circuit and hence the brightness of the lamp. The dimmer switch contains a variable resistor. Moving the knob or slider changes the length of resistance wire in the circuit.

The thickness of the wire also affects its resistance and hence the range of light levels available.

✸ Planning your investigation

These are the things you will need to consider when planning your investigation. (You can develop your plan in groups of two or three.)

1. How will you measure the current?

2. How will you measure the length of wire included in the circuit?

3. What do you need to keep the same to make it a fair test? How will you keep it/them the same? Choose suitable value(s). (You may need to try this out when planning your investigation.)

Remember!
You will not be using mains voltage to carry out your investigation. You will use a low-voltage power supply or a battery.

4. How many different lengths of wire will you need before you can identify a trend?

5. Will you need to repeat your readings? If so, how many times?

6. Draw a circuit diagram.

7. You should carry out a risk assessment before you start the investigation. (What precautions should you take?)

8. Write the plan for the investigation.

Try to write the plan in a logical order and ask yourself if someone can perform the investigation following just your plan.

✴ Performing the investigation

Once your plan has been approved you can perform the investigation.

1. If you repeated any readings, all of these will need to be recorded as well as the average result.

Why do scientists repeat readings?

2. Record your results in a table like this. You may need to add extra rows or columns

How many sets of data do you need to identify a trend?

length of wire in cm	current in A	brightness of lamp	

What graph would you draw?

3. If you were to complete this as a GCSE controlled assessment, you would go on to plot a graph and evaluate the investigation.

What would the labels be on the axes?

4. Is there any way in which you could have improved on how you performed the investigation?

How would you use the graph to decide on the answer to the task?

5. What have you found out about how the length of wire affects the current in a circuit?

6. How is this linked to the brightness of the lamp?

7. Suppose a thicker wire is used in a dimmer switch. How would this affect the brightness of the lamp?

8. Extension activity: You could modify your investigation to see how the thickness of wire used affects the brightness of a lamp.

P4 Checklist

To achieve your forecast grade in the exam you'll need to revise

Use this checklist to see what you can do now. It gives you many of the important points you will need to know. Refer back to the relevant pages in this book if you're not sure and to see if there is anything else you need to know. Look across the three columns to see how you can progress.

Remember you'll need to be able to use these ideas in various ways, such as:

> interpreting pictures, diagrams and graphs
> applying ideas to new situations
> explaining ethical implications
> suggesting some benefits and risks to society
> drawing conclusions from evidence you've been given.

Look at pages 278–299 for more information about exams and how you'll be assessed.

To aim for a grade E	To aim for a grade C	To aim for a grade A
describe how insulating materials can become charged **state** that there are two kinds of charge, positive and negative **describe** how you get an electric shock from charged objects **describe** how you get an electric shock if you become charged and then earthed	**recognise** that like charges repel and unlike charges attract **understand** that electron transfer causes electrostatic effects **explain** how static electricity can be dangerous **explain** how static electricity can be a nuisance	**describe** static electricity in terms of electron movement **explain** how the chance of receiving an electric shock can be reduced **explain** how the problems of static electricity can be reduced
recall that electrostatics is useful for electrostatic dust precipitators **recall** that electrostatics is useful for spraying paint and crops **recall** that electrostatics is useful for restarting the heart	**explain** how static electricity can be useful for electrostatic dust precipitators **explain** how static electricity can be useful for paint spraying **explain** how electrostatics is useful for restarting the heart	**explain** how static electricity is used in electrostatic dust precipitators in terms of electron movement **explain** how static electricity is used in paint spraying in terms of electron movement
understand what causes the greenhouse effect **recall** examples of greenhouse gases **describe** how climate change is linked to global warming **describe** difficulties in measuring global warming **explain** why scientists should share their data	**describe** how different wavelengths of radiation behave **name** natural and man-made sources of greenhouse gases **explain** how both human and natural activity affect the weather **list** evidence for and against man-made global warming **distinguish** between opinion and evidence-based statements	**explain** the greenhouse effect in terms of infrared wavelengths **interpret** data on abundance and impact of greenhouse gases interpret data about global warming and climate change **explain** how scientists can agree about the greenhouse effect but disagree about the effects of human activity
describe how resistors are used to change the current in a circuit **recall** the colour code for live, neutral and earth wires **describe** reasons for the use of fuses and circuit-breakers **recognise** that 'double insulated' appliances do not need earthing	**explain** how variable resistors are used to change current **use** the equation $$\text{resistance} = \frac{\text{voltage}}{\text{current}}$$ **describe** the functions of the live, neutral and earth wires **explain** how a wire fuse reduces the risk of fire if a fault occurs **use** the equation power = voltage × current **explain** why 'double insulated' appliances do not need earthing	**manipulate** the equation $$\text{resistance} = \frac{\text{voltage}}{\text{current}}$$ **explain** reasons for the use of fuses and circuit-breakers **manipulate** the equation power = voltage × current to select a fuse

To aim for a grade E

recall that ultrasound is a longitudinal wave

recognise wavelength, compression and rarefaction

recognise that ultrasound can be used in medicine for diagnosis

To aim for a grade C

describe features of longitudinal waves

recall that the frequency of ultrasound is above the upper threshold of human hearing

recognise that ultrasound can be used in medicine for therapy

To aim for a grade A

compare the motion of particles in longitudinal and transverse physical waves

explain how ultrasound is used in medicine

explain why ultrasound is used instead of X-rays for some scans

---

recognise how the radioactivity of an object is measured

understand that radioactivity decreases with time

recall that radioactivity ionises materials

recall that radiation comes from the nucleus of the atom

recall the relative nature and **describe** radioactive decay

explain and use the concept of half-life

explain ionisation in terms of electron movement

describe radioactivity as coming from the nucleus of an unstable atom

recall that an alpha particle is a helium nucleus and a beta particle a high-speed electron

interpret graphical data of radioactive decay and calculate half-life

explain why alpha particles are such good ionisers

describe what happens to a nucleus during alpha decay

describe what happens to a nucleus during beta decay

construct and balance nuclear equations

---

recall the main sources of background radiation

recall industrial examples of the use of tracers

recall that alpha sources are used in some smoke detectors

recall that radioactivity can be used to date rocks

recall background radiation can come from man-made sources and waste products

describe how tracers are used in industry

explain how a smoke detector with an alpha source works

explain how radioactivity is used to date rocks

recall radiocarbon dating is used to find the age of old materials

evaluate the relative significance of sources of background radiation

explain why gamma radiation is used in an industrial tracer

explain how radiocarbon dating finds the age of old materials

---

compare the properties of X-rays and gamma rays

recall that medical radioisotopes are produced by placing materials in a nuclear reactor

describe uses of nuclear radiation in medicine

recall that only beta and gamma rays can pass through skin

describe the role of a radiographer

understand how an X-ray image is produced

describe how materials can become radioactive by absorbing extra neutrons

explain why gamma (and beta) emitters can be used as tracers

understand why medical tracers should not remain in the body for long periods

explain how X-rays and gamma rays are produced

explain how radioactive sources are used as tracers in medicine

explain how radioactive sources are used to treat cancer

---

describe the main stages in the production of electricity

describe the process that gives out energy in a nuclear reactor

recall that nuclear fission produces radioactive waste

describe the difference between fission and fusion

recall the 'cold fusion' controversy

describe how electricity is generated at a nuclear power station

understand how the decay of uranium starts a chain reaction

describe a nuclear bomb as an 'out of control' chain reaction

describe how nuclear fusion releases energy

explain why 'cold fusion' results have been shared

describe what happens to allow uranium to release energy

explain what is meant by a chain reaction

explain how to stop nuclear reactions going 'out of control'

understand the conditions needed for fusion to take place

explain why 'cold fusion' is still not accepted by most scientists

Foundation Tier

1 A bicycle frame is being painted in a factory.
The paint coming out of the nozzle of the spray gun is positively charged.

AO1 **(a)** The paint spreads out as it leaves the paint spray. Explain why this happens. [2]

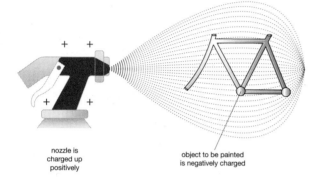

nozzle is charged up positively

object to be painted is negatively charged

AO1 **(b)** The bicycle frame is given a negative charge. Suggest why this is done. [1]

AO1 **(c)** State **two** advantages of this method of painting the bicycle frame. [2]

[Total: 5]

2 The diagram shows the wiring inside a plug attached to a kettle.

AO1 **(a)** Label the live wire with a letter L. [1]

AO1 **(b) (i)** Label the fuse with a letter F. [2]

AO2 **(ii)** How does the fuse reduce the risk of fire if the kettle develops a fault? [3]

[Total: 6]

AO1 **3 (a)** How are *ultrasound* waves different to sound waves? [1]

AO1 **(b)** Choose the correct answer, A or B, from each pair in the table to complete each sentence. [3]

	A	B
Ultrasound is a ……	longitudinal wave	transverse wave
A rarefaction is a…..	region of higher pressure	region of lower pressure
Frequency is the….	time for one complete wave	number of waves in a second

AO1 **(c)** Give one use of ultrasound. [1]

[Total: 5]

AO1 **4 (a)** Nuclear radiation ionises materials. What is meant by 'ionises'? [1]

AO1 **(b)** The answer to each question is alpha, beta or gamma.
(i) This particle is an electron. [1]
(ii) This radiation is similar to X-rays. [1]
(iii) This particle is a helium nucleus. [1]
(iv) This particle cannot pass through skin. [1]

[Total: 5]

AO3 **5** The table gives the properties of some radioisotopes.

radioisotope		radiation emitted	half-life
americium-241	solid	alpha	432 years
polonium-210	solid	alpha	138 days
carbon-14	solid	beta	5700 years
strontium-90	solid	beta	28 years
cobalt-60	solid	gamma	5 years

Choose the best radioisotope for the following applications. Give reasons for your choice.
• to detect a leak in an oil pipeline buried deep underground
• to estimate the age of a bone found in an archaeological dig
• to diagnose a girl with breathing problems believed to have damaged lungs
• as the source of radiation in a smoke alarm.

The quality of written communication ✐ will be assessed in your answer to this question. [6]

AO1 **6 (a)** What is the difference between fission and fusion? [2]

AO1 **(b)** Fusion happens in stars at extremely high temperatures. Scientists are trying to produce fusion on Earth. Two scientists claim to have achieved fusion at room temperature – 'cold fusion'.

Why do other scientists not believe their claim? [2]

[Total: 4]

AO1 recall the science   AO2 apply your knowledge   AO3 evaluate and analyse the evidence

✳ Worked Example – Foundation Tier

Nuclear radiation is used in hospitals in several ways.

(a) Gamma (and sometimes beta) emitters can be used as tracers. A tiny amount of a radioisotope is put into the body and its movement tracked with a radiation detector.

(i) Why cannot a source of alpha radiation be used as a tracer? [2]

Alpha radiation is easily stopped so wouldn't be able to get out of the body.

(ii) Why must the source used as a tracer have a short half-life? [2]

So it doesn't stay in the body for too long.

(b) Gamma emitters are also used to treat cancer by destroying cancerous cells. The photograph shows a man undergoing radiation treatment for lung cancer.

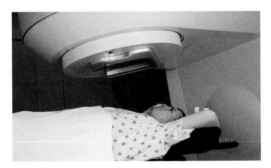

(i) Why cannot a source of alpha radiation be used to destroy cancerous cells? [1]

Alpha radiation would be stopped by the patient's skin.

(ii) How does the gamma source used here differ from that used as a tracer? [2]

It needs to be much more powerful if it is to kill cancer cells. A tracer needs to be a very weak source of radiation so it doesn't harm the patient.

(iii) Gamma radiation can damage healthy as well as cancerous tissue. What can be done to reduce damage to healthy tissue? [2]

Cover healthy tissue in lead so no gamma rays can reach it.

How to raise your grade!
Take note of these comments – they will help you to raise your grade.

This is true and makes alpha radiation useless as a tracer as it cannot be detected outside the body. Alpha radiation trapped in the body is also a serious health risk. The answer is 'getting there' but needs to give more detail. 1/2

This is correct but has missed the important part that it is emitting harmful radiation while in the body. 1/2

This is correct. 1/1

Two good points have been made here. An alternative point that could be mentioned is that a long half-life is needed in this case as opposed to the short half-life of the tracer. 2/2

Lead shielding is used. In addition the radiation beam can be focused on the tumour. But gamma rays are bound to reach some healthy tissue. 1/2

This student has scored 6 marks out of a possible 9. This is below the standard of Grade C. With more care the student could have achieved a Grade C.

Higher Tier

1 (a) Priya rubs a polythene rod with a duster. It becomes negatively charged.

AO1 **(i)** Explain this in terms of the movement of electrons. [1]

AO1 **(ii)** What, if anything, happens to the dust? [1]

AO2 **(b)** Priya hangs two small plastic balls side by side. She touches each ball with the charged polythene rod. Explain what happens. [2]

AO2 **(c)** Priya touches one ball with the polythene rod. She puts the ball against the wall. It sticks to the wall. Explain why this happens in terms of the movement of electrons. [3]

[Total: 7]

AO1 **2 (a)** An electric toaster has a power of 900 W. Mains voltage is 230 V. Calculate the current in the toaster when it is working normally. [3]
AO2

AO2 **(b)** 5 A and 13 A fuses are available. Which fuse should be used in the plug attached to the toaster? [1]

[Total: 4]

AO2 **3** Ultrasound is often used as an alternative to X-rays. Give two advantages of ultrasound compared to X-rays. [2]

AO1 **4 (a)** What is meant by half-life? [1]

(b) Cobalt-60 is a radioisotope frequently used in the treatment of cancers. The graph shows how the activity of cobalt-60 changes with time.

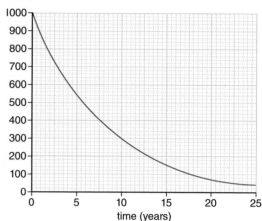

AO2 **(i)** Use the graph to find the half-life of cobalt-60. (Show how you got your answer). [2]

AO2 **(ii)** Cobalt-60 decays to a stable isotope of nickel, emitting a β-particle and γ-radiation. Complete the nuclear equation which represents this.

$$^{60}_{27}\text{Co} \; — \; ^{??}_{??}\text{Ni} + \; ^{0}_{-1}\text{e} \; (+\gamma)$$ [2]

[Total: 5]

AO3 **5** It is proposed to build a new nuclear power station in Greenshire. Local residents oppose the scheme. Give two points in favour and two against building the new power station. [4]

[Total: 4]

AO3 **6** Explain how body scans using ultrasound can produce an image of organs inside the body. Use the data in the table to support your answer.

medium	density in kg/m³	speed of ultrasound in m/s
air	1.3	330
water	1000	1500
blood	1060	1570
fat	1025	1540
bone (varies)	1400–1900	4080
xenon-133	gamma	5 days

The quality of written communication will be assessed in your answer to this question. [6]

AO1 recall the science   AO2 apply your knowledge   AO3 evaluate and analyse the evidence

✳ Worked Example – Higher Tier

In a nuclear reactor an atom of uranium-235 splits into two nearly equal parts when it is hit by a neutron, releasing energy and more neutrons.

Explain how this reaction is used to generate electricity in a nuclear power station.

The quality of your written communication ✎ will be assessed in your answer to this question. [6]

The extra <u>nutrons</u> released when a uranium-235 atom splits make a chain reaction because each <u>nutron</u> can go on to split another uranium-235 <u>nuclus</u>. This means more and more energy is released without the chain reaction you wouldn't get enough energy to produce large amounts of electricity.

The rate of reaction has to be controlled or it would be very dangerous, like a nuclear bomb. Boron rods are used to absorb some of the <u>nutrons</u>.

The energy is used to boil water – the steam turns a turbine to make electricity.

These longer 6 mark answers usually have marks awarded for the Quality of Written Communication shown by this symbol ✎ so answers need planning, and care is needed with spelling, punctuation and grammar. (See the example banded mark scheme on page 297.)

For the most part the information is relevant and presented in a structured and coherent format. Specialist terms are used for the most part appropriately. There are occasional errors in grammar, punctuation and spelling.

This student has scored 3 marks out of a possible 6. This is below the standard of Grade A. With a little more care the student could have achieved a Grade A.

How to raise your grade!
Take note of these comments – they will help you to raise your grade.

⬇

This paragraph is correct but a sketch to illustrate the meaning of a chain reaction may enhance the answer. The use of a graphite moderator to slow down the neutrons so that there is a greater probability of each neutron splitting another uranium nucleus has not been mentioned. There are careless spelling mistakes

It should be mentioned that the boron control rods are raised or lowered in the reactor to increase or decrease the number of neutrons absorbed. This alters the number of fissions that can occur and hence the amount of energy released.

This outlines the subsequent stages involved in producing electricity but the final step has been omitted. The rotating turbine must be connected to a generator in order to produce electricity. 3/6

When the evidence doesn't add up

Sometimes people use what sound like scientific words and ideas to sell you things or persuade you to think in a certain way. Some of these claims are valid, and some are not. The activities on these pages are based on the work of Dr Ben Goldacre and will help you to question some of the scientific claims you meet. Read more about the work of Ben at www.badscience.net.

Brown goo

You may have seen adverts for a foot spa that can remove toxins from your body. They are sometimes used in beauty salons or you might even buy one to use at home. The basin is filled with water, a sachet of special salts is added and then it is plugged in. You put your feet in to soak and the water turns brown!

It looks impressive, but is that because toxins have left your body through your feet?

Now, the advertisers of these products would tell us that we are being 'detoxed' and that horrible chemicals, toxins, which have accumulated in our bodies are at long last being released. It's perhaps not surprising that people are keen to be cleansed. However the talk doesn't match the facts. The chemicals in the water didn't come from your body which (as you know) is quite capable of getting rid of substances it doesn't need without using special equipment.

We are learning to:

> use primary and secondary evidence to investigate scientific claims
> apply scientific concepts to evaluate 'health products'
> explore the implications of these evaluations

✳ CAN YOU DETOX VIA YOUR FEET?

Read the leaflet – it sounds scientific but is it? Think about what you have learnt in science.

> Human metabolism is complex with the 'building blocks' of molecules being reshaped into new arrangements. The same molecule can be a waste product or a valued ingredient, depending on when and where it is in the body. There is no such thing as a 'detox system' in any medical textbook. Sometimes the body does need to dispose of waste but it does so by well-known ways.

> Electrolysis occurs when a direct electric current is passed through a liquid containing mobile ions, resulting in a chemical reaction at the electrodes.

Can you come up with a hypothesis about what's going on? How would you prove it? Ben came up with a good idea and gave his Barbie™ a foot bath – you might get a chance to replicate his experiment. Can you predict what might happen?

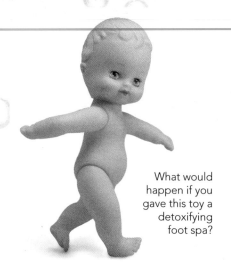

What would happen if you gave this toy a detoxifying foot spa?

Collins Detox Foot Bath

Before

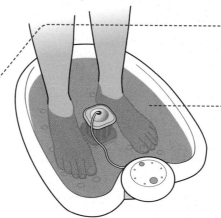

After 30 minutes

This looks like a serious piece of equipment.

This brown water looks horrible but is it brown because of toxins from the body?

This explanation sounds scientific, but is it?

The patented Collins Detox Foot Bath stimulates the active release of tingling ions that surge back and forth around your feet generating a flow of both negative and positive energy. This refreshes and renews the tissues, cleansing your body of accumulated toxins, readjusting the balance of energy at a bio-cellular level and removing excretory residues.

The centrally located micro-voltaic electrodes cause the flow of bi-polar ions producing an energy field that carries essential nutrients and life-giving oxygen. The release of toxins takes places through the myriad of microscopic pores in the soles of your feet. Graduated colour changes in the water present conclusive evidence of the beneficial effects.

The many enthusiastic users report a range of exhilarating effects including a heightened sense of awareness, improved circulation and relief of arthritic pain. The results are personal to each user as their toxin levels and combinations vary, but all report positive outcomes. One recent example of enthusiastic feedback said "The colour of the water shocked me in the realisation of what had accumulated in my body but the lightness I felt lasted for days!"

The people who tested it were impressed but did they enjoy the effects of detox or a relaxing foot bath?

✴ DETOX SELLS!

Words like 'toxins' and 'detoxification' (the removal of toxins) are sometimes used to promote products and techniques. Nobody likes to think of toxins accumulating in their body but we must consider whether there's any scientific basis for these ideas.

> Can you think of other products that claim to 'detox'?

> Why do you think that 'detoxing' can be used to sell these products?

> These treatments could all be said to be a little theatrical. How does this help to convince people that they're effective?

When the evidence doesn't add up

Sometimes people use what sound like scientific words and ideas to sell you things or persuade you to think in a certain way. Some of these claims are valid, and some are not. The activities on these pages are based on the work of Dr Ben Goldacre and will help you to question some of the scientific claims you meet. Read more about the work of Ben at www.badscience.net.

Bad news

In science you learn about ideas that scientists have developed by collecting evidence from experiments; you are also learning to collect and evaluate evidence yourself. You can use this outside of the laboratory to weigh up information you come across every day. Let's look at this example about how data can be used to support a story for a newspaper.

When data are produced you might think that there's only one way they can be used, and only one meaning that can be supported. This isn't always true.

We are learning to:

> understand how data can be used to make a good news story

> understand how science reports may be distorted to make headlines

> consider why science reports may be represented in various different ways

✳ GOOD ADVICE?

If a woman wants to be sexually active but doesn't want to get pregnant, one of the contraceptive methods available to her is to use a contraceptive implant. There are a number of factors to take into account; one of the most important ones is, of course, 'how well does it work?' Think about the headline on the right. What kind of questions might you ask that would reveal whether this method is, in fact, a failure?

600 pregnancies despite contraceptive implant

✳ STICKING TO THE NUMBERS

One of the questions we might want to consider is 'over what timescale?' Is this 600 over the last month, last year or since records began? In fact, the contraceptive implant had been available for ten years when this data was released, so it's 60 unintended pregnancies per year, on average. Still not ideal, but maybe not as disastrous as at first thought.

We might also want to know how widespread the use of the implant was. If the 60 pregnancies a year was out of say, 1,000 people, then that's not very good: it would mean that 6 out of every hundred women with an implant had got pregnant over a year.

If it was out of 100,000 then that means 6 out of every 10,000 women got pregnant over a year, so this method of contraception would compare well with other methods.

In fact, around 1.3 million implants have been used over the last ten years, and each lasts for three years. This works out as 1.4 unwanted pregnancies for every 10,000 women using the method per year if we assume that each implant lasts for the full 3 years.

Making the headlines

Four students are talking about this story.

Jo says

I think the journalists were doing a good job here to tell people about the fact that 600 women who thought that they couldn't get pregnant, then did. They got hold of the facts and then reported them.

Adam says

The journalists didn't write this up very well. Most of the people reading this story would be women who would be wondering if this method of contraception was one that they should use. The headline suggests that it's not safe and it is. Well, most of the time.

Will says

Journalists have to be responsible. If this story frightens women off one of the safest methods of contraception they've let people down.

Emma says

The main job of journalists is to be entertaining. Boring stories don't get read. '600 women using contraceptive get pregnant' makes you read the story. '0.014% of women using contraceptive get pregnant' looks boring.

- Look at these comments. Who do you think is right?

- Do you think the main purpose of a journalist is:
  - To be informative, even if it's sometimes boring?
  - To be engaging, even if it may sometimes give a false impression?

- If you had been the journalist assigned to this story, what headline would you have used?

✱ NUMBERS IN THE REAL WORLD

A useful way of presenting data like this is to use what's called the natural frequency. Out of a set number, this indicates how many will have a changed outcome as a result of this. In this case it's 1.4 out of 10,000. The figure of 600 isn't wrong, neither is the 0.014%, but 1.4 in 10,000 puts it in a simple form that people can make sense of and use to assess the likely impact on them.

When the evidence doesn't add up

Sometimes people use what sound like scientific words and ideas to sell you things or persuade you to think in a certain way. Some of these claims are valid, and some are not. The activities on these pages are based on the work of Dr Ben Goldacre and will help you to question some of the scientific claims you meet. Read more about the work of Ben at www.badscience.net.

MMR – don't die of ignorance

Autism is a condition which affects between one and two people in every thousand, affecting neural development and causing restricted and repetitive behaviour. It affects social behaviour and language. It is usually diagnosed from the age of three onwards.

In Britain, as in many countries, the majority of children are vaccinated against measles, mumps and rubella using a combined vaccine (MMR) between the ages of one and two. In 1998 a British doctor wrote a report on twelve children who had been vaccinated with the MMR vaccine and were subsequently diagnosed as autistic. The result of this report was that media interest was raised, many anti-MMR stories appeared and there was a significant fall in the number of children who were given the MMR vaccine.

Consider these questions:

> Does the fact that the children in the report were diagnosed with autism after being given the MMR vaccine prove that the vaccine caused the autism?

> At the time of the report being written well over 90% of children had the MMR vaccine. Why should it not be a surprise if some of those children are diagnosed with autism?

> What kind of survey would have helped to identify whether the MMR vaccine caused autism?

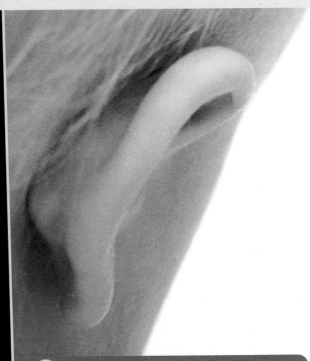

✳ THE RISE OF MEASLES

As the number of MMR vaccinations fell, the number of measles cases rose. Measles is a very dangerous disease that even in developed countries kills one in every 3000 people and causes pneumonia in one in 20.

It was subsequently established beyond reasonable doubt that there is no causal link between MMR vaccination and autism. The doctor had a commercial interest in the alleged link and was subsequently struck off. The scare affected no other countries. MMR vaccination rates in Britain are rising again. Doctors are still not sure why some children develop autism; its causes are unknown.

Consider these points of view:

"The doctor who wrote the report was right to alert people to his concerns and suggest that more research should be carried out."

"The media got hold of the story and turned it into a huge scare. It's their fault."

"There was never any evidence to prove a link. Thousands of children have caught diseases that could otherwise have been avoided."

✱ THE RISE OF MEASLES

One of the things this story illustrates is what can happen when you look at only a very small sample and the importance of working with large-scale surveys wherever possible. Such a study was carried out in Denmark: the Madsen study. Because Denmark tracks patients and the care they receive on a central system they have been able to study the correlation between vaccination and illness. The data clearly shows that there is no correlation between MMR vaccination and the incidence of autism.

The study was based on data from over half a million children: over 440,000 had been vaccinated and there was no greater incidence of autism in children vaccinated than in those not vaccinated.

> Identify the features of this study that make its findings reliable.

> What might you say to someone who still wasn't convinced by this study and decided to 'play it safe' by not having their child vaccinated for MMR?

Cause and effect?

Sometimes it looks like something causes something else, perhaps because they both happen at the same time. But scientists need to be very careful before saying that one thing causes another.

• Often you need to use common sense and extra information to help decide if there is true causation. For example, cocks crow in the morning, but nobody thinks that cocks crowing causes the Sun to come up, because there's no conceivable mechanism for that, and it conflicts with everything we know about the Sun and the Earth. On the other hand, we can observe that when it gets warmer, people wear fewer clothes, and it seems reasonable to say that the warm weather causes people to wear less.

• Sometimes two things are correlated, but it's harder to say what causes what, and there might be a third factor causing both of the things that we are observing. Let's say, for example, that a study finds that there is a strong correlation between a child's IQ and their height: perhaps both height and IQ are themselves related, through a complex causal pathway, to something else, like general health, or diet, or social deprivation.

• Often, although things happen at the same time, there is no link at all. For example, Halley's Comet appears once every 76 years. Previous appearances have coincided with King Harold's defeat at the Battle of Hastings, Genghis Khan's invasion of Europe and both the birth and death of great American novelist Mark Twain, author of Tom Sawyer.

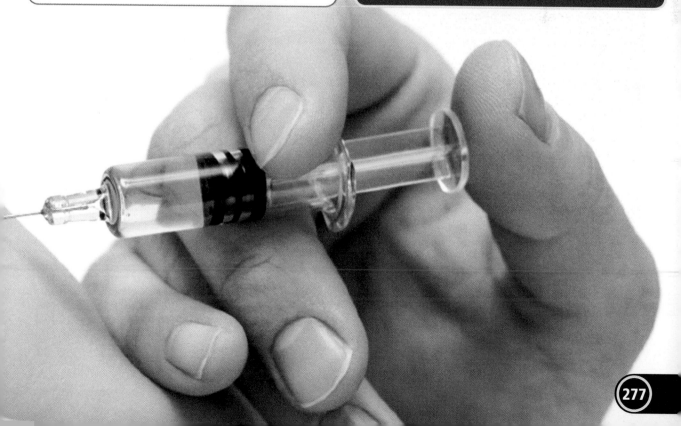

Carrying out controlled assessments in GCSE Additional Science

Introduction

As part of your GCSE Additional Science course, you will have to carry out a controlled assessment. This will be divided into three parts.

> Research and collecting secondary data

> Planning and collecting primary data

> Analysis and evaluation

The tasks will be set by OCR, the awarding body, and marked by the teachers in school. The marking will be checked to make sure standards are the same in every school.

Some of the work you do must be supervised and you will have to work on your own under examination conditions.

Some experimental work may be performed in groups and results shared.

Some research work may be done as part of a homework exercise.

As well as your scientific skills, the quality of written communication will also be assessed as part of the controlled assessment.

Controlled assessment is worth 25% of the marks for your GCSE. It's worth doing it well.

✺ Part 1: Research and collecting secondary data

You will have to plan and carry out research. The task will be given to you in the form of a handout. You will be allowed to research in class and/or as a homework activity.

Secondary data needs to be appropriate and you will have to select the information from a variety of sources to answer the questions you have been set.

Secondary data can be collected by a variety of different methods:

> survey

> questionnaire

> interview

> textbook

> newspapers and magazines

> internet search

Always make sure that you reference the material you use from any secondary source.

The work you do for your GCSE examination will need to be hand written or typed and printed out. You will take the work to a supervised lesson where you will use the information you have collected to answer specific questions in an answer booklet. Your research will be retained. It will be needed when you complete the analysis and evaluation part of your controlled assessment.

Your research may also be needed when you are planning your experiment.

Assessment tip

Make sure that the data you look at is relevant and appropriate. The data is more likely to be reliable if the same results and conclusions are obtained by a number of different researchers.

Assessment tip

As part of your research, you may need to design your own survey or questionnaire as well as collecting the data.

Choose the search terms you use on the internet carefully.

Assessment tip

When referencing, make sure you record:
> book title, author and page
> newspaper or magazine title, date, author and page
> full website address, author (if possible).

Definition

Secondary data are measurements/observations made by anyone other than you.

Part 2: Planning and collecting primary data

As part of your GCSE Additional Science course, you will develop practical skills which will help you to plan and collect primary data from a science experiment. Your experimental work will be divided into several parts:

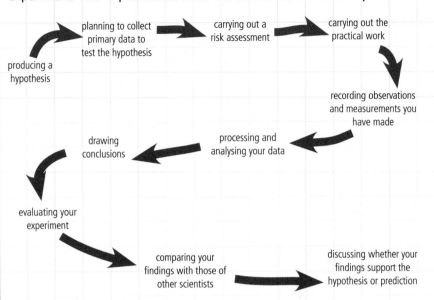

A scientific investigation usually begins with a scientist testing an idea, answering a question, or trying to solve a problem.

You first have to plan how you will carry out the investigation.

Your planning will involve testing a **hypothesis**. For example, you might observe that plants grow faster in a heated greenhouse than an unheated one.

So your hypothesis might be 'the rate of photosynthesis increases because the temperature increases'.

In Science, you will be given a hypothesis to test. In Additional Science, or if you're doing Separate Sciences, you will have to produce a hypothesis.

To formulate a hypothesis you may have to research some of the background science.

First of all, use your lesson notes and your textbook. The topic you've been given to investigate will relate to the science you've learned in class.

Also make use of the Internet, but make sure that your Internet search is closely focused on the topic you're investigating.

✔ The search terms you use on the Internet are very important. 'Investigating photosynthesis' is a better search term than just 'photosynthesis', as it's more likely to provide links to websites that are more relevant to your investigation.

✔ The information on websites also varies in its reliability. Free encyclopaedias often contain information that hasn't been written by experts. Some question and answer websites might appear to give you the exact answer to your question, but be aware that they may sometimes be incorrect.

✔ Most GCSE Science websites are more reliable, but if in doubt, use other information sources to verify the information.

If you do have to produce a hypothesis, you can use your lesson notes, the research you have already done and textbooks.

Definition

A **hypothesis** is a possible explanation that someone suggests to explain some scientific observations.

Assessment tip

When you're formulating a hypothesis, it's important that it's testable. In other words, you must be able to test the hypothesis in the school lab.

Assessment tip

In the planning stage, scientific research is important if you are going to obtain higher marks.

Example 1

Investigation: Plan and research an investigation into the effect of temperature on the change in height of a plant over 2 weeks.

Your hypothesis might be 'when I increase the temperature, the percentage increase in the height of the plant will be greater.'

You should be able to justify the hypothesis by some facts you have found. For example 'growing lettuces in greenhouses halves the time it takes for them to be ready to sell.'

✳ Choosing a method and suitable apparatus

As part of your planning, you must choose a suitable way of carrying out the investigation.

You will have to choose suitable techniques, equipment and technology, if this is appropriate. How do you make this choice?

You will have already carried out the techniques you need to use during the course of practical work in class (although you may need to modify these to fit in with the context of your investigation). For most of the experimental work you do, there will be a choice of techniques available. You must select the technique:

✔ that is most appropriate to the context of your investigation, and

✔ that will enable you to collect valid data, for example if you are measuring the effects of light intensity on photosynthesis, you may decide to use an LED (light-emitting diode) at different distances from the plant, rather than a light bulb. The light bulb produces more heat, and temperature is another independent variable in photosynthesis.

Your choice of equipment, too, will be influenced by measurements you need to make. For example:

✔ you might use a one-mark or graduated pipette to measure out the volume of liquid for a titration, but

✔ you may use a measuring cylinder or beaker when adding a volume of acid to a reaction mixture, so that the volume of acid is in excess to that required to dissolve, for example, the calcium carbonate.

In science, the measurements you make as part of your investigation should be as precise as you can, or need to, make them. To achieve this, you should use:

✔ the most appropriate measuring instrument

✔ the measuring instrument with the most appropriate size of divisions.

The smaller the divisions you work with, the more precise your measurements. For example:

✔ in an investigation on how your heart rate is affected by exercise, you might decide to investigate this after a 100 m run. You might measure out the 100 m distance using a trundle wheel, which is sufficiently precise for your investigation

✔ in an investigation on how light intensity is affected by distance, you would make your measurements of distance using a metre rule with millimetre divisions; clearly a trundle wheel would be too imprecise

Assessment tip

Technology, such as data-logging and other measuring and monitoring techniques, for example heart sensors, may help you to carry out your experiment.

Definition

The **resolution** of the equipment refers to the smallest change in a value that can be detected using a particular technique.

Assessment tip

Carrying out a preliminary investigation, along with the necessary research, may help you to select the appropriate technique to use.

in an investigation on plant growth, in which you measure the thickness of a plant stem, you would use a micrometer or Vernier callipers. In this instance, a metre rule would be too imprecise.

✳ Variables

In your investigation, you will work with independent and dependent variables.

The factors you choose, or are given, to investigate the effect of are called **independent variables**.

What you choose to measure, as affected by the independent variable, is called the **dependent variable**.

✳ Independent variables

In your practical work, you will be provided with an independent variable to test, or will have to choose one – or more – of these to test. Some examples are given in the table.

Investigation	Possible independent variables to test
activity of amylase enzyme	> temperature > sugar concentration
rate of a chemical reaction	> temperature > concentration of reactants
stopping distance of a moving object	> speed of the object > the surface on which it's moving

Independent variables can be **discrete** or **continuous**.

> When you are testing the effect of different disinfectants on bacteria you are looking at discrete variables.

> When you are testing the effect of a range of concentrations of the same disinfectant on the growth of bacteria you are looking at continuous variables.

Range

When working with an independent variable, you need to choose an appropriate **range** over which to investigate the variable.

You need to decide:

 which treatments you will test, and/or

✔ the upper and lower limits of the independent variables to investigate, if the variable is continuous.

Once you have defined the range to be tested, you also need to decide the appropriate intervals at which you will make measurements.

> **Definition**
>
> Variables that fall into a range of separate types are called **discrete variables**.

> **Definition**
>
> Variables that have a continuous range are called **continuous variables**.

> **Definition**
>
> The **range** defines the extent of the independent variables being tested.

The range you would test depends on:

✔ the nature of the test
✔ the context in which it is given
✔ practical considerations, and
✔ common sense.

Example 2

1 Investigation: Investigating the factors that affect how quickly household limescale removers work in removing limescale from an appliance

You may have to decide on which acids to use from a range you're provided with. You would choose a weak acid, or weak acids, to test, rather than a strong acid, such as sulfuric acid. This is because of safety reasons, but also because the acid might damage the appliance you were trying to clean. You would then have to select a range of concentrations of your chosen weak acid to test.

2 Investigation: How speed affects the stopping distance of a trolley in the lab

The range of speeds you would choose would clearly depend on the speeds you could produce in the lab.

Temperature

You might be trying to find out the best temperature to grow tomatoes.

The 'best' temperature is dependent on a number of variables that taken together would produce tomatoes as fast as possible whilst not being too costly.

You should limit your investigation to just one variable, temperature and then consider other variables such as fuel cost later.

✸ Dependent variables

The dependent variable may be clear from the problem you're investigating, for example the stopping distance of moving objects. But you may have to make a choice.

Example 3

1 Investigation: Measuring the rate of photosynthesis in a plant

There are several ways in which you could measure the rate of photosynthesis in a plant. These include:

> counting the number of bubbles of oxygen produced in a minute by a water plant such as *Elodea* or *Cabomba*

> measuring the volume of oxygen produced over several days by a water plant such as *Elodea* or *Cabomba*

> monitoring the concentration of carbon dioxide in a polythene bag enclosing a potted plant using a carbon dioxide sensor

> measuring the colour change of hydrogencarbonate indicator containing algae embedded in gel.

Assessment tip

Again, it's often best to carry out a trial run or preliminary investigation, or carry out research, to determine the range to be investigated.

Assessment tip

The value of the *depend*ent variable is likely to *depend* on the value of the independent variable. This is a good way of remembering the definition of a dependent variable.

2 Investigation: Measuring the rate of a chemical reaction

You could measure the rate of a chemical reaction in the following ways:

> the rate of formation of a product

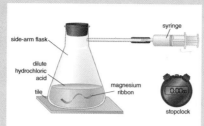

> the rate at which the reactant disappears

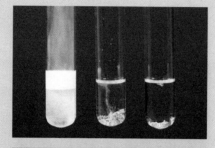

> a colour change

> a pH change.

✳ Control variables

The validity of your measurements depend on you measuring what you're supposed to be measuring.

Some of these variables may be difficult to control. For example, in an ecology investigation in the field, factors such as varying weather conditions are impossible to control.

Experimental controls

Experimental controls are often very important, particularly in biological investigations where you're testing the effect of a treatment.

Definition

Other variables that you're not investigating may also have an influence on your measurements. In most investigations, it's important that you investigate just one variable at a time. So other variables, apart from the one you're testing at the time, must be controlled, and kept constant, and not allowed to vary. These are called **control variables**.

Definition

An **experimental control** is used to find out whether the effect you obtain is from the treatment, or whether you get the same result in the absence of the treatment.

Example 4

Investigation: The effect of temperature on the growth of tomato plants.

The tomato plants grow most at 35 °C, but some plants at lower temperatures grow just as well. You need to be certain that the effect is caused by the temperature. There are lots of things that affect plant growth, so you should make sure these variables are controlled. These include the volume of water they receive, the soil that the plants are grown in, the nutrients present in the soil, and that the plants are as genetically similar as possible. Farmers often use f1 hybrid seeds as the plants are virtually genetically identical and will be ready to harvest at the same time.

Assessing and managing risk

Before you begin any practical work, you must assess and minimise the possible risks involved.

Before you carry out an investigation, you must identify the possible hazards. These can be grouped into biological hazards, chemical hazards and physical hazards.

Biological hazards include:
> microorganisms
> body fluids
> animals and plants.

Chemical hazards can be grouped into:
> irritant and harmful substances
> toxic
> oxidising agents
> corrosive
> harmful to the environment.

Physical hazards include:
> equipment
> objects
> radiation.

Scientists use an international series of symbols so that investigators can identify hazards.

Hazards pose risks to the person carrying out the investigation.

A risk posed by chlorine gas produced in the electrolysis of sodium chloride will be reduced if you increase the ventilation of the process, or devise a method to remove the gas so that workers cannot inhale it.

When you use hazardous materials, chemicals or equipment in the laboratory, you must use them in such a way as to keep the risks to absolute minimum. For example, one way is to wear eye protection when using hydrochloric acid.

Definition

A **hazard** is something that has the potential to cause harm. Even substances, organisms and equipment that we think of being harmless, used in the wrong way, may be hazardous.

Hazard symbols are used so that hazards can be identified

Definition

The **risk** is the likelihood of a hazard to cause harm in the circumstances it's being used in.

Assessment tip

When assessing risk, and suggesting control measures these should be specific to the hazard and risk, and not general. Hydrochloric acid is dangerous as it is 'corrosive, and skin and eye contact should be avoided 'will be given credit but wear 'eye protection' is too vague.

☀ Risk assessment

Before you begin an investigation, you must carry out a risk assessment. Your risk assessment must include:

✔ all relevant hazards (use the correct terms to describe each hazard, and make sure you include them all, even if you think they will pose minimal risk)

✔ risks associated with these hazards

✔ ways in which the risks can be minimised

✔ results of research into emergency procedures that you may have to take if something goes wrong.

You should also consider what to do at the end of the practical. For example, used agar plates should be left for a technician to sterilise; solutions of heavy metals should be collected in a bottle and disposed of safely.

Assessment tip

To make sure that your risk assessment is full and appropriate:

> remember that for a risk assessment for a chemical reaction, the risk assessment should be carried out for the products and the reactants

> when using chemicals, make sure the hazard and ways of minimising risk match the concentration of the chemical you're using; many acids, for instance, while being corrosive in higher concentrations, are harmful or irritant at low concentrations.

☀ Collecting primary data

✔ You should make sure that observations, if appropriate are recorded in detail. For example, it's worth recording the colour of your precipitate when making an insoluble salt, in addition to any other measurements you make.

✔ Measurements should be recorded in tables. Have one ready so that you can record your readings as you carry out the practical work.

✔ Think about the dependent variable and define this carefully in your column headings.

Definition

When you carry out an investigation, the data you collect are called **primary data.** The term 'data' is normally used to include your observations as well as measurements you might make.

✔ You should make sure that the table headings describe properly the type of measurements you've made, for example 'time taken for magnesium ribbon to dissolve'.

✔ It's also essential that you include units – your results are meaningless without these.

✔ The units should appear in the column head, and not be repeated in each row of the table.

☀ Repeatability and reproducibility of results

When making measurements, in most instances, it's essential that you carry out repeats.

These repeats are one way of checking your results.

Definition

One set of results from your investigation may not reflect what truly happens. Carrying out repeats enables you to identify any results that don't fit. These are called **outliers** or **anomalous results**.

Results will not be repeatable of course, if you allow the conditions the investigation is carried out in to change.

You need to make sure that you carry out sufficient repeats, but not too many. In a titration, for example, if you obtain two values that are within 0.1 cm³ of each other, carrying out any more will not improve the precision of your results.

This is particularly important when scientists are carrying out scientific research and make new discoveries.

Once you have planned your experiment and collected your primary data, your work will be retained. It will be needed when you complete the analysis and evaluation part of your controlled assessment.

Definition

If, when you carry out the same experiment several times, and get the same, or very similar results, we say the results are **repeatable**.

Definition

Taking more than one set of results will help to make sure your data is **precise**.

⚡ Part 3: Analysis and evaluation

Calculating the mean

Using your repeat measurements you can calculate the arithmetical mean (or just 'mean') of these data. Often, the mean is called the 'average.'

Here are the results of an investigation into the energy requirements of three different mp3 players. The students measured the energy using a joulemeter for ten seconds.

Mp3 player	Energy used in joules (J)			
	Trial 1	Trial 2	Trial 3	Mean
viking	5.5	5.3	5.7	5.5
anglo	4.5	4.6	4.9	4.7
saxon	3.2	4.5	4.7	4.6

You may also be required to use formulae when processing data. Sometimes, these will need rearranging to be able to make the calculation you need. Practise using and rearranging formulae as part of your preparation for assessment.

Significant figures

When calculating the mean, you should be aware of significant figures.

For example, for the set of data below:

18	13	17	15	14	16	15	14	13	18

The total for the data set is 153, and ten measurements have been made. The mean is 15, and not 15.3.

This is because each of the recorded values has two significant figures. The answer must therefore have two significant figures. An answer cannot have more significant figures than the number being multiplied or divided.

Using your data

When calculating means (and displaying data), you should be careful to look out for any data that don't fit in with the general pattern.

It might be the consequence of an error made in measurement. But sometimes outliers are genuine results. If you think an outlier has been introduced by careless practical work, you should ignore it when calculating the mean. But you should examine possible reasons carefully before just leaving it out.

Definition

The **reproducibility** of data is the ability of the results of an investigation to be reproduced by someone else, who may be in a different lab, carrying out the same work.

Definition

The **mean** is calculated by adding together all the measurements, and dividing by the number of measurements.

Definition

Significant figures are the number of digits in a number based on the precision of your measurements.

Definition

An **outlier** (or **anomalous result**) is a reading that is very different from the rest.

Displaying your data

Displaying your data – usually the means – makes it easy to pick out and show any patterns. And it also helps you to pick out any anomalous data.

It is likely that you will have recorded your results in tables, and you could also use additional tables to summarise your results. The most usual way of displaying data is to use graphs. The table will help you decide which type to use.

Type of graph	When you would use the graph	Example
Bar charts or bar graph	Where one of the variables is discrete	'The energy requirements of different mp3 players'
Line graph	Where independent and dependent variables are both continuous	'The volume of carbon dioxide produced by a range of different concentrations of hydrochloric acid'
Scatter graph	To show an association between two (or more) variables	'The association between length and breadth of a number of privet leaves' In scatter graphs, the points are plotted, but not usually joined

If it's possible from the data, join the points of a line graph using a straight line, or in some instances, a curve. In this way graphs can also help us to process data.

Assessment tip

Remember when drawing graphs, plot the independent variable on the *x*-axis, and the dependent variable on the *y*-axis.

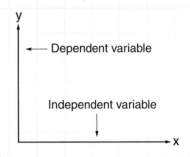

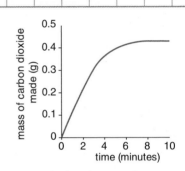

We can calculate the rate of production of carbon dioxide from the gradient of the graph

✵ Conclusions from differences in data sets

When comparing two (or more) sets of data, we often compare the values of two sets of means.

Example 5

Investigation: Comparing the braking distance of two tyres.

Two groups of students compared the braking distance of two tyres, labelled A and B. Their results are shown in the table.

Tyre	Braking distance in metres (m)										
	1	2	3	4	5	6	7	8	9	10	Mean
A	15	13	17	15	14	16	15	14	13	18	15
B	25	23	24	23	26	27	25	24	23	22	24

When the means are compared, it appears that tyre A will bring a vehicle to a stop in less distance than tyre B. The difference might have resulted from some other factor, or could be purely by chance.

Scientists use statistics to find the probability of any differences having occurred by chance. The lower this probability is, which is found out by statistical calculations, the more likely it is that tyre A is better at stopping a vehicle than tyre B.

Assessment tip

You have learnt about probability in your maths lessons.

Definition

If there is a relationship between dependent and independent variables that can be defined, we say there is a **correlation** between the variables.

✹ Drawing conclusions

Observing trends in data or graphs will help you to draw conclusions. You may obtain a linear relationship between two sets of variables, or the relationship might be more complex.

> ### Example 6
> **Conclusion:** The higher the concentration of acid, the shorter the time taken for the magnesium ribbon to dissolve.
>
> **Conclusion:** The higher the concentration of acid, the faster the rate of reaction.

When drawing conclusions, you should try to relate your findings to the science involved.

> In the first investigation in Example 6, your discussion should focus on the greater possibility/increased frequency of collisions between reacting particles as the concentration of the acid is increased.

> In the second investigation in Example 6, there's a clear scientific mechanism to link the rate of reaction to the concentration of acid.

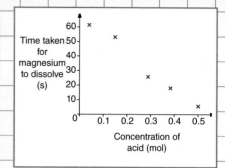

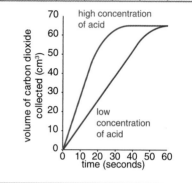

✹ Evaluating your investigation

Your conclusion will be based on your findings, but must take into consideration any uncertainty in these introduced by any possible sources of error. You should discuss where these have come from in your evaluation.

The two types of errors are:

✔ random error

✔ systematic error.

Random errors can occur when the instrument you're using to measure lacks sufficient sensitivity to indicate differences in readings. It can also occur when it's difficult to make a measurement. If two investigators measure the height of a plant, for example, they might choose different points on the compost, and the tip of the growing point to make their measurements.

Systematic errors are either consistently too high or too low. One reason could be down to the way you are making a reading, for example taking a burette reading at the wrong point on the meniscus. Another could be the result of an instrument being incorrectly calibrated, or not being calibrated.

> ## Definition
> **Error** is a difference between a measurement you make, and its true value.

> ## Definition
> With **random error**, measurements vary in an unpredictable way.

> ## Definition
> With **systematic error**, readings vary in a controlled way.

> ## Assessment tip
> A pH meter must be calibrated before use using buffers of known pH.

✳ Accuracy and precision

When evaluating your investigation, you should mention accuracy and precision. But if you use these terms, it's important that you understand what they mean, and that you use them correctly.

The terms accuracy and precision can be illustrated using shots at a target.

The shots are precise but not accurate.

The shots are precise and accurate.

The shots are not precise and not accurate.

Definition

When making measurements:
> the **accuracy** of the measurement is how close it is to the true value

> **precision** is how closely a series of measurements agree with each other.

✳ Improving your investigation

When evaluating your investigation, you should discuss how your investigation could be improved. This could be by:

✔ confirming your data. For example, you could make more repeats, or more frequent readings, or 'fine-tune' the range you chose to investigate, or refine your technique in some other way

✔ improving the accuracy and precision of your data, by using more precise measuring equipment.

✳ Using secondary data

As part of controlled assessment, you will be expected to compare your data – primary data – with **secondary data** you have collected.

The secondary data you collected earlier and the primary data you collected from your experiment will now be returned to you. You will be provided with an answer booklet to complete. This will help you to organise the data you have collected and answer specific questions about the topic you have been investigating.

You should review secondary data and evaluate it. Scientific studies are sometimes influenced by the **bias** of the experimenter.

✔ One kind of bias is having a strong opinion related to the investigation, and perhaps selecting only the results that fit with a hypothesis or prediction.

✔ Or the bias could be unintentional. In fields of science that are not yet fully understood, experimenters may try to fit their findings to current knowledge and thinking.

There have been other instances where the 'findings' of experimenters have been influenced by organisations that supplied the funding for the research.

You must fully reference any secondary data you have used, using one of the accepted referencing methods.

Assessment tip

Make sure you relate your conclusions to the hypothesis you are investigating. Do the results confirm or reject the hypothesis. Quote some results to back up your statement. e.g. 'My results at 35°C and 65°C show that over a 30°C change in temperature the time taken to produce 50 cm³ of carbon dioxide halved'.

Do the data support the hypothesis?

You need to discuss, in detail, whether all, or which of your primary, and the secondary data you have collected, support your original hypothesis. They may, or may not.

You should communicate your points clearly, using the appropriate scientific terms, and checking carefully your use of spelling, punctuation and grammar. You will be assessed on this written communication as well as your science.

If your data do not completely match the hypothesis, it may be possible to modify the hypothesis or suggest an alternative one. You should suggest any further investigations that can be carried out to support your original hypothesis or the modified version.

It is important to remember, however, that if your investigation does support the hypothesis, it can improve the confidence you have in your conclusions and scientific explanations, but it can't prove your explanations are correct.

 ## Referencing methods

The two main conventions for writing a reference are the:

✔ Harvard system
✔ Vancouver system.

In your text, the Harvard system refers to the authors of the reference, for example 'Smith and Jones (1978)'.

The Vancouver system refers to the number of the numbered reference in your text, for example '... the reason for this hypothesis is unknown.[5]'.

Though the Harvard system is usually preferred by scientists, it is more straightforward for you to use the Vancouver system.

Harvard system

In your references list a book reference should be written:

> Author(s) (year of publication). *Title of Book*, publisher, publisher location.

The references are listed in alphabetical order according to the authors.

Vancouver system

In your references list a book reference should be written:

> 1 Author(s). *Title of Book*. Publisher, publisher location: year of publication.

The references are number in the order in which they are cited in the text.

Assessment tip

Remember to write out the URL of a website in full. You should also quote the date when you looked at the website.

How to be successful in your GCSE Additional Science written examination

Introduction

OCR uses assessments to test how good your understanding of scientific ideas is, how well you can apply your understanding to new situations and how well you can analyse and interpret information you've been given. The assessments are opportunities to show how well you can do these.

To be successful in exams you need to:

✔ have a good knowledge and understanding of science

✔ be able to apply this knowledge and understanding to familiar and new situations, and

✔ be able to interpret and evaluate evidence that you've just been given.

You need to be able to do these things under exam conditions.

✳ The language of the external assessment

When working through an assessment paper, make sure that you:

✔ re-read a question enough times until you understand exactly what the examiner is looking for

✔ make sure that you highlight key words in a question. In some instances, you will be given key words to include in your answer

✔ look at how many marks are allocated for each part of a question. In general, you need to write at least as many separate points in your answer as there are marks.

✳ What verbs are used in the question?

A good technique is to see which verbs are used in the wording of the question and to use these to gauge the type of response you need to give. The table lists some of the common verbs found in questions, the types of responses expected and then gives an example.

Verb used in question	Response expected in answer	Example question
write down; state; give; identify	These are usually more straightforward types of question in which you're asked to give a definition, make a list of examples, or the best answer from a series of options	'Write down three types of microorganism that cause disease' 'State one difference and one similarity between radio waves and gamma rays'
calculate	Use maths to solve a numerical problem	'Calculate the percentage of carbon in copper carbonate $(CuCO_3)$'

estimate	Use maths to solve a numerical problem, but you do not have to work out the exact answer	'Estimate from the graph the speed of the vehicle after 3 minutes'
describe	Use words (or diagrams) to show the characteristics, properties or features of, or build an image of something	'Describe how meiosis halves the number of chromosomes in a cell to make egg or sperm cells'
suggest	Come up with an idea to explain information you're given, usually in a new or unfamiliar context	'Suggest why tyres with different tread patterns will have different braking distances'
demonstrate; show how	Use words to make something evident using reasoning	'Show how enzyme activity changes with temperature'
compare	Look for similarities and differences	'Compare aerobic and anaerobic respiration'
explain	To offer a reason for, or make understandable, information you're given	'Explain how carbon-14 dating could be used to estimate the age of the remains of an animal'
evaluate	To examine and make a judgement about an investigation or information you're given	'Evaluate the benefits of using a circuit breaker instead of a fuse in an electrical circuit'

What is the style of the question?

Try to get used to answering questions that have been written in lots of different styles before you sit the exam. Work through past papers, or specimen papers, to get a feel for these. The types of questions in your assessment fit the three assessment objectives shown in the table.

Assessment objective	Your answer should show that you can...
AO1 Recall the science	Recall, select and communicate your knowledge and understanding of science
AO2 Apply your knowledge	Apply skills, knowledge and understanding of science in practical and other contexts
AO3 Evaluate and analyse the evidence	Analyse and evaluate evidence, make reasoned judgements and draw conclusions based on evidence

Assessment tip

Of course you must revise the subject material adequately. But it's as important that you are familiar with the different question styles used in the exam paper, as well as the question content.

☀ How to answer questions on: AO1 Recall the science

These questions, or parts of questions, test your ability to recall your knowledge of a topic. There are several types of this style of question:

✔ Describe a process
✔ Explain a concept
✔ Complete sentences, tables or diagrams
✔ Tick the correct statements
✔ Use lines to link a term with its definition or correct statement

Example 7

What is meant by the term *exothermic reaction*?
Tick (✓) **one** box.

☐ a reaction that gives out heat energy
☐ a reaction that takes in energy from the surroundings
☐ a reaction that can go in either direction.

☀ How to answer questions on: AO1 Recall the science in practical techniques

You may be asked to recall how to carry out certain practical techniques; either ones that you have carried out before, or techniques that scientists use.

To revise for these types of questions, make sure that you have learnt definitions and scientific terms. Produce a glossary of these, or key facts cards, to make them easier to remember. Make sure your key facts cards also cover important practical techniques, including equipment, where appropriate.

Example 8

Describe how to find the work done when an object of 15N is moved 1 metre.

Assessment tip

Don't forget that mind maps – either drawn by you or by using a computer program – are very helpful when revising key points.

How to answer questions on: AO2 Apply skills, knowledge and understanding

Some questions require you to apply basic knowledge and understanding in your answers.

You may be presented with a topic that's familiar to you, but you should also expect questions in your Science exam to be set in an unfamiliar context.

Questions may be presented as:

✔ experimental investigations

✔ data for you to interpret

✔ a short paragraph or article.

The information required for you to answer the question might be in the question itself, but for later stages of the question, you may be asked to draw on your knowledge and understanding of the subject material in the question.

Practice will help you to become familiar with contexts that examiners use and question styles. But you will not be able to predict many of the contexts used. This is deliberate; being able to apply your knowledge and understanding to different and unfamiliar situations is a skill the examiner tests.

Practise doing questions where you are tested on being able to apply your scientific knowledge and your ability to understand new situations that may not be familiar. In this way, when this type of question comes up in your exam, you will be able to tackle it successfully.

> **Assessment tip**
>
> Work through the Preparing for Assessment: Applying your knowledge tasks in this book as practice.

Example 9

The force of gravity acts on all objects falling towards the Earth. Two identical packages are dropped out of an aircraft at the same time. Both have parachutes. One package's parachute opens, the other doesn't.

Suggest which package will hit the ground first. Explain your answer.

How to answer questions on: AO2 Apply skills, knowledge and understanding in practical investigations

Some opportunities to demonstrate your application of skills, knowledge and understanding will be based on practical investigations. You may have carried out some of these investigations, but others will be new to you, and based on data obtained by scientists. You will be expected to describe patterns in data from graphs you are given or that you will have to draw from given data.

Again, you will have to apply your scientific knowledge and understanding to answer the question.

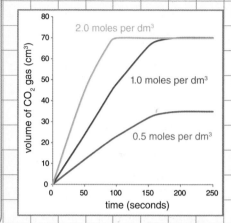

Example 10

Look at the graph showing the volume of gas collected when 10 g of calcium carbonate is reacted with three different concentrations of hydrochloric acid.

a What is the maximum volume of gas that can be produced using 1 mole per dm^3 of hydrochloric acid?

b Explain why this volume of gas is produced quicker when using 2 mole per dm^3 of hydrochloric acid?

c Suggest why 0.5 mole per dm^3 of hydrochloric acid does not produce this volume of gas.

 ## How to answer questions on: AO3 Analysing and evaluating evidence

For these types of questions, you will analyse and evaluate scientific evidence or data given to you in the question. It's likely that you won't be familiar with the material.

When describing patterns and trends in the data, make sure you:

✔ explain a pattern or trend in as much detail as you can

✔ mention anomalies where appropriate

You must also be able to evaluate the information you're given. This is one of the hardest skills. Think about the validity of the scientific data: did the technique(s) used in any practical investigation allow the collection of accurate and precise data?

Your critical evaluation of scientific data in class, along with the practical work and controlled assessment work, will help you to develop the evaluation skills required for these types of questions.

Assessment tip

Remember, when carrying out any calculations, you should include your working at each stage. You may get credit for getting the process correct, even if your final answer is wrong.

Example 11

1 The table shows the properties of some metals.

metal	melting point in °C	density in g/cm³	relative electrical conductivity	cost per tonne in £
aluminium	660	2.7	38	2491
copper	1083	8.9	60	9048
silver	962	10.5	63	1 125 276
zinc	420	7.1	17	2260

Pylon wires are made from metal.

Which metal would be most suitable for using to make pylon wires?

Use information about each of the metals in the table to explain your answer.

✳ The quality of your written communication

Scientists need good communication skills to present and discuss their findings. You will be expected to demonstrate these skills in the exam. Questions will [end] with the sentence: The quality of your written communication will be assessed in your answer to this question.

✔ You must also try to make sure that your spelling, punctuation and grammar are accurate, so that it's clear what you mean in your answer. Again, examiners can't award marks for answers where the meaning isn't clear.

✔ Make sure your language is concise. When describing and explaining science, use correct scientific vocabulary.

Practise answering some longer 6 mark questions. These will examine your quality of written communication as well as your knowledge and understanding of science. Look at how marks are awarded in mark schemes provided by the awarding body. You'll find these in the specimen question papers, and past papers.

You will also need to remember the writing and communication skills you've developed in English lessons. For example, make sure that you understand how to construct a good sentence using connectives.

Assessment tip

You will be assessed on the way in which you communicate science ideas.

Assessment tip

When answering questions, you must make sure that your writing is legible. An examiner can't award marks for answers that he or she can't read.

All long answer, 6 mark, questions will have the same general mark scheme which will be amplified and made specific to the question by examiners. Three levels of answer will gain credit.

Level 3

All information in answer is relevant, clear, organised and presented in a structured and coherent format. Specialist terms are used appropriately. Few, if any, errors in grammar, punctuation and spelling. (5–6 marks)

Level 2

For the most part the information is relevant and presented in a structured and coherent format. Specialist terms are used for the most part appropriately. There are occasional errors in grammar, punctuation and spelling. (3–4 marks)

Level 1

Answer may be simplistic. There may be limited use of specialist terms. Errors of grammar, punctuation and spelling prevent communication of the science. (1–2 marks)

Level 0

Insufficient or irrelevant science. Answer not worthy of credit. (0 marks)

✳ Revising for your Science exam

You should revise in the way that suits you best. But it's important that you plan your revision carefully, and it's best to start well before the date of the exams. Take the time to prepare a revision timetable and try to stick to it. Use this during the lead up to the exams and between each exam.

When revising:

- ✔ find a quiet and comfortable space in the house where you won't be disturbed. It's best if it's well ventilated and has plenty of light
- ✔ take regular breaks. Some evidence suggests that revision is most effective when you revise in 30 to 40 minute slots. If you get bogged down at any point, take a break and go back to it later when you're feeling fresh. Try not to revise when you are feeling tired. If you do feel tired, take a break
- ✔ use your school notes, textbook and possibly a revision guide. But also make sure that you spend some time using past papers to familiarise yourself with the exam format
- ✔ produce summaries of each [topic] or [module]
- ✔ draw mind maps covering the key information on a [topic] or [module]

Assessment tip

Try to make your revision timetable as specific as possible – don't just say 'science on Monday, and Thursday', but list the [modules] that you'll cover on those days.

✔ set up revision cards containing condensed versions of your notes

✔ ask yourself questions, and try to predict questions, as you're revising topics or [modules]

✔ test yourself as you're going along. Try to draw key labelled diagrams, and try some questions under timed conditions

✔ prioritise your revision of topics. You might want to allocate more time to revising the topics you find most difficult.

Assessment tip

Start your revision well before the date of the exams, produce a revision timetable, and use the revision strategies that suit your style of learning. Above all, revision should be an active process.

✳ How do I use my time effectively in the exam?

Timing is important when you sit an exam. Don't spend so long on some questions that you leave insufficient time to answer others. For example, in a 60-mark question paper, lasting one hour, you will have, on average, one minute per question.

If you're unsure about certain questions, complete the ones you're able to do first, then go back to the ones you're less sure of.

If you have time, go back and check your answers at the end of the exam.

✳ What will my exam look like?

Your science exam consists of two papers.

Paper 1 contains three sections and lasts 1 hour 15 minutes.
There are 25 marks for each section.
You should spend about 25 minutes answering each section.
The questions in each section will test objectives AO1, AO2 and AO3 using structured questions.

Paper 2 contains four sections and lasts 1 hour 30 minutes.
The first three sections will be similar to the Paper 1 sections.
Section D contains a ten mark data response question which primarily assesses objective AO3. You will be required to analyse and evaluate evidence, make reasoned judgements and draw conclusions based on evidence.

You should spend about 15 minutes answering this section.

✳ On exam day

A little bit of nervousness before your exam can be a good thing, but try not to let it affect your performance. When you turn over the exam paper keep calm. Look at the paper and get it clear in your head exactly what is required from each question. Read each question carefully. Don't rush.

If you read a question and think that you have not covered the topic, keep calm – it could be that the information needed to answer the question is in the question itself or the examiner may be asking you to apply your knowledge to a new situation.

Finally, good luck!

✳ Mathematical skills

You will be allowed to use a calculator in all assessments.

These are the maths skills that you need, to complete all the assessments successfully.

You should understand:

✔ the relationship between units, for example, between a gram, kilogram and tonne

✔ compound measures such as speed

✔ when and how to use estimation

✔ the symbols $=$ $<$ $>$ $\sim$

✔ direct proportion and simple ratios

✔ the idea of probability.

You should be able to:

✔ give answers to an appropriate number of significant figures

✔ substitute values into formulae and equations using appropriate units

✔ select suitable scales for the axes of graphs

✔ plot and draw line graphs, bar charts, pie charts, scatter graphs and histograms

✔ extract and interpret information from charts, graphs and tables.

You should be able to calculate:

✔ using decimals, fractions, percentages and number powers, such as 10^3

✔ arithmetic means

✔ areas, perimeters and volumes of simple shapes

In addition, if you are a higher tier candidate, you should be able to:

✔ **change the subject of an equation**

and should be able to use:

✔ **numbers written in standard form**

✔ **calculations involving negative powers, such as 10^{-1}**

✔ **inverse proportion**

✔ **percentiles and deciles.**

✳ Some key physics equations

With the written papers, there will be an equation sheet. Below are some of the key equations found on the sheet; it will help if you practise using them.

$$\text{average speed} = \frac{\text{distance}}{\text{time}}$$

distance = average speed × time $\qquad s = \dfrac{(u + v)t}{2}$

$$\text{acceleration} = \frac{\text{change in speed}}{\text{time taken}} \qquad a = \frac{(v - u)}{t}$$

force = mass × acceleration $\qquad F = ma$

stopping distance = thinking distance + braking distance

work done = force × distance

weight = mass × gravitational field strength

$$\text{power} = \frac{\text{work done}}{\text{time}}$$

power = force × speed

$KE = \frac{1}{2}\, mv^2$

momentum = mass × velocity

$$\text{force} = \frac{\text{change in momentum}}{\text{time}}$$

GPE = mgh

$$\text{resistance} = \frac{\text{voltage}}{\text{current}}$$

power = voltage × current

Glossary

A

ABS braking system known as advance braking system which helps to control a skidding car

accelerate an object accelerates if it speeds up

acceleration a measurement of how quickly the speed of a moving object changes (if speed is in m/s the acceleration is in m/s²)

acid rain rain water which is made more acidic by pollutant gases

acrosome part of the sperm that contains enzymes

activation energy the energy needed to start a chemical reaction

active site the place on an enzyme where the substrate molecule binds

active transport in active transport, cells use energy to transport substances through cell membranes against a concentration gradient

activity average number of nuclei that decay every second

aerobic respiration respiration that involves oxygen

air bags cushions which inflate with gas to protect people in a vehicle accident

air resistance the force exerted by air to any object passing through it

alkali metals very reactive metals in group 1 of the periodic table, e.g. sodium

alkalis substances which produce OH⁻ ions in water

allotropes different forms of the same element

alpha particles radioactive particles which are helium nuclei – helium atoms without the electrons (they have a positive charge)

alternating current or voltage an electric current that is not a one-way flow

amino acids small molecules from which proteins are built

ammeter meter used in an electric circuit for measuring current

ampere (A) the unit used to measure electrical current, often abbreviated to amp

amplitude wave in which the vibrations are in the same direction as the direction in which the wave travels

anaerobic respiration respiration without using oxygen

aquifer underground layer of permeable rock or loose materials (gravel or silt) where groundwater is stored

arteries blood vessels that carry blood away from the heart

asexual reproduction reproduction involving only one parent

atom the basic 'building block' of an element which cannot be chemically broken down

atom economy a way of measuring the amount of atoms that are wasted or lost when a chemical is made

atomic number the number of protons found in the nucleus of an atom

ATP molecule used to store energy in the body

attract move towards, for example, unlike charges attract

average speed total distance travelled divided by the total time taken for a journey

B

background radiation ionising radiation from space and rocks, especially granite, that is around us all the time but is at a very low level

bacteria single-celled micro-organisms which can either be free-living organisms or parasites (they sometimes invade the body and cause disease)

balanced equation chemical equation where the number of atoms on each side of the equation balance each other

balanced forces forces acting in opposite directions that are equal in size

barium chloride testing chemical for sulfates in water

batch process a process used to make small fixed amounts of substances, like medicines, with a clear start and finish

battery two or more electrical cells joined together

becquerels (Bq) unit of activity or count rate; 1Bq = 1 count per second

best-fit straight line or curve best suited to experimental results when plotted on a graph

beta particles particles given off by some radioactive materials (they have a negative charge)

biodiversity range of different living organisms in a habitat

biofuels fuels made from plants – these can be burned in power stations

biological catalyst molecules in the body that speed up chemical reactions

biological control a natural predator is released to reduce the number of pests infesting a crop

boiling point temperature at which the bulk of a liquid turns to vapour

boron control rods rods that are raised or lowered in a nuclear reactor to control the rate of fission

braking distance distance travelled while a car is braking

buckminsterfullerene a very stable sphere of 60 carbon atoms joined by covalent bonds. An allotrope of carbon

C

cancer life-threatening condition where body cells divide uncontrollably

capillaries small blood vessels that join arteries to veins

carbon a very important element, carbon is present in all living things and forms a huge range of compounds with other elements

carbon-14 radioactive isotope of carbon

carbon dioxide (CO_2) a greenhouse gas which is emitted into the atmosphere as a by-product of combustion

carotene plant pigment involved in photosynthesis

catalyst a chemical that speeds up a reaction but is not changed or used up by the reaction

cathode the negative electrode in a circuit or battery

cell differentiation when cells become specialised

chain reaction a reaction where the products cause the reaction to go further or faster, e.g. in nuclear fission

charge(s) a property of matter charge exists in two forms, positive and negative, which attract each other

chemical properties the characteristic reactions of substances

chlorination addition of chlorine to water supplies to kill micro-organisms

chlorophyll pigment found in plants which is used in photosynthesis (gives green plants their colour)

chloroplast a cell structure found in green plants that contains chlorophyll

chromatography a method for splitting up a substance to identify compounds and check for purity

chromosomes thread-like structures in the cell nucleus that carry genetic information

circuit breakers resettable fuses

clone genetically identical copy

close packed metal ions structure of a metal

'cold fusion' attempts to produce fusion at normal room temperature that have not been validated since other scientists could not reproduce the results

collagen protein used for support in animal cells

collision frequency the number of successful collisions between reacting particles that happen in one second.

combustion process where fuels react with oxygen to produce heat

community all the plants and animals living in an ecosystem, e.g. a garden

compost dead and decaying plant material

compound two or more elements which are chemically joined together, e.g. H_2O

compressions particles pushed together, increasing pressure

computer modelling using a computer to 'model' situations to see how they are likely to work out if you do different things

concentration the amount of chemical dissolved in a certain volume of solution

conducting gel applied to a patient's chest before using a defibrillator to ensure good electrical contact

conductors materials which transfer thermal energy easily; electrical conductors allow electricity to flow through them

conservation of energy principle stating that energy cannot be created or destroyed, but can be altered from one form to another

conservation of mass the total mass of reactants equals the total mass of products formed

continuous process a process for making a large amount of chemicals which runs all the time if new materials are added and wastes removed

cosmic rays radiation from space that contributes to background radiation

count rate average number of nuclei that decay every second

covalent bonds bonds between atoms where some of the electrons are shared

crash barrier barrier used to prevent vehicles crossing from one carriageway to the other, causing a head-on collision

crash testing deliberately crashing vehicles and analysing results to improve car safety

crenation when red blood cells shrink in concentrated solutions, they look partly deflated

crop rotation system of growing crops in sequence

crop sprayer charges fertiliser, pesticides, etc. to give wider, even coverage

cross-sectional area the area displaced by a moving object

cruise control system that automatically controls the speed of a vehicle

crumple zones areas of a car that absorb the energy of a crash to protect the centre part of the vehicle

current flow of electrons in an electric circuit

curved line line of changing gradient

D

decay   to rot

decelerates   an object decelerates if it slows down

deceleration   a measurement of how quickly the speed of a moving object decreases

deciduous   type of tree that loses its leaves in winter

decomposer   an organism that breaks down dead organic matter

defibrillator   machine which gives the heart an electric shock to start it beating regularly

delocalised electrons   electrons which are free to move away through a collection of ions – as in a metal

denatured   an enzyme is denatured if its shape changes so that the substrate cannot fit into the active site

density   the density of a substance is found by dividing its mass by its volume

detritivore   an organism that eats dead material, e.g. an earthworm

detritus   the dead and semi-decayed remains of living things

diagnostic   process for identifying the nature or cause of a medical problem

diesel oil   fuel for diesel engines, traditionally obtained from oil but other forms such as biodiesel are being developed

diffuse   when particles diffuse they spread out

diploid   cells that have two copies of each chromosome

direct current   an electric current that flows in one direction only

displacement reaction   chemical reaction where one element displaces or 'pushes out' another element from a compound

distance–time graph   a plot of the distance moved against the time taken for a journey

distillation   the process of evaporation followed by condensation

DNA   molecule found in all body cells in the nucleus – its sequence determines how our bodies are made (e.g. do we have straight or curly hair), and gives each one of us a unique genetic code

DNA bases   four chemicals that are found in DNA, they make up the base sequence and are given the letters A, T, G and C

dot and cross model   a drawn model representing the number of electrons in the outside shell of bonding atoms or ions

double circulatory system   where the blood is pumped to the lungs then returned to the heart before being pumped round the body

double covalent bond   covalent bond where each atom shares two electrons with the other atom

double-insulated   an electrical device in which there are at least two layers of insulation between the user and the electrical wires

drag   energy losses caused by the continual pushing of an object against the air or a liquid

dummies   used in crash testing to learn what would happen to the occupants of a car in a crash

E

earth wire   the third wire in a mains cable which connects the case of an appliance to the ground so that the case cannot become charged and cause an electric shock

earthed (electrically)   connected to the ground (at 0V)

echoes   reflection of sound (or ultrasound)

efficient   a process in which losses are minimised

electric cars   cars running on solar power or batteries

electric windows   windows that can be opened or closed at the push of a button

electrical conductivity   the measurement of the ability to conduct electricity

electrical conductors   materials which let electricity pass through them

electromagnet   a magnet which is magnetic only when a current is switched on

electromagnetic waves   a group of waves that carry different amounts of energy – they range from low frequency radio waves to high frequency gamma rays

electron gain   gaining of electrons to form negative ions

electron loss   losing of electrons to form positive ions

electron shells   the orbit around the nucleus likely to contain the electron

electronic structure   the number of electrons in sequence that occupy the shells, e.g. the 11 electrons of sodium are in sequence 2.8.1

electrons   small particles within an atom that orbit the nucleus (they have a negative charge)

electrostatic attraction   attraction between opposite charges, e.g. between Na^+ and Cl^-

electrostatic dust precipitators   charged plates inside factory chimneys remove dust particles from smoke

(electrostatic) paint sprayer   charges paint droplets to give even coverage

elements   substances made out of only one type of atom

endothermic reaction   chemical reaction which takes in heat

energy   the ability to 'do work' – the human body needs energy to function

enriched uranium uranium containing more of the U-235 isotope than occurs naturally

environment the surroundings of an object

enzymes biological catalysts that increase the speed of a chemical reaction

escape lane rough-surfaced uphill path adjacent to a steep downhill road enabling vehicles with braking problems to stop safely

estimated calculated approximately the value of something

evaporation when a liquid changes to a gas, it evaporates

exhaust gases gases discharged into the atmosphere as a result of combustion of fuels

exothermic reaction chemical reaction in which heat is given out

explosion a very fast reaction making large volumes of gas

extrapolation making an estimate beyond the range of results

F

fertilisation when a sperm fuses (joins with) an egg

fertiliser chemical put on soil to increase soil fertility and allow better growth of crop plants

filament a very fine wire, typically in an old-style incandescent lamp, that emits heat and light when a current passes through it

filtration the process of filtering river or ground water to purify it for drinking water

fission splitting apart, especially of large radioactive nuclei such as uranium

flaccid floppy

flame test test where a chemical burns in a Bunsen flame with a characteristic colour – tests for metal ions

force a push or pull which is able to change the velocity or shape of a body

fossil fuels fuels such as coal, oil and gas

free-fall a body falling through the atmosphere without an open parachute

frequency the number of waves passing a set point per second

friction energy losses caused by two or more objects rubbing against each other

fuel consumption the distance travelled by a given amount of fuel, e.g. in km/100 litres

fuel rods rods of enriched uranium produced to provide fuel for nuclear power stations

fullerenes cage-like carbon molecules containing many carbon atoms, e.g. buckyballs

fungicide chemical used to kill fungi

funicular railway inclined plane or cliff railway

fuse(s) a special component in an electric circuit containing a thin wire which is designed to melt if too much current flows through it, breaking the circuit

fusion the joining together of small nuclei, such as hydrogen isotopes, at very high temperatures with the release of energy

fusion bombs hydrogen bombs or H-bombs based on fusion reactions

G

gametes the male and female sex cells (sperm and eggs)

gamma rays ionising electromagnetic waves that are radioactive and dangerous to human health – but useful in killing cancer cells

Geiger counter a device used to detect some types of radiation

Geiger-Muller tube a device used to detect some types of radiation

gel in ultrasound scanning, placed on the skin so that nearly all the ultrasound passes into the body and is not reflected by the skin

gene section of DNA that codes for a particular characteristic

gene pool the different genes available within a species

gene therapy medical procedure where a virus is used to 'carry' a gene into the nucleus of a cell (this is a new treatment for genetic diseases)

generator device that converts rotational kinetic energy to electrical energy

genetic engineering transfer of genes from one organism to another

giant ionic lattice sodium chloride forms a lattice, also called a giant ionic structure

giant molecular structures a large regular three-dimensional covalently bonded structure containing more that one non-metal element

gradient rate of change of two quantities on a graph; change in y / change in x

granite mineral containing low levels of uranium

graphite a type of carbon used as a moderator in a nuclear power station

gravitational field strength the force of attraction between two masses

gravitational potential energy the energy a body has because of its position in a gravitational field, e.g. an object

gravity an attractive force between objects (dependent on their mass)

303

greenhouse gas   any of the gases whose absorption of solar radiation is responsible for the greenhouse effect, e.g. carbon dioxide, methane

group   within the periodic table the vertical columns are called groups

group 1 metals   metals in the group 1 of the periodic table, e.g. lithium, sodium and potassium

group 7 elements   non-metals in group 7 of the periodic table, e.g. fluorine, bromine and iodine

H

Haber process   industrial process for making ammonia

habitat   where an organism lives, e.g. the worm's habitat is the soil

haemoglobin   chemical found in red blood cells which carries oxygen

half-life   average time taken for half the nuclei in a radioactive sample to decay

halogens   reactive non-metals in group 7 of the periodic table , e.g. chlorine

haploid   cells that have only one copy of each chromosome

hardness   hardness of solid material as tested by the scratch test

heat conductor   a material that conducts heat

helium   second element in periodic table; an alpha particle is a helium nucleus

herbicide   chemical used to kill weeds

hertz (Hz)   units for measuring wave frequency

hybrid cars   cars powered by electric batteries which also have fuel engines

hydroponics   growing plants in mineral solutions without the need for soil

i

impact   collision between two moving objects or a moving object and a stationary object

inbreeding   breeding closely related animals

insecticide   a chemical that can kill an insect

insoluble salt   salt which is not soluble in water (forms a precipitate)

instantaneous speed   the speed of a moving object at one particular moment

insulator   a material that transfers thermal energy only very slowly

insulin   hormone made by the pancreas which controls the level of glucose in the blood

intensive farming   farming that uses a lot of artificial fertilisers and energy to produce a high yield per farm worker

intermolecular force   force between molecules

interpolation   making an estimate beyond the range of results in a range

iodine   radioactive isotopes of iodine are used in diagnosing and treating thyroid cancer

ionic bond   a chemical bond between two ions of opposite charges

ionic equation   an equation representing the formation of ions by the transfer of electrons

ionises   adds or removes electrons from an atom leaving it charged

ions   charged particles (can be positive or negative)

isotopes   atoms with the same number of protons but different numbers of neutrons

J

joule   unit of work done and energy

joule meter   meter used to measure electrical energy.

K

kilogram (kg)   unit of mass

kinetic energy   the energy that moving objects have

kite diagram   method of displaying results from a transect line

L

lead   heaviest element having a stable isotope; all isotopes of the elements above it in the periodic table are unstable

limewater   calcium hydroxide particles in water – this clear liquid turns milky in the presence of carbon dioxide

limiting reactant   chemical used up in a reaction that limits the amount of product formed

linear   a line of constant gradient on a graph

live (wire)   carries a high voltage into and around the house

longitudinal (wave)   wave in which the vibrations are in the same direction as the direction in which the wave travels

lubricating   oiling

lustrous   shiny

lysis   to split apart

M

magnitude   size of something

mass   describes the amount of something; it is measured in kilograms (kg)

mass number   number of protons and neutrons in a nucleus

meiosis   cell division that results in haploid cells

melting point   the temperature at which a solid turns into a liquid

meristem   tips of roots and shoots where cell division and elongation takes place

messenger RNA   copy of a section of DNA used to carry the gene code to the ribosomes

metabolic rate   amount of energy the body needs

metal halide   a compound of a halogen and a metal, e.g. potassium bromide

metallic bonding   the bonding between close-packed metal ions due to delocalised electrons

metallic properties   the physical properties specific to a metal, such as lustre and electrical conductivity

metals   solid substances that are usually lustrous, conduct electricity and form ions by losing electrons

microbes   tiny microscopic organisms

minerals   natural solid materials with a fixed chemical composition and structure, rocks are made of collections of minerals; mineral nutrients in our diet are things like calcium and iron, they are simple chemicals needed for health

mitochondria   structures in a cell where respiration takes place

mitosis   cell division that results in genetically identical diploid cells

moderator   material used to slow down neutrons in a nuclear power station

molecular formula   the formula of a chemical using symbols in the periodic table, e.g. methane has a molecular formula of CH_4

molecule   two or more atoms which have been chemically combined

molten liquid   a liquid that has just melted, usually referring to rock, ores, metals or salts with very high melting points

momentum   the product of mass and velocity

multicellular organism   organisms made up of many specialised cells

mutation   where the DNA within cells have been altered (this happens in cancer)

N

nanometre   units used to measure very small things (one billionth of a metre)

nanoparticles   very small particles on the nanoscale

nanotube   carbon atoms formed into a very tiny tube

negative ion   an ion made by an atom gaining electrons

net force   same as resultant force

neutral   a neutral substance has a pH of 7

neutral (wire)   provides a return path for the current in a mains supply to a local electricity substation

neutrons   small particle which does not have a charge found in the nucleus of an atom

newtons   unit of force (abbreviated to N)

nickel catalyst   a catalyst used in the hardening of margarine

nitrate residue   unwanted residues sometimes found in water contaminated by farm run-off

non-metals   substances that are dull solids, liquids or gases that do not conduct electricity and form ions by gaining electrons

non-renewable   something which is used up at a faster rate than it can be replaced e.g. fossil fuels

nuclear equation   equation showing changes to the nuclei in a nuclear reaction

nuclear power stations   power stations using the energy produced by nuclear fission to generate heat

nuclear transfer   type of cloning that involves taking a nucleus from a body cell and placing it into an egg cell

nucleons   protons and neutrons (both found in the nucleus)

nucleus   central part of an atom that contains protons and neutrons

O

ohms   units used to measure resistance to the flow of electricity

optimum temperature   the temperature range that produces the best reaction rate

organic food   food produced by organic farming using no artificial fertilisers or pesticides

osmosis   when solutions of different concentrations are separated by a semi-permeable membrane, water molecules pass through the membrane moving from the dilute solution to the more concentrated one to reduce the difference

oxidation   the process of electron loss

oxygen debt   the debt for oxygen that builds up in the body when demand for oxygen is greater than supply

P

paddle shift controls controls attached to the steering wheel of a car so that the driver can use them without taking their eyes off the road

paddles charged plates in a defibrillator that are placed on the patient's chest

palisade cells tightly packed together cells found on the upper side of a leaf

parallel circuit electric circuit formed by more than one loop so that the electrons can go through different paths

partially-permeable membrane a membrane that allows some small molecules to pass through but not larger molecules

percentage yield comparing the amount of useful product made to the amount expected

period a row in the periodic table

periodic table a table of all the chemical elements based on their atomic number

pesticide residue unwanted residues sometimes found in water contaminated by local pesticide use

petrol volatile mixture of hydrocarbons used as a fuel

pharmaceuticals medical drugs

phloem specialised transporting cells which form tubules in plants to carry sugars from leaves to other parts of the plant

photocopier uses electrostatics to copy documents

photosynthesis process carried out by green plants where sunlight, carbon dioxide and water are used to produce glucose and oxygen

physical property property that can be measured without changing the chemical composition of a substance, e.g. hardness

pitch whether a sound is high or low on a musical scale

plantlets small plants formed by strawberries during asexual reproduction plasma yellow liquid found in blood

plasma yellow liquid found in blood

plasmolysis the shrinking of a plant cell due to loss of water, the cell membrane pulls away

platelets cell fragments which help in blood clotting

pollutants unwanted residues found that can sometimes cause damage

pollute contaminate or destroy the environment

pollution contaminating or destroying the environment as a result of human activities

population group of organisms of the same species

positive ion an ion made by an atom losing electrons

potential difference another word for voltage (a measure of the energy carried by the electric current)

power rate of transfer of energy;
electric power = voltage × current

power station facility that generates electricity on a large scale

power transmission transmission of electricity

precipitate solid formed in a solution during a chemical reaction

precipitation reaction chemical test in which a solid precipitate is formed – tests for metal ions

pressure wave vibrating particles in a longitudinal wave creating pressure variations

primary safety features help to prevent a crash, e.g. ABS brakes, traction control

product molecules produced at the end of a chemical reaction

protons small positive particles found in the nucleus of an atom

R

radioactive waste waste produced by radioactive materials used at nuclear power stations, research centres and some hospitals

radiocarbon dating method of dating some old artefacts using carbon-14

radiographer a technician who works in a hospital radiography department, possibly taking x-rays or treating some types of cancer with radiation

radioisotope isotope of an element that is radioactive

radiotherapy using ionizing radiation to kill cancer cells in the body

random having no regular pattern

rarefactions particles further apart than usual, decreasing pressure

rate of reaction the speed with which a chemical reaction takes place

ratemeter a device that measures the amount of radiation detected by a Geiger-Muller tube

reactants chemicals which are reacting together in a chemical reaction

reaction time the time it takes for a driver to step on the brake after seeing an obstacle

recharging battery being charged with a flow of electric current

red blood cells blood cells which are adapted to carry oxygen

reduction the process of electron gain

refined the refining process turns crude oil into usable forms such as petrol

reflected radiation rebounding off a surface

relative atomic mass the mass of an atom compared to 1/12 of a carbon atom

relative formula mass the sum of the relative atomic masses in a compound

relative velocity vector difference between the velocities of two objects

renewable energy that can be replenished at the same rate that it's used up e.g. biofuels

repel move away, for example, like charges repel

reservoir a water resource where large volumes of water are held

resistance measurement of how hard it is for an electric current to flow through a material

respiration process occurring in living things where oxygen is used to release the energy in foods

respiratory quotient (RQ) equation used to determine the substrate used in respiration

resultant force the combined effect of forces acting on an object

rheostat a variable resistor

ribosome structures in a cell where protein synthesis takes place

S

safety cage a car's rigid frame that protects occupants in a roll-over accident

saprophyte an organism that breaks down dead organic matter, usually used to refer to fungi

sea water water containing high levels of dissolved salts making it undrinkable

seatbelts harness worn by occupants of motor vehicles to prevent them from being thrown about in a collision

secondary safety features protect occupants in the event of a crash, e.g. crumple zones, air bags, seat belts

sedimentation a process during water purification where small solid particles are allowed to settle

selective breeding process of breeding organisms with the desired characteristics

series circuit circuit formed by a single loop of electrical conductors

shock occurs when a person comes into contact with an electrical energy source so that electrical energy flows through a portion of the body

side impact beams bars in the side of a car to lessen the amount of bodywork distortion inside the car

silver nitrate a chemical used for testing halide ions in water

single covalent bond bond between hydrogen atoms where each atom shares its electron with the other

smoke detector device to detect smoke, some forms of which contain a source of alpha radiation

solar energy energy from the Sun

solar-powered energy provided by the Sun

soluble a soluble substance can dissolve in a liquid, e.g. sugar is soluble in water

solution when a solute dissolves in a solvent, a solution forms

sound energy anything making a noise gives out sound energy

sparks type of electrostatic discharge briefly producing light and sound

speed how fast an object travels: speed = distance ÷ time

speed camera device used to measure the speed of a moving vehicle

speed–time graph a plot of how the speed of an object varies with time

spongy mesophyll cells found in the middle of a leaf with an irregular shape and large air spaces between them

stable electronic structure an achieved structure where the outer electron shell of an atom is full

stable (nucleus) (nucleus) is not radioactive; it will not decay

stem cells unspecialised body cells (found in bone marrow) that can develop into other, specialised, cells that the body needs, e.g. blood cells

sterilise killing all the organisms in an area, usually used to mean killing micro-organisms

stomata (*singular* stoma) small holes in the surface of leaves which allow gases in and out of leaves

stopping distance sum of the thinking and braking distances

straight line line of constant gradient

streamlining shaping an object to reduce resistance to motion

sub-atomic particles particles that make up an atom, e.g. protons, neutrons and electrons

substrate molecules at the start of a chemical reaction

superconductors materials that conduct electricity with little or no resistance

T

temperature coefficient (Q_{10}) equation used to calculate the effect of temperature on the rate of an enzyme-controlled reaction

tensile strength a force which stretches something

terminal speed or velocity the top speed reached when drag matches the driving force

therapy treatment of a medical problem

thermal decomposition the breaking down of a compound into two or more products on heating

thermal energy another name for heat energy

thinking distance distance travelled while the driver reacts before braking

thinking time time for a driver to react before braking

thyroid gland gland at the base of the neck which makes the hormone thyroxin

tissue culture process that uses small sections of tissue to clone plants

toxic a toxic substance is one which is poisonous, e.g. toxic waste

tracers a radioactive, radiation-emitting substance used to follow movement of a particular chemical, e.g. in nuclear medicine, tracking the path of an underground pipe, etc.

traction control helps limit tyre slip in acceleration on slippery surfaces

transect line across an area to sample organisms

transition element an element in the middle section of the periodic table, between the group 1 and 2 block and the group 3 to group 0 block

transmitted radiation passing through an object

transverse (wave) wave in which the vibrations are at right angles to the direction in which the wave travels

tread pattern on part of tyre that comes in contact with road surface to provide traction

trials tests to find if something works and is safe

tumour abnormal mass of tissue that is often cancerous

turbine device for generating electricity – the turbine moves through a magnetic field and electricity is generated

turgid plant cells which are full of water with their walls bowed out and pushing against neighbouring cells

turgor pressure the pressure exerted on the cell membrane by the cell wall when the cell is fully inflated

U

ultrasound high-pitched sounds which are too high for detection by human ears

unbalanced (forces) forces acting in opposite directions that are unequal in size

unicellular organism organisms made of only one cell

unstable (nucleus) liable to decay

uranium radioactive element with a very long half-life used in nuclear power stations

V

vacuum space containing hardly any particles

Van de Graaff generator a machine which uses a moving belt to accumulate very high charges on a hollow metal globe

variable resistor a resistor whose resistance can change

vascular bundle group of xylem and phloem cells

veins blood vessels that carry blood back to the heart

velocity how fast an object is travelling in a certain direction: velocity = displacement ÷ time

voltage a measure of the energy carried by an electric current (also called the potential difference)

voltmeter instrument used to measure voltage or potential difference

volts (V) units used to measure voltage

W

water conservation the acts of reducing water consumption through planned choice, e.g. hosepipe bans and water metering

water resources places form where water is extracted or where it is stored, e.g. aquifers, reservoirs or lakes

watt (W) a unit of power, 1 watt equals 1 joule of energy being transferred per second

wave oscillatory motion

wavelength (λ) distance between two wave peaks

weight the force of gravity acting on a body

white blood cells blood cells which defend against disease

work work is done when a force moves

work done work done is the product of the force and distance moved in the direction of the force

X

xanthophylls plant pigments involved in photosynthesis

xylem cells specialised for transporting water through a plant; xylem cells have thick walls, no cytoplasm and are dead, their end walls break down and they form a continuous tube

x-rays ionising electromagnetic waves used in x-ray photography (where x-rays are used to generate pictures of bones)

Index

Index

and weight   198–9, 212, 213, 215
work done   198–9
fossil fuels   204, 205, 206, 207
Franklin, Rosalind   11
free-fall   198, 212, 213, 215
frequency   242–3, 245, 246
friction   197, 206, 207, 212, 214, 217, 230, 232
Frozen Ark, The   10
fuels
biofuels   204, 205, 207
cars   200–1, 204, 205, 206–7
combustion   122–3, 132–3, 201
consumption   200–1, 205, 206–7
efficiency   206–7
energy comparisons   122–3
kinetic energy   204
fullerenes   130–1
fungi   80, 81, 82–3
funicular railway   218
fuses   240, 241
fusion   262–3

G

gametes   24–5, 39
gamma radiation   248, 250, 253, 256, 257, 258–9
gases
collision theory   103
explosions   106–7
products of reactions   100, 104, 108, 109, 113
see also individual gases
Geiger counter   248, 252, 253
gene pool   37
gene therapy   39
genes
cell division   22–5
cloning   40–3
disorders   26–7, 39
function   10–15
gametes   24–5
mutations   14, 15, 26–7
proteins   12–15
selective breeding   36–7
switched on/off   15
transferring   38–9
genetic code   12, 13, 14
genetic disorders   26–7, 39
genetic engineering   38–9
genetic modification (GM)   38–9
glucose
photosynthesis   58–9
respiration   18–21

GM (genetic modification)   38–9
gold   142, 169
gradient of graphs   99, 104–5, 188–9
granite   252, 254
graphite   128–9, 130, 261
gravitation potential energy   216–19, 220–1
gravitational field strength   198, 199, 215, 217
gravity   198, 199, 212, 215, 216–21
greenhouse gases   201, 205, 206, 207
greenhouses, growing in   60–1
group 1 elements   152, 154–7, 161
group 7 elements   152, 160–3
group 8 elements   153
group numbers (periodic table)   152–3
growth
animals   32, 33, 34–5
cell division   22, 23, 32
measuring   34–5, 60
plants   32, 34–5, 42–3, 58–65, 76–9
guard cells   63, 64, 67, 73, 75

H

Haber process   158–59
habitats   54, 55, 57
haemoglobin   14, 15, 29
haemophilia   15, 26–7
half-life   249, 253, 254, 255, 258–9
halogens   160–3
haploid cells   24, 25
hearing, human   242–3
heart   20, 28, 29, 30–1
defibrillators   234, 236–2
heat
combustion   122–3, 132–3
conduction   165, 169
fusion and fission   260–3
thermal decomposition   166–7
thermal energy, falling objects   216, 217
helium   250, 251, 263
hertz (Hz)   242–3
Hindenburg disaster   106
homologous pairs, chromosomes   23, 25
human cloning   41
human tissue, growing   22
hybrid cars   206, 207
hydrochloric acid   98–9, 100, 101, 104–5, 106

hydrogen
car fuel   119
covalent bonds   151
nuclear fusion   262, 263
photosynthesis   59
product of reactions   98–101, 104–5, 106, 109, 154–5
relative atomic mass   110, 111
hydrogen bombs   262
hydroponics   84, 85
hydroxides   154–5, 166–7
hypothesis   279, 290

I

Ice Maiden   255
identification keys   56
inbreeding   37
independent variables   281–2
inheritance   24, 26–7
insecticides   84–5
insects, collecting   56
instantaneous speed   187
insulators   230, 232, 233, 240, 241
insulin   12, 14, 38, 39, 41
intensive farming   84–5
intermolecular forces   151, 153
interpolation   105
iodides   162, 163, 174, 175
iodine   160, 161, 162, 163
iodine-123   259
ionic bonds   146–9
ionisation   248, 250–1, 252, 253, 257
ions
alkali metals   157
bonds   146–9
halogens   161, 163
positive and negative   146–7, 148, 149, 157, 169, 250
testing water   174–5
transition metals   167
iron   164, 165, 166, 167, 168
isotopes
definition   143
radioactive   248–9, 252–9

J

JET project   262, 263

Internet research

The Internet is a great resource to use when you are working through your GCSE Science course.

Below are some tips to make the most of it.

1   Make sure that you get information at the right level for you by typing in the following words and phrases after your search: 'GCSE', 'KS4', 'KS3', 'for kids', 'easy', or 'simple'.

2   Use OR, AND, NOT and NEAR to narrow down your search.

> Use the word OR between two words to search for one or the other word.

> Use the word AND between two words to search for both words.

> Use the word NOT, for example, 'York NOT New York' to make sure that you do not get unwanted results (hits).

> Use the word NEAR, for example, 'London NEAR Art' to bring up pages where the two words appear very close to each other.

3   Be careful when you search for phrases. If you search for a whole phrase, for example, A Room with a View, you may get a lot of search results matching some or all of the words. If you put the phrase in quote marks, 'A Room with a View' it will only bring search results that have that whole phrase and so bring you more pages about the book or film and less about flats to rent!

4   For keyword searches, use several words and try to be specific. A search for 'asthma' will bring up thousands of results. But, a search for 'causes of asthma' or 'treatment of asthma' will bring more specific and fewer returns. Similarly, if you are looking for information on cats, for example, be as specific as you can by using the breed name.

5   Most search engines list their hits in a ranked order so that results that contain all your listed words (and so most closely match your request) will appear first. This means the first few pages of results will always be the most relevant.

6   Avoid using lots of smaller words such as A or THE unless it is particularly relevant to your search. Choose your words carefully and leave out any unnecessary extras.

7   If your request is country-specific, you can narrow your search by adding the country. For example, if you want to visit some historic houses and you live in the UK, search 'historic houses UK' otherwise it will search the world. With some search engines you can click on a 'web' or 'pages from the UK only' option.

8   Use a plus sign (+) before a word to force it into the search. That way only hits with that word will come up.

Group

	1	2

1		
1 H		
hydrogen		

7	9
3 Li	4 Be
lithium	beryllium

23	24
11 Na	12 Mg
sodium	magnesium

39	40	45	48	51	52	55	56	59	59	64	65
19 K	20 Ca	21 Sc	22 Ti	23 V	24 Cr	25 Mn	26 Fe	27 Co	28 Ni	29 Cu	30 Zn
potassium	calcium	scandium	titanium	vanadium	chromium	manganese	iron	cobalt	nickel	copper	zinc

85	88	89	91	93	96	99	101	103	106	108	112
37 Rb	38 Sr	39 Y	40 Zr	41 Nb	42 Mo	43 Tc	44 Ru	45 Rh	46 Pd	47 Ag	48 Cd
rubidium	strontium	yttrium	zirconium	niobium	molybdenum	technetium	ruthenium	rhodium	palladium	silver	cadmium

133	137	139	178	181	184	186	190	192	195	197	201
55 Cs	56 Ba	57 La	72 Hf	73 Ta	74 W	75 Re	76 Os	77 Ir	78 Pt	79 Au	80 Hg
caesium	barium	lanthanum	hafnium	tantalum	tungsten	rhenium	osmium	iridium	platinum	gold	mercury

223	226	227
87 Fr	88 Ra	89 Ac
francium	radium	actinium

Group

3	4	5	6	7	0

					4
					2 He
					helium

11	12	14	16	19	20
5 B	6 C	7 N	8 O	9 F	10 Ne
boron	carbon	nitrogen	oxygen	fluorine	neon

27	28	31	32	35	40
13 Al	14 Si	15 P	16 S	17 Cl	18 Ar
aluminium	silicon	phosphorus	sulfur	chlorine	argon

70	73	75	79	80	84
31 Ga	32 Ge	33 As	34 Se	35 Br	36 Kr
gallium	germanium	arsenic	selenium	bromine	krypton

115	119	122	128	127	131
49 In	50 Sn	51 Sb	52 Te	53 I	54 Xe
indium	tin	antimony	tellurium	iodine	xenon

204	207	209	210	210	222
81 Tl	82 Pb	83 Bi	84 Po	85 At	86 Rn
thallium	lead	bismuth	polonium	astatine	radon

Modern periodic table. You need to remember the symbols for the highlighted elements.

Acknowledgements

The publishers gratefully acknowledge the following for permission to reproduce images. Every effort has been made to trace copyright holders. However, any cases where this has not been possible, or for any inadvertent omission, the publishers will gladly rectify at the first opportunity.

cover & p.1 Sovereign, ISM/Science Photo Library, p.8t Eye of Science/Science Photo Library, p.8u alxhar/Shutterstock, p.8l Power and Syred/Science Photo Library, p.8b Ria Novosti/Science Photo Library, p.9t Dr Gopal Murti/Science Photo Library, p.9u marema/Shutterstock, p.9l Benjamin Albiach Galan/Shutterstock, p.9b U.S. National Institute of Health/Science Photo Library, p.10t A. Barrington Brown/Science Photo Library, p.10b CNRI/Science Photo Library, p.11 Martin McCarthy/iStockphoto, p.12tl Antagain/iStockphoto, p.12tr geopaul/iStockphoto, p.12bl Martin McCarthy/iStockphoto, p.12br Martin McCarthy/iStockphoto, p.13 dra_schwartz/iStockphoto, p.14t Eye of Science/Science Photo Library, p.14b Science Photo Library, p.15 Equinox Graphics/Science Photo Library, p.16 Rosenfeld Images Ltd/Science Photo Library, p.18t Andrew Lambert Photography/Science Photo Library, p.18b redmal/iStockphoto, p.21 sportgraphic/Shutterstock, p.22t BBC Photo Library, p.22b CNRI/Science Photo Library, p.24l David M. Phillips/Science Photo Library, p.24c D. Phillips/Science Photo Library, p.24b Yuri Arcurs/Alamy, p.26 Mary Evans Picture Library/Alamy, p.28t Sebastian Kaulitzki/Shutterstock, p.28l Dr. Richard Kessel & Dr. Gene Shih, Visuals Unlimited/Science Photo Library, p.28r Aaliya Landholt/Shutterstock, p.29 bioraven/Shutterstock, p.30r Bo Veisland, MI&I/Science Photo Library, p.30l CIOT/Science Photo Library, p.32 Catalin Petolea/Shutterstock, p.33t Sebastian Kaulitzki/Shutterstock, p.33b Sven Hoppe/Shutterstock, p.34l Smit/Shutterstock, p.34bl Phototake Inc./Alamy, p.34br Sinclair Stammers/Science Photo Library, p.36l asharkyu/Shutterstock, p.36c Tomas Sereda/Shutterstock, p.36r Chris Turner/Shutterstock, p.37t Joe Gough /Shutterstock, p.37c Len Green/Shutterstock, p.37b Eric Isselée/Shutterstock, p.38t Ryan Carter/iStockphoto, p.38l zentilia/Shutterstock, p.40t 2005 Seoul National University/Getty Images, p.40l Monkey Business Images/Shutterstock, p.40r ZanyZeus/Shutterstock, p.41 Jeremy Sutton Hibbert/Rex Features, p.42 ra3rn/Shutterstock, p.43t Natali Glado/Shutterstock, p.43b Peter Menzel/Science Photo Library, p.44bl Jim Jurica/iStockphoto, p.44tr tBoyan/iStockphoto, p.48 alohaspirit/iStockphoto, p.49l Jaroslaw Wojcik/iStockphoto, p.49r Viorika Prikhodko/iStockphoto, p.50 sgame/iStockphoto, p.52t J.C. Revy, ISM/Science Photo Library, p.52u Pakhnyushcha/Shutterstock, p.52l Darryl Sleath/Shutterstock, p.52b esemelwe/iStockphoto, p.53t Jubal Harshaw/Shutterstock, p.53u Colin Bell, p.53l David Cook /blueshiftstudios/Alamy, p.53b Dario Sabljak/Shutterstock, p.54t Rick Wylie/iStockphoto, p.54l Colin Bell, p.54r Colin Bell, p.55 Panagiotis Milonas/iStockphoto, p.56t Philippe Psaila/Science Photo Library, p.56b Martyn F. Chillmaid/Science Photo Library, p.57 Astrid & Hanns-Frieder Michler/Science Photo Library, p.58 Amy Mikler/Alamy, p.59 Emilio Ereza/Alamy, p.60l Matthew Noble/Alamy, p.60r BasPhoto/Shutterstock, p.62t Stock Connection Blue/Alamy, p.62b Colin Bell, p.63 Sidney Moulds/Science Photo Library, p.64t Colin Bell, p.64b Ivonne Wierink/Shutterstock, p.67 Fernando Alonso Herrero/iStockphoto, p.68t David Cook/blueshiftstudios/Alamy, p.68l liubomir/Shutterstock, p.68r Colin Bell, p.73 Jubal Harshaw/Shutterstock, p.74 LazarevaEl/Shutterstock, p.75l Bas Meelker/Shutterstock, p.75r Jogn Clegg/Science Photo Library, p.76c Paul Yates/Shutterstock, p.76b Jill Fromer/iStockphoto, p.80 Colin Bell, p.81 Rikard Stadler/Shutterstock, p.82 Filipe B. Varela/Shutterstock, p.84t David Bagnall/Alamy, p.84b Khoo Si Lin/Shutterstock, p.85 Dario Sabljak/Shutterstock, p.86 Vladimir Chernyanskiy/Shutterstock, p.87t Asianet-Pakistan/Alamy, p.87b Art_man/Shutterstock, p.96t Hiob/iStockphoto, p.96u Martyn F. Chillmaid/Science Photo Library, p.96l ppart/Shutterstock, p.96b dirkr/iStockphoto, p.97t Ashok Rodrigues/iStockphoto, p.97u Sergio Ponomarev/Shutterstock, p.97l Dean Kerr/Shutterstock, p.97b Martin McCarthy/iStockphoto, p.98 Dmitri Melnik/Shutterstock, p.102 Galyna Andrushko/Shutterstock, p.106t CSU Archv/Everett/Rex Features, p.106b Martyn F. Chillmaid/Science Photo Library, p.114t Topham Picturepoint/TopFoto, p.114b Tui De Roy/Getty Images, p.116 BostjanT/iStockphoto, p.118l bajars/Shutterstock, p.118r Andrew Lambert/Science Photo Library, p.119 Mona Makela/Shutterstock, p.120t Olga Utlyakova/Shutterstock, p.120b Akimoto Hiroyuki/Shutterstock, p.121 dinadesign/Shutterstock, p.124 Mustafa Arican/iStockphoto, p.125 Science Photo Library, p.126 Steven Paul Pepper/Shutterstock, p.127 Maximilian Stock Ltd/Science Photo Library, p.128t marema/Shutterstock, p.128b Elzbieta Sekowska/Shutterstock, p.130t Martin Mccarthy/iStockphoto, p.130c Oliver Hoffmann /Shutterstock, p.130b Lijuan Guo/Shutterstock, p.131 Andrew Lambert Photography/Science Photo Library, p.132 Minerva Studio/Shutterstock, p.140t Victor Soares/Shutterstock, p.140u Magdalena Jankowska/iStockphoto, p.140l Andraž Cerar/Shutterstock, p.140b Zeljko Radojko/Shutterstock, p.141t Smit/Shutterstock, p.141u Sherri Camp/iStockphoto, p.141l Darren Falkenberg/iStockphoto, p.141b somchaij/Shutterstock, p.142 Philippe Plailly/Science Photo Library, p.144l David Guyler/Shutterstock, p.144r Glowimages RM/Alamy, p.146 Irina Opachevsky/iStockphoto, p.148t Vlue/Shutterstock, p.148b T.W. van Urk/iStockphoto, p.149 Vasilyev/Shutterstock, p.150t Elle1/Shutterstock, p.150b westphalia/iStockphoto, p.154t Martyn F. Chillmaid/Science Photo Library, p.154l Andrew Lambert Photography/Science Photo Library, p.154c Andrew Lambert Photography/Science Photo Library, p.154r Andrew Lambert Photography/Science Photo Library, p.154b Dr P. Marazzi/Science Photo Library, p.155 Charles D. Winters/Science Photo Library, p.156l David Taylor/Science Photo Library, p.156c Andrew Lambert Photography/Science Photo Library, p.156r Andrew Lambert Photography/Science Photo Library, p.156b Dieter Melhorn/Alamy, p.158t Mary Evans Picture Library/Alamy, p.158b Charles D. Winters/Science Photo Library, p.160t Tim Abramowitz/iStockphoto, p.160l John Wollwerth/Shutterstock, p.160tr Ambient Ideas/Shutterstock, p.160br Dmitry Naumov/Shutterstock, p.160b Andrew Lambert Photography/Science Photo Library, p.161l Andrew Lambert Photography/Science Photo Library, p.161r MarcelClemens/Shutterstock, p.161b Karin Hildebrand Lau/Shutterstock, p.162l Andrew Lambert Photography/Science Photo Library, p.162a Andrew Lambert Photography/Science Photo Library, p.162b Charles D. Winters/Science Photo Library, p.162c Andrew Lambert Photography/Science Photo Library, p.164 Martyn F. Chillmaid/Science Photo Library, p.165 Martyn F. Chillmaid/Science Photo Library, p.166t Andrew Lambert Photography/Science Photo Library, p.166c Charles D. Winters/Science Photo Library, p.166b Andrew Lambert Photography/Science Photo Library, p.168t Anticipator/Wikimedia Commons, p.168c Tom Curtis/Shutterstock, p.168b Terekhov Igor/Shutterstock, p.169l Yuri Samsonov/Shutterstock, p.169c Lukich/Shutterstock, p.169r Stuart Monk/Shutterstock, p.169b Science Photo Library, p.170t G. Muller, Struers GmbH/Science Photo Library, p.170b Andrea Leone/Shutterstock, p.172 Kodda/Shutterstock, p.173 John Kasawa/Shutterstock, p.174l Andrew Lambert

Photography/Science Photo Library, p.174r Andrew Lambert Photography/Science Photo Library, p.176 Andrew Whittle/iStockphoto, p.184t Ahmad Faizal Yahya/Shutterstock, p.184u Ljupco Smokovski/Shutterstock, p.184l Piotr Sikora/Shutterstock, p.184b suzz/Shutterstock, p.185t Richard Watson/iStockphoto, p.185u Christian Waadt/iStockphoto, p.185l Kim D. French/Shutterstock, p.185b Skyhobo/iStockphoto, p.186t stocknshares/iStockphoto, p.186b Paul White, Transport Infrastructures/Alamy, p.187 Stephen Aaron Rees/Shutterstock, p.188 Adrian Phillips/Shutterstock, p.190 Chen Wei Seng/Shutterstock, p.191 Action Plus Photo Library, p.192 NASA, p.193t hfng/Shutterstock, p.193b Charles Stirling/Alamy, p.194t Christopher Elwell/Shutterstock, p.194l Rudolf Stricker/Wikimedia Commons, p.194r Rudolf Stricker/Wikimedia Commons, p.195l Steve Stock/Alamy, p.195r Jaggat/Shutterstock, p.197l daseaford/Shutterstock, p.197r seraficus/iStockphoto, p.198t snowturtle/Shutterstock, p.198b AVAVA/Shutterstock, p.199l Arena Creative/Shutterstock, p.199r Monkey Business Images/Shutterstock, p.200t Motoring Picture Library/Alamy, p.200c Max Earey/Shutterstock, p.200b Matthew Brown/iStockphoto, p.201 Baloncici/Shutterstock, p.202t image100/Alamy, p.202b Nik Taylor Sport/Alamy, p.204t Max Earey/Shutterstock, p.204c photobar/Shutterstock, p.204r MC_PP/Shutterstock, p.204b Yegor Korzh/Shutterstock, p.205 Justin Kase z12z/Alamy, p.206t Robert Conley/Alamy, p.206c Alvey & Towers Picture Library/Alamy, p.206b Idealink Photography/Alamy, p.208 TRL Ltd./Science Photo Library, p.209 Wfmillar/Wikimedia Commons, p.210t Lexan/Shutterstock, p.210b Shout/Alamy, p.211 Corbis Flirt/Alamy, p.212t Germanskydiver/Shutterstock, p.212cr erashov/Shutterstock, p.212br NASA, p.212bl Ljupco Smokovski/Shutterstock, p.213 Dusan Bartolovic/Shutterstock, p.214 Hank Morgan/Science Photo Library, p.216t Edwin Verin/Shutterstock, p.216b urmoments/Shutterstock, p.217l Noam Armonn/Shutterstock, p.217r Germanskydiver/Shutterstock, p.218t Brian North/Alamy, p.218l Neil Roy Johnson/Shutterstock, p.218r Jacqueline Abromeit/Shutterstock , p.219 Angela Weiss/Getty Images, p.224 Aspen Photo/Shutterstock, p.228t Bill Frische/Shutterstock, p.228u Adrian Britton/Shutterstock, p.228l jovannig/Shutterstock, p.228b VR Photos/Shutterstock, p.229t muratseyit/iStockphoto, p.229u Bork/Shutterstock, p.229l Thomas Reekie/iStockphoto, p.229b ermejoncqc/Shutterstock, p.230t Petr Mašek/Shutterstock, p.230b Charles D. Winters/Science Photo Library, p.231 sciencephotos/Alamy, p.232t Catherine Wessel/Corbis, p.232c MrTwister/Shutterstock, p.232b Artem Samokhvalov/Shutterstock, p.233t SDBIndustry/Alamy, p.233b Chris Pearsall/Alamy, p.234t Image Source/Alamy, p.234b Arve Bettum/Shutterstock, p.235 Leslie Garland Picture Library/Alamy, p.236t jannoon028/Shutterstock, p.236b Scott Camazine/Alamy, p.237t Dario Sabljak/Shutterstock, p.237l mjl84/Shutterstock, p.237r B Brown/Shutterstock, p.238t Kosmonaut/Shutterstock, p.238c Heintje Joseph T. Lee/Shutterstock, p.238b Trevor Clifford Photography/Science Photo Library, p.240 ra3rn/Shutterstock, p.242 Kirsanov/Shutterstock, p.244t Chad Ehlers/Alamy, p.244b BSIP, Boucharlat/Science Photo Library, p.245r nutech21/Shutterstock, p.245l Ruth Jenkinson/Midirs/Science Photo Library, p.246 Hank Morgan/Science Photo Library, p.247 Robyn Mackenzie/Shutterstock, p.248t Science Source/Science Photo Library, p.248b Andrew Lambert Photography/Science Photo Library, p.249 Astrid & Hanns-Frieder Michler/Science Photo Library, p.250t C. Powell, P.Fowler & D. Perkins/Science Photo Library, p.250b Science Photo Library, p.251 N. Feather/Science Photo Library, p.252t Maria Gioberti/Shutterstock, p.252c Dr Morley Read/Science Photo Library, p.252b Tanikewak/Shutterstock, p.254t Davidyoung/Shutterstock, p.254b Larry Miller/Science Photo Library, p.255t Kenneth V. Pilon/Shutterstock, p.255c Daniel Baránek/Wikimedia Commons, p.255b Ria Novosti/Science Photo Library, p.256 Monkey Business Images/Shutterstock, p.257 jannoon028/Shutterstock, p.258t Simon Fraser/NCCT, Freeman Trust, Newcastle-Upon-Tyne/Science Photo Library, p.258b Phototake Inc./Alamy, p.260t US Department of Energy/Science Photo Library, p.260b Patrick Landmann/Science Photo Library, p.262t Philippe Plailly/Eurelios/Science Photo Library, p.262b Los Alamos National Laboratory/Science Photo Library, p.263 Jerry Mason/Science Photo Library, p.264 Andrew Lambert Photography/Science Photo Library, p.269 Mark Kostich/iStockphoto, p.271 Michael Utech/iStockphoto, p.272-273 Tischenko Irina/Shutterstock, p.272 Kletr/Shutterstock, p.274-275 Diego Cervo/Shutterstock, p.276-277 Dmitry Naumov/Shutterstock, p.283t Andrew Lambert Photography/Science Photo Library, p.283c Pedro Salaverría/Shutterstock, p.283b Shawn Hempel/Shutterstock, p.288 Martyn F. Chillmaid/Science Photo Library.